Aeromonas

Edited by

Joerg Graf

Department of Molecular and Cell Biology
University of Connecticut
Storrs, CT
USA

Caister Academic Press

Caister Academic Press
Norfolk, UK

www.caister.com

British Library Cataloguing-in-Publication Data
A catalogue record for this book is available from the British Library

ISBN (hardback): 978-1-908230-56-0
ISBN (ebook): 978-1-908230-57-7

Cover design adapted from Figures 8.10 and 9.2.

Contents

Current Books of Interest

Advanced Vaccine Research Methods for the Decade of Vaccines	2015
Antifungals: From Genomics to Resistance and the Development of Novel Agents	2015
Bacteria-Plant Interactions: Advanced Research and Future Trends	2015
Antibiotics: Current Innovations and Future Trends	2015
Leishmania: Current Biology and Control	2015
Acanthamoeba: Biology and Pathogenesis (2nd edition)	2015
Microarrays: Current Technology, Innovations and Applications	2014
Metagenomics of the Microbial Nitrogen Cycle: Theory, Methods and Applications	2014
Pathogenic *Neisseria*: Genomics, Molecular Biology and Disease Intervention	2014
Proteomics: Targeted Technology, Innovations and Applications	2014
Biofuels: From Microbes to Molecules	2014
Human Pathogenic Fungi: Molecular Biology and Pathogenic Mechanisms	2014
Applied RNAi: From Fundamental Research to Therapeutic Applications	2014
Halophiles: Genetics and Genomes	2014
Molecular Diagnostics: Current Research and Applications	2014
Phage Therapy: Current Research and Applications	2014
Bioinformatics and Data Analysis in Microbiology	2014
The Cell Biology of Cyanobacteria	2014
Pathogenic *Escherichia coli*: Molecular and Cellular Microbiology	2014
Campylobacter Ecology and Evolution	2014
Burkholderia: From Genomes to Function	2014
Myxobacteria: Genomics, Cellular and Molecular Biology	2014
Next-generation Sequencing: Current Technologies and Applications	2014
Omics in Soil Science	2014
Applications of Molecular Microbiological Methods	2014
Mollicutes: Molecular Biology and Pathogenesis	2014
Genome Analysis: Current Procedures and Applications	2014
Bacterial Toxins: Genetics, Cellular Biology and Practical Applications	2013
Bacterial Membranes: Structural and Molecular Biology	2014
Cold-Adapted Microorganisms	2013
Fusarium: Genomics, Molecular and Cellular Biology	2013
Prions: Current Progress in Advanced Research	2013

Full details at www.caister.com

Contributors

Brian Austin
Institute of Aquaculture
University of Stirling
Stirling
UK

brian.austin@stir.ac.uk

Roxana Beaz-Hidalgo
Unit of Microbiology
Department of Basic Health Sciences
School of Medicine and Health Sciences
IISPV
Universitat Rovira i Virgili
Sant Llorenç
Reus
Spain

roxana.beaz@urv.cat

Ashok K. Chopra
Department of Microbiology & Immunology
University of Texas Medical Branch
Galveston, TX
USA

achopra@utmb.edu

Maria José Figueras
Departament de Ciències Mèdiques Bàsiques
Facultat de Medicina i Ciències de la Salut
Universitat Rovira i Virgili
Sant Llorenç
Reus
Spain

mariajose.figueras@urv.cat

Joerg Graf
Department of Molecular and Cell Biology
University of Connecticut
Storrs, CT
USA

joerg.graf@uconn.edu

Christopher J. Grim
Center for Food Safety and Applied Nutrition
U.S. Food and Drug Administration
Laurel, MD
USA

christopher.grim@fda.hhs.gov

Geert Huys
Laboratory of Microbiology & BCCM/LMG
Bacteria Collection
Faculty of Sciences
Ghent University
Gent
Belgium

geert.huys@ugent.be

Elena V. Kozlova
Department of Microbiology & Immunology
University of Texas Medical Branch
Galveston, TX
USA

evkozlov@utmb.edu

Brigitte Lamy
Laboratoire de Bactériologie-Virologie
Equipe Pathogènes et Environnements
Université Montpellier
Montpellier
France

brigitte_lamy@yahoo.fr

Antonio Martínez-Murcia
Area de Microbiología
Universidad Miguel Hernández
Orihuela
Alicante
Spain

ammurcia@umh.es

Jennifer Parker
Department of Infection and Immunity
University of Sheffield
Sheffield
UK

jennifer.parker@sheffield.ac.uk

Duraisamy Ponnusamy
Department of Microbiology & Immunology
University of Texas Medical Branch
Galveston, TX
USA

duponnus@utmb.edu

Jonathan Shaw
Department of Infection and Immunity
University of Sheffield
Sheffield
UK

j.g.shaw@sheffield.ac.uk

Preface

The conception of this book and the groundwork were done by Amy Horneman. Amy has been a driving force in the *Aeromonas* field, which is reflected by her being known to many of us in the community as '*Aeromonas*' Amy. It was natural for her to be the editor of this book. Unfortunately for us in the *Aeromonas* research community, her new position as the Director of Microbiology and Molecular Diagnostics at the Veterans Administration Maryland Health Care System prevented her from editing this book. However, Amy had already convinced Horizon Press of the importance of *Aeromonas.* She laid out a plan identifying the major topics and got leading researchers in this field to commit to write chapters. So I would like to express my thanks to Amy for starting this work.

As the reader will discover in this book about *Aeromonas,* this genus of bacteria encompasses a very interesting group of bacteria that have something of interest to many different microbiologists. These bacteria are capable of causing disease in humans and fish, which can be food borne or acquired from the environment. In contrast, these bacteria are also beneficial symbionts in other animals. In addition, it has proven difficult to identify *Aeromonas* strains at the species level using standard biochemical tests, which complicates many different aspects. The sheer number of species contained within this genus is ever increasing. We now know of more than one dozen *Aeromonas* species and can be confident that more species are yet to be described. Perhaps a better knowledge of the taxonomy will aid in identifying the virulent strains. Over recent years there has been *tremendous* progress in understanding the mechanisms by which *Aeromonas* causes disease and the surface properties that can lay the groundwork for vaccine development.

I would like to thank all of the authors for contributing high-quality manuscripts that cover each of the topic areas in a comprehensive fashion. I would also like to thank Hugh Griffin, Melanie Woodward and other staff members from Horizon Press with editing the manuscripts and helping in the production process.

While science moves at a frantic pace and books are static in time, I hope that this book will provide the reader with an excellent overview of *Aeromonas* that will serve them as a valuable resource for many years.

Joerg Graf
University of Connecticut

Introduction and Overview

Joerg Graf

1

As the reader will discover, the genus *Aeromonas* is a fascinating group of organisms for a variety of reasons, including the complicated taxonomy, the controversy of causing infections in humans, and the wide range of hosts colonized. The history of bacteria now considered to be *Aeromonas* dates back to 1891, when these bacteria were linked to 'red leg' disease in frogs. It was over 50 years later when *Proteus hydrophila,* which later was reclassified to *Aeromonas hydrophila,* was described by Stanier. During the 1980s DNA–DNA hybridization studies lead to the description of the main hybridization groups that included the species that are the most important fish and human pathogens, *A. salmonicida, A. hydrophila, A. veronii* and *A. caviae.* The advance of rRNA gene sequencing led to separation of the aeromonads from the Vibrionaceae as their own family. Over the years the interest in *Aeromonas* has increased as reflected by the increase in publications involving *Aeromonas* (Fig. 1.1).

In Chapter 2, Huys presents a brief history of the *Aeromonas* taxonomy followed by a description of how different technologies affected our understanding of the *Aeromonas* taxonomy as the field moved from DNA–DNA hybridization coupled with biochemical tests to more molecular approaches such as amplified fragment length polymorphism and the sequencing of multiple housekeeping genes. He also comments on several taxonomic controversies regarding species identification and makes general recommendations for future species descriptions.

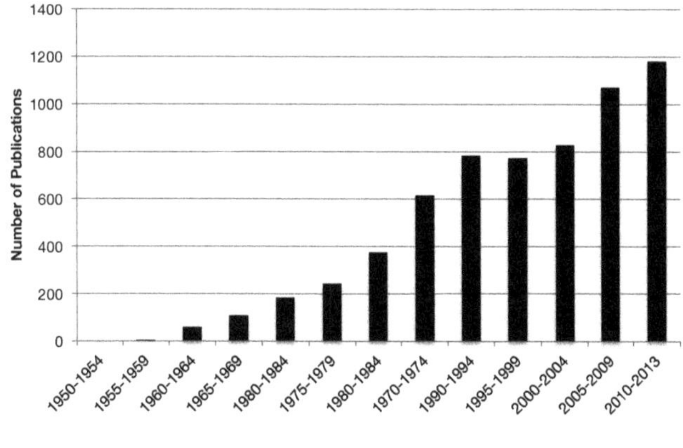

Figure 1.1 Number of *Aeromonas* publications. The number of citations returned with *Aeromonas* as a search term for the indicated time frame is shown.

Perhaps the best-known diseases caused by aeromonads involve fish and are traditionally thought to be caused by non-motile (*A. salmonicida*) or motile (*A. hydrophila*) *Aeromonas* strains. However as Austin describes the current state it becomes clear that an increasing number of other *Aeromonas* spp. that have been shown to be able to cause disease in fish. This chapter covers relevant taxonomy, diagnosis, ecology, virulence factors and disease control.

The role of *Aeromonas* spp. in causing disease in humans is perhaps the most controversial one, especially when it comes to the digestive tract illnesses such as diarrhoea. Figueras and Beaz-Hidalgo present the current evidence for *Aeromonas* being a human pathogen focuses not just on gastrointestinal illnesses but also other infections that *Aeromonas* isolates have been associated with injury or wound infections, nosocomial infections, septicaemia, respiratory tract infections and peritonitis.

While most research focuses on pathogenic associations of *Aeromonas*, some members of this genus are found in a number of different animals that are colonized in a benign or even beneficial manner. Graf describes the best studied example of this is the digestive tract symbiosis of *A. veronii* and the medicinal leech in Chapter 3. In addition, it was shown that *Aeromonas* can induce the normal development of the digestive tract in gnotobiotic zebrafish. In addition, *Aeromonas* spp. have been reported in mosquitoes and vampire bats, but the role in those organisms is less well understood.

Surface structures play an important role in adherence and are also an important target for developing vaccines. Parker and Shaw review *Aeromonas* surface structures such as polar and lateral flagella, pili and fimbriae, lipopolysaccharides, outer membrane proteins, capsule and the S-layer. One interesting feature of *Aeromonas* is that some of the surface proteins are also glycosylated, which is touched upon by the authors.

A description of the virulence factors of *A. hydrophila* is provided by Kozlova, Ponnusamy, and. Chopra. Using molecular genetic investigations of the strain SSU, insight into the molecular requirements for causing disease have been revealed. In this chapter the importance of type III secretion, type VI secretion and quorum sensing is discussed. In addition, the role of the secreted effector molecules or other virulence factors is mentioned.

Finally, but not least, *Aeromonas* in water and food is described in a chapter by Grim. As *Aeromonas* species are wide spread in water understanding the ecology, prevalence and factors influencing their abundance is important. Both results from classical culture-dependent studies and 16S rRNA gene studies are presented.

2 Taxonomy

Geert Huys

Abstract

Since its description by Kluyver and van Niel in 1936, the taxonomic structure of the genus *Aeromonas* has been drastically reshaped each time new technological advances were made in bacterial systematics. Modern *Aeromonas* taxonomy started off at the end of the 1970s essentially relying on physiological and biochemical characterization and DNA–DNA hybridizations, the latter still being considered the 'golden standard' for delineation of bacterial species. The original 'four-species concept' encompassing *Aeromonas hydrophila, Aeromonas caviae, Aeromonas sobria* and *Aeromonas salmonicida* was soon expanded with multiple DNA hybridization groups (HGs), most of which were later given the species status. The introduction of 16S rRNA gene sequence analysis and amplified fragment length polymorphism (AFLP) fingerprinting in the 1990s allowed to characterize the phylogenetic and genotypic diversity of larger sets of *Aeromonas* isolates, and facilitated the description of several new *Aeromonas* species or the synonymization of existing taxa. Next, the availability of bacterial whole-genome sequences allowed to evaluate single-copy protein-encoding housekeeping genes such as *gyrB* and *rpoD* as alternative molecular markers in *Aeromonas* taxonomy. Compared to the 16S rRNA gene, these markers display a higher taxonomic resolution and can be combined in a multilocus sequence approach to construct a stable phylogenetic framework to rapidly and reliably recognize new *Aeromonas* taxa and thus avoid new nomenclatural problems.

Following the first reference to an organism which would later be recognized as a motile aeromonad (Sanarelli, 1891), the taxonomy of the genus *Aeromonas* has undergone major changes. The main purpose of this chapter is to reconstruct and discuss the major steps in this process. Following a short description of the genus as commonly found in renowned taxonomic manuals, past and current views on the phylogenic position of *Aeromonas* in the Gammaproteobacteria and the case-specific lack of taxonomic discrimination of the 16S rRNA gene at species level are discussed. Next, the gradual expansion of the genus *Aeromonas* with new species is presented in a more or less chronological order that follows the path of the various technological advances witnessed by bacterial systematics since the 1970s. Finally, a critical overview is given of the methods which helped to shape our present view on *Aeromonas* taxonomy and have proven their use for reliable assignment of unknown isolates to known taxa (i.e. identification).

Short description of the genus

This description is based on the *Aeromonadaceae* chapter in the second edition of *Bergey's Manual of Systematic Bacteriology* (Martin-Carnahan and Joseph, 2005). The reader is also referred to this chapter for additional reading on *Aeromonas* taxonomy, together with the taxonomy chapter (Carnahan and Altwegg, 1996) in the previous *Aeromonas* monograph, the review of Janda and Abbott (2010) and the forthcoming chapter on the *Aeromonadaceae* in the fourth edition of *The Prokaryotes.*

Members of the genus *Aeromonas* are facultatively anaerobic, Gram-negative bacteria that are widespread in still and streaming aquatic biotopes. Cells occur as rigid rods with rounded ends but can also have a coccoid shape; most species, except *A. salmonicida* and *A. media,* are motile by a single polar flagellum. Cells exist singly, in pairs, or in short chains. Their size range is 0.3–1.0 µm in diameter and 1.0–3.5 µm in length. *Aeromonas* strains are oxidase- and catalase-positive, and reduce nitrates to nitrites. Metabolism of glucose is both respiratory and fermentative. Carbohydrates are broken down with the production of acid or acid and gas. Aeromonads are chemoorganotrophic, capable of using a wide range of sugars and organic acids as their source of carbon. They do not require Na for growth, and are resistant to the vibriostatic agent 0/129 (2,4-diamino-6,7-diiso-propylpteridine) and to most penicillins. Growth can occur between pH 4.5 and 9.0. Two major groups are known to exist, i.e. the motile, mesophilic aeromonads (optimum growth temperature 28 to 30°C), and the non-motile, psychrophilic aeromonads (optimum growth temperature 22–25°C). The G + C content ranges from 57 to 63 mol%. The type species is *Aeromonas hydrophila.*

Phylogeny

Phylogenetic position of the genus

In the first edition of *Bergey's Manual of Systematic Bacteriology,* the genus *Aeromonas* was assigned to the *Vibrionaceae* family primarily on the basis of phenotypically expressed properties (Popoff, 1984). At the time, this classification contradicted a previous DNA hybridization study by Staley and Colwell (1973) in which the genomic relatedness between selected reference strains of *Aeromonas* and *Vibrio* was reported to be relatively low (i.e. <10%); this important finding thus suggested that a significant evolutionary distance exists between both aforementioned genera. During the 1980s, the influential work of Woese (reviewed in 1987) and De Ley (reviewed in 1992) demonstrated that comparison of ribosomal RNA (rRNA) cistrons is a highly informative method to study phylogenetic relationships within the Proteobacteria. These new insights triggered the generation of 16S rRNA catalogues, 5S rRNA sequences, and DNA–rRNA hybridization data on the basis of which Colwell and co-workers (1986) concluded that the genus *Aeromonas* represents an evolutionary line that is sufficiently different from the *Vibrionaceae* and the *Enterobacteriaceae* to justify its exclusion from these two families. As a result, these authors proposed to assign the genus *Aeromonas* in a new family within rRNA superfamily I *sensu* De Ley (1992), the *Aeromonadaceae.* The allocation to a separate family was supported by subsequent rDNA sequencing (Martínez-Murcia *et al.,* 1992a) and rRNA sequencing (Kita-Tsukamoto *et al.,* 1993; Ruimy *et al.,* 1994) studies.

In the second edition of Bergey's Manual of Systematic Bacteriology, the *Aeromonadaceae* family is phylogenetically placed in the order *Aeromonadales* (Martin-Carnahan & Joseph,

2005) within the phylum Gammaproteobacteria. Since its first description by Colwell and colleagues (1986), several new genera have been situated in the *Aeromonadaceae* (Fig. 2.1). Based on the 16S rRNA gene tree of the All-Species Living Tree Project (Yarza *et al.*, 2008, 2010) and J.P. Euzéby's List of Prokaryotic names with Standing in Nomenclature (http://www.bacterio.cict.fr/), the family currently contains the type genus *Aeromonas* (Stanier, 1943), *Tolumonas* (Fischer-Romero *et al.*, 1996), *Oceanimonas* (Brown *et al.*, 2001), *Oceanisphaera* (Romanenko *et al.*, 2003) and *Zobellella* (Lin and Shieh, 2006).

Inter-specific phylogeny

Martínez-Murcia and colleagues (1992a) reported on the taxonomic value of small-subunit rRNA sequencing in *Aeromonas* beyond the genus level. Their results showed that the 16S rDNA sequences of 10 *Aeromonas* type strains exhibited a very high level of similarity ranging from 98 to 100%. When comparing with previously published DNA reassociation data, the authors found several cases for which the phylogenetic interrelationships of the corresponding species completely disagreed with DNA–DNA hybridization results. For instance, the ribosomal sequences of the type strains of *A. caviae* (HG4) and *A. trota* (HG13) differed by only one nucleotide (i.e. 99.9% sequence similarity) (Martínez-Murcia *et al.*, 1992a), whereas Carnahan and co-workers (1991c) determined that the DNA relatedness between these two strains was as low as 30%. A similar phenomenon was observed with HG2 and HG3, and HG1 and *A. media*, respectively. Conversely, reference strains of *A. veronii* HG8 and HG10 showed identical 16S rDNA sequences in agreement with the very high genotypic similarity between these two taxa (Hickman-Brenner *et al.*, 1987). Not surprisingly, the occasional lack of congruence between 16S rRNA gene sequence phylogeny and species

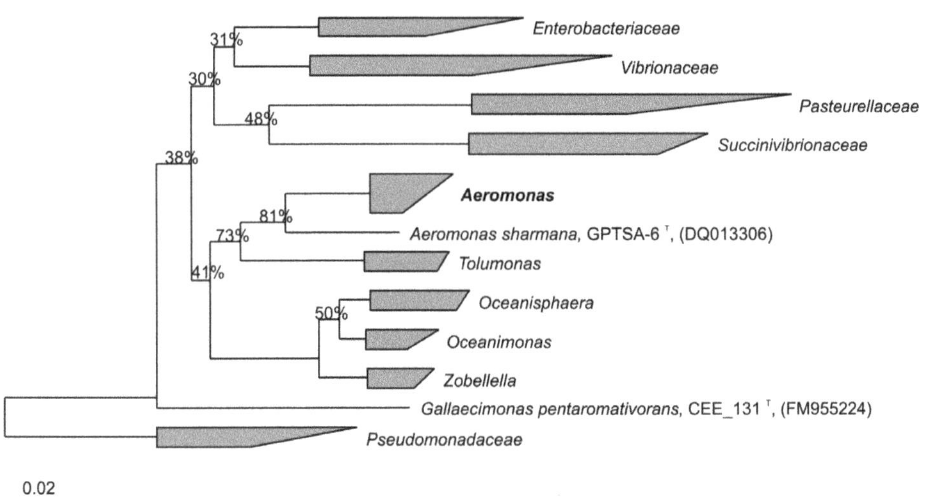

Figure 2.1 Phylogenetic reconstruction of the family Aeromonadaceae based on 16S rRNA and created using the neighbour-joining algorithm with the Jukes-Cantor correction and 100 runs bootstrapping. Percentages on nodes indicate bootstrap values. The sequence datasets and alignments were used according to the All-Species Living Tree Project (LTP) database (Yarza *et al.*, 2010; http://www.arb-silva.de/projects/living-tree). Representative sequences from closely related families were used as outgroups. Scale bar indicates estimated sequence divergence.

delineation based on DNA–DNA hybridizations has fuelled several taxonomic discussions. For example, Martínez-Murcia (1999) revealed discrepancies between 16S rDNA sequence analysis and DNA–DNA hybridization results (Huys *et al.*, 1997b) on the taxonomic position of *Aeromonas* HG11. Likewise, genotypic and phylogenetic discrepancies have been reported for the discrimination between *A. salmonicida* and *A. bestiarum* (Martínez-Murcia *et al.*, 2005). Along the same lines, Fox and colleagues (1992) also reported cases of discongruence between 16S rRNA sequences and DNA–DNA hybridization data for *Bacillus* spp. The latter authors concluded that 16S rRNA sequences could be routinely used to establish relationships between genera and well-resolved species, but may fail to recognize very recently diverged species.

In a Letter to the Editor, Sneath (1993) suggested that the observed discrepancies between phylogenetic and DNA–DNA hybridization data in *Aeromonas* may be a typical example of the lack of sensitivity displayed by 16S rRNA sequence analysis at the species level (Fox *et al.*, 1992). In addition, Sneath (1993) also presented evidence for the occurrence of hybrid events in the form of rare recombinations in ribosomal gene sequences of some *Aeromonas* species. In comparison, Eardly and colleagues (1996) reported a similar type of segment-dependent polymorphic site partitioning as described by Sneath (1993) among species of *Rhizobium* and *Agrobacterium*. Further evidence for the occurrence of intragenomic heterogeneity in the 16S rRNA gene sequences of some *Aeromonas* species suggested that this gene should be used with caution for use in *Aeromonas* phylogeny and identification (Morandi *et al.*, 2005). A phylogenetic reconstruction of the genus *Aeromonas* based on 16S rRNA sequences is shown in Fig. 2.2.

Systematics

Historical aspects

In the years following Sanarelli's first description (1891) of *Bacillus hydrophilus fuscus* as what is now considered a motile *Aeromonas* strain, this organism was allocated to many different bacterial genera. As reviewed by Carnahan and Altwegg (1996), early designations of mesophilic *Aeromonas* spp. include members of the genera *Bacillus, Bacterium, Aerobacter, Achromobacter, Pseudomonas, Proteus,* and *Vibrio*. Chester (1901) emended the name *Bacillus hydrophilus fuscus* Sanarelli 1891 and renamed it *Bacterium hydrophilum*, meaning a bacterium that was 'water loving'. The description of the non-motile aeromonads, on the other hand, was initiated by the work of Emmerich and Wiebel (1894). They reported that an organism named *Bacillus der Forellenseuche* was the cause of epizoonosis in trout, for which Chester (1897) proposed the species name *Bacterium salmonicida*. In contrast to the motile aeromonads, the latter bacterium was able to produce a brown diffusible pigment on trypticase soy agar and did not grow at 37°C.

In 1936, Kluyver and van Niel proposed the genus name *Aeromonas*, meaning 'gas-producing unit', in reference to the suspected similarity of the organism's metabolism with the fermentative characteristics of the genera *Aerobacter* and *Aerobacillus*. Originally, the type species and sole member situated in the new genus was *Aeromonas liquefaciens* Beijerinck. In the seventh edition of *Bergey's Manual of Determinative Bacteriology*, Snieszko (1957) described four species in the genus *Aeromonas* which then still resided in the family

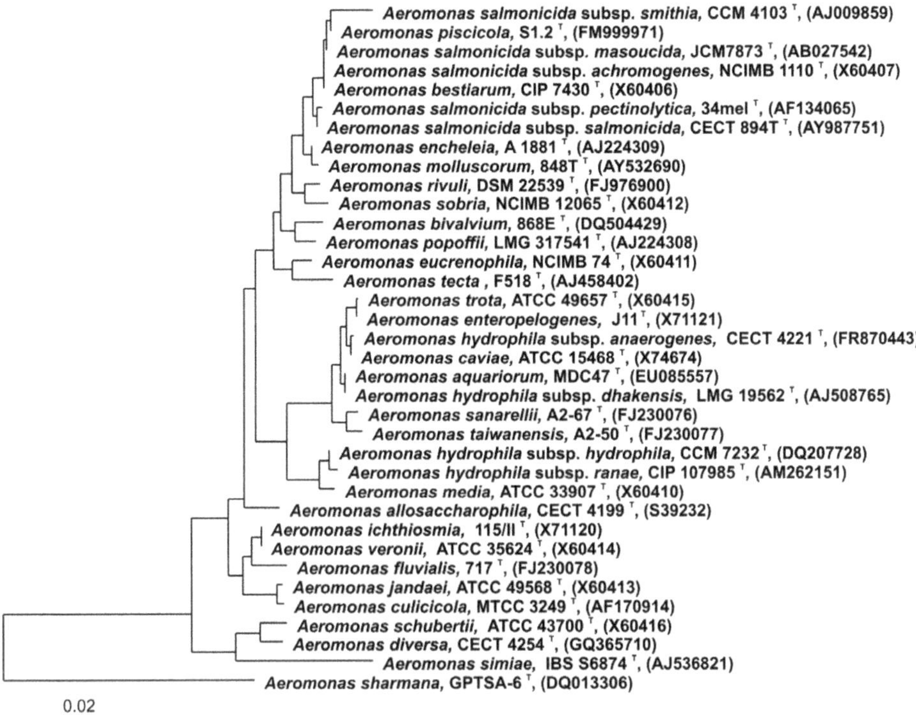

Figure 2.2 Phylogenetic reconstruction of the genus *Aeromonas* and the relative position of *A. sharmana* as extracted from Figure 2.1. The tree is based on 16S rRNA and created using the neighbour-joining algorithm with the Jukes-Cantor correction and 100 runs bootstrapping. The sequence datasets and alignments were used according to the All-Species Living Tree Project (LTP) database (Yarza *et al.*, 2010; http://www.arb-silva.de/projects/living-tree). Scale bar indicates estimated sequence divergence.

Pseudomonadaceae: the three motile species *A. hydrophila, A. punctata,* and *A. liquefaciens,* and the non-motile species *A. salmonicida.*

During the 1960s, the taxonomic structure of the genus *Aeromonas* changed continuously as a result of several phenotypic studies and this was often leading to conflicting conclusions. Eddy (1960) suggested that Snieszko's description of the genus *Aeromonas* in *Bergey's Manual* (1957) should be redefined towards a new, three-species-concept including *A. liquefaciens, A. formicans,* and *A. salmonicida.* Likewise, the proposal of Ewing and colleagues (1961) to classify all motile aeromonads in the species *A. hydrophila* and *A. shigelloides* encountered much resistance. First, Habs and Schubert (1962) suggested to remove the species *A. shigelloides* from the genus *Aeromonas* and reclassify it as *Plesiomonas shigelloides.* Until now, this species remains the only member of the genus *Plesiomonas* (Farmer III *et al.,* 1992). Secondly, Eddy (1962) and Eddy and Carpenter (1964) suggested that the motile *Aeromonas* strains rather belong to the species *A. punctata* and *A. caviae.* These authors argued that *A. punctata,* being a legitimate designation for both *A. liquefaciens* and *A. hydrophila,* should be the type species of the genus *Aeromonas,* and that *A. caviae* was a better name for *A. formicans.* Eventually, Schubert (1968) concluded that *A. liquefaciens* could no longer serve as the type species of the genus *Aeromonas* and, in agreement with the findings of Eddy and colleagues, conserved this position for *A. punctata*. Smith (1963) proposed to transfer the non-motile *Aeromonas* strains to the new genus *Necromonas* that also comprised the non-pigment-producing species *N. achromogenes*. However, Smith's proposal was not acknowledged by other *Aeromonas* taxonomists.

With the publication of the eighth edition of *Bergey's Manual of Determinative Bacteriology* (1974), another new *Aeromonas* taxonomy was described. In this edition, Schubert compiled all his previously reported findings (1967a,b, 1969), including the division of the genus *Aeromonas* into aerogenic and anaerogenic aeromonads based on their ability to produce gas from glucose. The genus now comprised the motile species *A. hydrophila* (with the subspecies *hydrophila, anaerogenes* and *proteolytica*) and *A. punctata* (with the subspecies *punctata* and *caviae*), and the non-motile species *A. salmonicida* (with the subspecies *salmonicida, achromogenes,* and *masoucida*). In addition, the genus *Aeromonas* was now considered a member of the *Vibrionaceae*, as suggested in an earlier study by Véron (1966).

The foundations of modern *Aeromonas* taxonomy

In 1976, Popoff and Véron performed a taxonomic study on 68 strains of the *A. hydrophila–A. punctata* group in an attempt to improve Schubert's (1974) identification scheme that, according to the former two authors, gave unsatisfactory results with the classification of new motile *Aeromonas* isolates in the *hydrophila–punctata* complex. Based on the numerical analysis of 203 phenotypic characters, the aeromonads under study were classified in two major classes that were actually considered motile *Aeromonas* species. Most strains (62%) belonged to the species *A. hydrophila*, which could be further divided in *A. hydrophila* biovar *hydrophila* and *A. hydrophila* biovar *anaerogenes*, the latter comprising all anaerogenic aeromonads. The remaining 26 strains constituted a second group that did not correspond to any of the previously described species, and these strains were allocated to a new species named *A. sobria*. In the same study, it was argued that *A. punctata* was a later and thus illegitimate synonym for *A. hydrophila*. As a result, the current *Aeromonas* taxonomy still recognizes the latter species as the type species of this genus. Finally, Popoff and Véron also indicated that the halophilic bacterium *A. hydrophila* subsp. *proteolytica* (Schubert, 1974) should be excluded from the genus *Aeromonas*, a suggestion that was readily confirmed at a later stage by transferring this subspecies to the genus *Vibrio* as *V. proteolyticus* (Baumann *et al.*, 1980).

In the years following the phenotypic work of Popoff and Véron, several researchers to further clarified the taxonomic relationships among the existing *Aeromonas* species by means of DNA–DNA hybridizations. McInnes and colleagues (1979) determined in their study that the two main phenotypic groups in the genus *Aeromonas* also corresponded to two legitimate genotypic groups: a diverse group of motile, mesophilic aeromonads and a more homogeneous group of non-motile, psychrophilic aeromonads. Subsequently, Popoff and associates (1981) found at least seven DNA hybridization groups among a collection of 55 motile *Aeromonas* strains. In this respect, the genotypic delineation of three groups in *A. hydrophila* (formerly *A. hydrophila* biovar hydrophila), two groups in the newly proposed *A. caviae* (formerly *A. hydrophila* biovar anaerogenes), and two groups in *A. sobria* by Popoff and co-workers (1981) are now regarded as being the most essential contributions to modern *Aeromonas* taxonomy. These results were adopted by Popoff in the first edition of (1984), in which the author already confronted the reader with two new taxonomic *Bergey's*

Manual of Systematic Bacteriology challenges. First, is was obvious that the individual DNA hybridization groups respectively situated in *A. hydrophila, A. caviae,* and *A. sobria* could not be established as new *Aeromonas* species since, at that time, they were phenotypically indistinguishable from one another. A second problem mentioned by Popoff concerned the names of the three *A. salmonicida* subspecies, i.e. subsp. *salmonicida, achromogenes,* and *masoucida* (Schubert, 1969, 1974), who were still present in the Approved Lists of Bacterial Names. However, according to the DNA hybridization data of McInnes and colleagues (1979), these three taxa were not sufficiently diverse to warrant their assignation to separate subspecies.

Naming of Popoff's DNA hybridization groups and first emended descriptions and synonyms

Inspired by the pioneering work of Popoff and colleagues (1981), several studies were launched in the 1980s and early 1990s to examine the precise taxonomic position of the various DNA hybridization groups (HGs) in the phenotypically and genotypically heterogeneous species *A. hydrophila, A. caviae* and *A. sobria.* In order to find diagnostic markers useful for taxonomic differentiation of HGs in *Aeromonas,* most of these studies relied on a combination of physiological and biochemical characterization with DNA–DNA hybridizations. Based on the guidelines proposed by Wayne and associates (1987), the definition of a HG implies that all constituting strains share ≥7 0% DNA relatedness with ≤ 5% divergence between the related sequences.

According to the chronological order of their first reporting in literature, the HGs originally delineated by Popoff and co-workers (1981) were referred to as HG1, HG2, HG3 (phenotypically resembling *A. hydrophila*), HG4, HG5, HG6 (phenotypically resembling *A. caviae*), HG7 and HG8 (phenotypically resembling *A. sobria*). The pioneering work of Popoff and colleagues resulted in the general usage of the terms phenotypic species or phenospecies (i.e. taxa delineated on the basis of phenotypic characterization) and genospecies, genomospecies, genomic species, or HGs (i.e. taxa delineated by DNA–DNA hybridization) in modern *Aeromonas* taxonomy. As a result, the terms species and phenospecies were long considered interchangeable, whereas HGs were also referred to as geno(mo)(mic)species in *Aeromonas* literature from the 1980s onwards. However, it should be clearly stressed that this apparent duality in *Aeromonas* nomenclature is only relevant in a specific number of cases where referring to only the (pheno)species or the HG designation of a given *Aeromonas* isolate does not fully specify the taxon in question. For example, a group of aeromonads phenotypically identified as belonging to the species *A. hydrophila* may contain members of HG1 and/or HG3. Given the significant difference in the potential clinical relevance of both taxa, an extended identification of *A. hydrophila* isolates down to the HG level may in these cases be an important requirement. Conversely, isolates that were genotypically assigned to HG3 without determining their physiological or biochemical features can belong to either *A. hydrophila* or *A. salmonicida.* Regarding the numbering of *Aeromonas* HGs, a lack of consensus has occasionally been noticed in subsequent species descriptions as one HG designation has been given to several species. For instance, whereas most taxonomists had referred to *A. trota* as *Aeromonas* HG13, researchers at the CDC considered *Aeromonas* Group 501 to be HG13 and have placed *A. trota* in HG14. Alternatively, Esteve and co-workers (1995b) proposed *A. allosaccharophila* to be HG14 although this designation was already assigned to *A. trota* or *Aeromonas* Group 501.

In Popoff's original *A. hydrophila* complex (including HGs 1–3), HG1 harbours the type strain of *A. hydrophila* (ATCC 7966^{T}) and therefore corresponds to *A. hydrophila sensu stricto*. As further discussed in 'New taxa recognized by 16S rRNA gene sequencing and AFLP fingerprinting', two groups of phenotypically atypical *A. hydrophila* HG1 strains were later proposed as subspecies of *A. hydrophila*, i.e. *A. hydrophila* subsp. *dhakensis* (Huys *et al.*, 2002b) and *A. hydrophila* subsp. *ranae* (Huys *et al.*, 2003). With the description of *A. bestiarum* (Ali *et al.*, 1996), formerly referred to as *A. hydrophila* HG2, a significant contribution was made to solve the structural and nomenclatural problems observed in the *A. hydrophila* complex (comprising HG1, HG2 and HG3), thus leaving only the motile strains in HG3 as the sole unnamed taxon in this group. Three phenotypical tests, including utilization of urocanic acid and DL-lactate, and acid from D-sorbitol, were highlighted as being relevant discriminatory tests within the *A. hydrophila* complex (Ali *et al.*, 1996; Hänninnen, 1994). At the first International Workshop on *Aeromonas* and *Plesiomonas* in 1986, Farmer III and colleagues showed that the type strain of the non-motile species *A. salmonicida* was genomically highly related to the reference strain of the motile *A. hydrophila* strains in HG3. This important finding led to the conclusion that the three subspecies of *A. salmonicida* (i.e. subsp. *salmonicida*, subsp. *achromogenes* and subsp. *masoucida*) should be included in HG3, despite the striking phenotypic differences with the other, motile members of HG3. Since, a fourth and a fifth subspecies named *A. salmonicida* subsp. *smithia* (Austin *et al.*, 1989) and *A. salmonicida* subsp. *pectinolytica* (Pavan *et al.*, 2000), respectively, have been described. It has been suggested that the motile strains in HG3 phenotypically resembling *A. hydrophila* might constitute another new subspecies in *A. salmonicida* (Altwegg *et al.*, 1990), but until a taxonomic proposal is made this group remains unnamed.

In the original *A. caviae* complex (including HGs 4–6), members of HG4 are considered as *A. caviae sensu stricto* and includes the type strain ATCC 15468^{T}. In 1983, Allen and associates described a new *Aeromonas* taxon isolated from the water of a trout farm which possessed some typical characteristics of *A. salmonicida*, i.e. absence of motility and the production of a brown diffusible pigment. In spite of the fact that no DNA relatedness data were included, the authors concluded from a further comparison of phenotypic and genotypic traits that this group of strains deserved the status of a new *Aeromonas* species, named *A. media*. Later, it was found that the type strain of *A. media* resided in HG5 together with representatives of *A. caviae* (Hickman-Brenner *et al.*, 1988a). In fact, new DNA hybridization data revealed that HG5 should be split up in HG5A (phenotypically *A. caviae*) and HG5B (comprising strains that resemble *A. caviae* or *A. media*). The suggestions made by Altwegg and co-workers (1990) to consider HGs 5A and 5B as two subspecies of *A. media*, and their proposal to create two separate biogroups in HG5B, have not been validated to date. In 1988, Schubert and Hegazi demonstrated that the type strain of the aerogenic subspecies *A. punctata* subsp. *punctata* (Schubert, 1974) and six other aquatic *Aeromonas* isolates recovered from unpolluted surface and groundwater belonged to a new species named *A. eucrenophila*. It is noteworthy that the type strain of *A. eucrenophila* had been included before in the DNA hybridization study of Popoff and associates (1981) as strain 546. The latter workers considered this strain to be the sole representative of a new HG in the *A. caviae* complex next to HGs 4 and 5, later referred to as HG6 (Hickman-Brenner *et al.*, 1988a). Schubert and Hegazi (1988) determined that *A. eucrenophila* was the sole aerogenic member of *A. caviae* group. Based on chemotaxonomic and genotypic evidence, Huys and co-workers (1996a) later discovered a significant degree of taxonomic heterogeneity among the original *A. eucrenophila* strains leading to the definition of subgroups I and II within this species. As confirmed by a DNA reassociation study and phenotypic characterization (Huys *et al.*, 1997b), *A. eucrenophila* subgroup I (including Popoff's strain 546) represented the taxonomic core of this species and an emended description of the species was provided. Subgroup II of *A. eucrenophila* was found to belong to the previously described species *A. encheleia* (Esteve *et al.*, 1995a) and was consequently also given an extended description (Huys *et al.*, 1997b).

In the *A. sobria* complex (HGs 7 and 8) as first described by Popoff and co-workers (1981), HG7 originally harboured only two strains. In contrast, the nomenclatural history of HG8 is much less straightforward. In 1987, Hickman-Brenner and associates proposed the name *A. veronii* for a group of human isolates formerly referred to as Enteric Group 77 at the Centers for Disease Control and Prevention (CDC), Atlanta, Ga. (USA). All strains of this group uniformly displayed the unique feature of producing ornithine decarboxylase (ODC^+). This was seen as an indication for the discovery of a new *Aeromonas* species previously not recognized by Popoff and co-workers (1981). DNA–DNA hybridization data further strengthened the notion of a new species, and resulted in the description of *A. veronii* as *Aeromonas* HG10 (Kuijper *et al.*, 1989a). The order of HG numbering took into account that Fanning and associates (1985) previously discovered a third HG in *A. sobria*, i.e. HG9, that had not been recognized before by Popoff and co-workers (1981). Interestingly, two ODC^+ strains of Enteric Group 77 were not included in the description of *A. veronii* because their respective levels of DNA relatedness to the type strain were below 70% at optimal renaturation conditions. One of these two strains, i.e. strain ATCC 35941, was later allocated to *Aeromonas* HG11 (Hickman-Brenner *et al.*, 1988a), whereas the other strain (ATCC 35942) now resides in the species *Aeromonas allosaccharophila* (Martínez-Murcia *et al.*, 1992b; see 'New taxa recognized by 16S rRNA gene sequencing and AFLP fingerprinting'). The taxonomic position of *Aeromonas* HG11 was investigated in two polyphasic studies (Huys *et al.*, 1996a, 1997b) from which it was decided that this taxon should be included in *Aeromonas encheleia*. Furthermore, Hickman-Brenner and colleagues (1987) stated in an Addendum in Proof that representatives of *A. sobria* HG8 exhibited high levels of DNA relatedness to the type strain of *A. veronii* but differed phenotypically from the latter species in a number of characteristics including the lack of ODC production. As a result, HG8 was considered a biogroup of *A. veronii* for which the name *A. veronii* biovar sobria was proposed and, accordingly, HG10 is now referred to as *A. veronii* biovar veronii (Joseph *et al.*, 1991).

Hickman-Brenner and co-workers (1988b) proposed the name *A. schubertii* for a group of human isolates formerly referred to as Enteric Group 501 at the CDC (Hickman-Brenner *et al.*, 1988a). A negative reaction in the fermentation of D-mannitol was found to be a useful phenotypic marker to differentiate *A. schubertii* from all other known *Aeromonas* species. In the genotypic classification scheme of the genus *Aeromonas*, the new species was added as HG12 (Altwegg *et al.*, 1990). In an Addendum in Proof, Hickman-Brenner and co-workers (1988b) reported on the existence of two *A. schubertii*-like isolates from human origin that probably had diverged somewhat from this species. These two strains were placed in *Aeromonas* Group 501. Phylogenetic data have supported the notion that *Aeromonas* Group 501 may represent a new *Aeromonas* species closely related to *A. schubertii* (Martínez-Murcia, 1999). Finally, Miñana-Galbis and co-workers (2010a) concluded from the results of their polyphasic taxonomic study to include both strains in a new *Aeromonas* species for which the name *A. diversa* was proposed.

In 1990, two new species were proposed in the genus *Aeromonas*, i.e. *A. enteropelogenes* (Schubert *et al.*, 1990a) and *A. ichthiosmia* (Schubert *et al.*, 1990b). In the following years, the descriptions of these two species were the subject of taxonomic disputes based on the fact that both studies included too few *Aeromonas* species for comparative phenotypic and DNA–DNA hybridization research. In their study of 16S rRNA sequences, Collins and co-workers (1993) reported a few years later that *A. ichthiosmia* and *A. enteropelogenes* were in fact identical to *A. veronii* and *A. trota*, respectively. The latter species was described by Carnahan and colleagues (1991c) only a couple of months following the publication of the *A. enteropelogenes* proposal (Schubert *et al.*, 1990a), and remarkably, both species were composed of human stool isolates that primarily originated from Southern Asia and that were negative for aesculin hydrolysis and the Voges-Proskauer reaction. In addition, a most remarkable feature of *A. trota* that had not been observed with any other *Aeromonas* species before was its susceptibility to ampicillin (Carnahan *et al.*, 1991c). For the first time, the use of ampicillin-containing screening media for aeromonads was somewhat questioned. In support of previous phylogenetic findings (Collins *et al.*, 1993), Carnahan (1993) remarked that *A. ichthiosmia* and *A. veronii*, and *A. enteropelogenes* and *A. trota* shared highly similar phenotypic profiles, respectively. Many years later, Huys and co-workers (2001b) published DNA–DNA hybridization data showing that the type strains of *A. veronii* and *A. ichthiosmia* exhibited a genomic relatedness of 84–96%, whereas extended phenotypic analysis clearly allocated the latter taxon as a member of *A. veronii* biovar sobria (HG8). Given the fact that *A. veronii* (Hickman-Brenner *et al.*, 1987) was the first validated of the two species, it was proposed to reject the species name *A. ichthiosmia* as a junior synonym of *A. veronii* (Huys *et al.*, 2001b). In a subsequent study, Huys and co-workers (2002a) likewise demonstrated that *A. trota* and *A. enteropelogenes* should be considered synonyms on the basis of 86–98% DNA homology and identical phenotypic profiles observed among their respective type strains. A Request for an Opinion was proposed to decide whether the later validated but most cited species name *A. trota* (Carnahan *et al.*, 1991c) can obtain priority over the first proposed but rarely referred name of *A. enteropelogenes* (Schubert *et al.*, 1990a).

Next to the description of *A. trota*, one other new species was added to the genus *Aeromonas* in 1991. Based on the preliminary findings reported by Fanning and co-workers (1985), Carnahan and colleagues (1991a) proposed the name *A. jandaei* for the rare and unnamed *Aeromonas* HG9 that mainly consisted of human isolates. A negative reaction in the fermentation of sucrose was mentioned as a useful biochemical character to distinguish *A. jandaei* from other *A. sobria*-like strains. More than a decade later, the description of *A. jandaei* was emended following a taxonomic study of sucrose-positive *A. jandaei*-like isolates from faeces, water and eels (Esteve *et al.*, 2003). The latter authors proposed to also include sucrose-positive clinical and environmental strains as well as environmental isolates that are pathogenic for fish and humans in *A. jandaei* and reported new species-specific traits such as the ability to grow at 4–42°C, acid production from glycerol but not from lactose, D-melibiose or D-raffinose, the use of D-gluconate, L-glutamate or L-proline but not L-lactate, L-alanine, L-arabinose or L-arginine, hydrolytic activity against casein, elastin, starch and lecithin and the inability to lyse arbutin.

Four strains isolated from European eels were found to constitute a new species for which the name *A. encheleia* was proposed (Esteve *et al.*, 1995a). The new species displayed

0–50% genomic relatedness with type and reference strains of all recognized *Aeromonas* taxa except *A. hydrophila* HG2 (later named *A. bestiarum*) which was not included in the DNA–DNA hybridization study. Members of *A. encheleia* were generally not pathogenic for eels or mice. As a result of polyphasic taxonomic studies (Huys *et al.*, 1996a, 1997b), the taxonomic core of *A. encheleia* was later extended with strains belonging to *A. eucrenophila* subgroup II and with the two members of *Aeromonas* HG11 i.e. the *A. veronii*-like clinical isolate ATCC 35941 (Hickman-Brenner *et al.*, 1988a) and the environmental strain CDC 3136–78. This finding was in contrast to the results of Esteve and colleagues (1995a) who reported that DNA homology values between *A. encheleia* and HG11 strain ATCC 35941 were not higher than 37%, whereas Huys and co-workers (1997b) reported genomic relatedness between the two taxa ranging from 83% to 98%. The inclusion of HG11 in *A. encheleia* proposed by the latter authors was further supported by cellular fatty acid analysis, whole-cell protein profiling, ribotyping and AFLP fingerprinting (Huys *et al.*, 1996a). Only in phenotypic analysis, the two strains of HG11 displayed aberrant behaviour and not clearly fitted the biochemical profile of the other *A. encheleia* strains. Phylogenetic analysis by Martínez-Murcia (1999) later revealed that the two HG11 strains exhibited a species-specific sequence including unique nucleotides at two positions, which seemingly represents another inconsistency between DNA hybridization data and 16S rRNA sequencing in *Aeromonas*.

New taxa recognized by 16S rRNA gene sequencing and AFLP fingerprinting

The introduction of molecular methods such as 16S rRNA gene sequence analysis and amplified fragment length polymorphism (AFLP) fingerprinting in bacterial taxonomy during the early 1990ies allowed to discover potentially new *Aeromonas* taxa at a much faster rate than was the case using DNA–DNA hybridizations.

The description of the species *A. allosaccharophila* (Martínez-Murcia *et al.*, 1992b) was the first species proposal in *Aeromonas* to be based on 16S rRNA gene sequence analysis and phenotypic characterization but lacking DNA–DNA hybridization data. The new species comprised two atypical isolates from diseased elvers and one human stool isolate, i.e. strain ATCC 35942, that was previously described as *A. veronii*-like by Hickman-Brenner and co-workers (1987). Since *A. allosaccharophila* only comprised three strains which did not clearly exhibit species-specific biochemical features, phenotypic identification of this species was initially problematic. Later, Esteve and associates (1995b) demonstrated in a comparative DNA–DNA hybridization study that *A. allosaccharophila* represented a distinct genomic species. However, the latter finding was contradicted by new DNA homology data reported by Huys and colleagues (2001b) showing that the type strains of *A. allosaccharophila* and *A. veronii* were 78–82% related whereas Esteve and colleagues (1995b) reported 0% DNA homology between these two strains. Previously, Huys and co-workers (1996b) had already shown that both species cannot be separated using AFLP fingerprinting. The latter authors suggested that additional DNA–DNA hybridization studies should be carried out to solve this apparent contradiction. From additional evidence based on partial sequencing of two housekeeping genes, it was argued that although the phylogenetic tree based on *rpoD* gene sequences showed a borderline relationship to *A. veronii*, the *gyrB* gene sequence results provided evidence that supports the original proposal of *A. allosaccharophila* (Saavedra *et al.*, 2006). However, sequence analysis of another housekeeping gene, *dnaJ*, indicated a close enough relationship between both taxa to consider *A. allosaccharophila* a heterotypic synonym of *A. veronii* (Nhung *et al.*, 2007).

Based on clustering analysis of AFLP fingerprints, Huys and colleagues had previously located a HG2-like group in the *A. hydrophila* complex (Huys *et al.*, 1996c) comprising isolates from Flemish and Scottish drinking water production plants and reservoirs. This group was later described as *A. popoffii* (Huys *et al.*, 1997c) and could be biochemically differentiated from the closely related species *A. hydrophila sensu stricto* and *A. bestiarum* of the *A. hydrophila* complex by negative reactions in the tests for D-sucrose fermentation and lysine decarboxylation. In support of the species description, analysis of 16S rRNA sequences likewise indicated that *A. popoffii* belongs to a new phylogenetic line in the genus *Aeromonas* (Martínez-Murcia, 1999) and can be differentiated from other members of this genus based on a signature region within its 16S rDNA sequence (Demarta *et al.*, 1999).

Using AFLP for initial genotypic delineation, two groups of HG1-like strains were proposed as subspecies of *A. hydrophila sensu stricto*. First, Huys and colleagues (2002b) described a group of mainly clinical *Aeromonas* isolates originating from children with diarrhoea in Bangladesh that were classified as phenotypically atypical members of *A. hydrophila* HG1. Whereas DNA–DNA hybridization levels did not indicate that these isolates deserved species status, the lack of acid production from L-arabinose and the ability to utilize urocanic acid clearly discriminated these isolates from *A. hydrophila* HG1. Taken together with its separate clustering in AFLP, the diarrhoeal isolates were assigned to the new subspecies *A. hydrophila* subsp. *dhakensis*. In the same paper (Huys *et al.*, 2002b), an extended description was given of *A. hydrophila* subsp. *hydrophila* (Chester 1901) Stanier 1943 (formerly *A. hydrophila sensu stricto*). One year later, another new subspecies was proposed in *A. hydrophila* HG1 for a set of isolates recovered from the internal organs of septicaemic farmed frogs (*Rana rugulosa*) in Thailand. The new taxon, for which the name *A. hydrophila* subsp. *ranae* was proposed (Huys *et al.*, 2003), could be biochemically differentiated from *A. hydrophila* subsp. *hydrophila* and subsp. *dhakensis* using assimilation of L-glycine and isobutyrate as sole carbon source, acid production from salicin and D-sucrose, and aesculin hydrolysis as diagnostic tests.

Phylogenetic analysis based on 16S rRNA gene sequences of two isolates from the faeces of two healthy monkeys (*Macaca fascicularis*) from Mauritius kept in quarantine formed the basis of the delineation of *A. simiae* (Harf-Monteil *et al.*, 2004). DNA–DNA hybridizations indicated highest relatedness of the two isolates to *A. trota* (18%) and *A. schubertii* (16%). Phenotypically, *A. simiae* was shown to differ from most previously described mesophilic *Aeromonas* species by their lack of haemolysis on sheep-blood agar and inability to produce indole, gas from glucose or acid from mannitol.

A numerical taxonomy study of *Aeromonas* isolates from environmental and clinical samples in Spain (Miñana-Galbis *et al.*, 2002) revealed two groups of isolates from bivalve molluscs (phena VI and VII) which were later subjected to a polyphasic taxonomic study. The study of phenon VI concentrated on five isolates that showed sufficiently low relatedness with the other *Aeromonas* HGs in DNA–DNA hybridization experiments and AFLP clustering analysis to justify their assignment to a new species, *A. molluscorum* (Miñana-Galbis *et al.*, 2004). In 16S rRNA gene sequence analysis, the new species exhibited highest similarity to *A. encheleia*. The authors reported negative reactions for indole production, lysine decarboxylase, gas from glucose and starch hydrolysis as useful phenotypic characters to differentiate *A. molluscorum* from other *Aeromonas* species. Likewise, polyphasic characterization of two strains of phenon VII using DNA–DNA hybridizations, AFLP and 16S rRNA gene sequencing led to the conclusion that this group belongs to a new species, for which the name *A. bivalvium* was proposed (Miñana-Galbis *et al.*, 2007). Phenotypic differentiation of *A. bivalvium* from other mesophilic *Aeromonas* species relied on tests for gas from glucose, lysine decarboxylase, Voges-Proskauer reaction, acid from L-arabinose, hydrolysis of aesculin and utilization of L-lactate.

Novel taxa and reclassifications based on housekeeping gene sequence analysis

In search of alternative gene markers for the 16S rRNA gene, several housekeeping genes or a combination thereof have been proposed. Sequence analysis of single or multiple genes have been used as a basis for the phylogenetic delineation of several novel *Aeromonas* species. For a review of the housekeeping genes used in *Aeromonas* taxonomy, the reader is referred to 'Genomic methods' of this chapter.

Three isolates recovered from the midgut of mosquitoes (*Culex quinquefasciatus* and *Aedes aegyptii*) were assigned to the novel species *A. culicicola* on the basis of biochemical characterization, DNA–DNA hybridization, and 16S rRNA gene sequencing (Pidiyar *et al.*, 2002). Highest DNA reassociation values for the type strain MTCC 3249^T were reported with the type strains of *A. sobria* (61%) and *A. jandaei* (54%), whereas phylogenetic 16S rDNA-based positioning revealed an extremely high similarity (>99.9%) between MTCC 3249^T and the type strain of *A. jandaei*. Soon after its description, however, inclusion of *A. culicicola* in *gyrB* gene sequence analysis demonstrated that the latter species grouped closely with *A. veronii* biovar *sobria* and biovar *veronii* (Pidiyar *et al.*, 2003; Yáñez *et al.*, 2003). Likewise, several other findings questioned the status of *A. culicicola* as a separate species in the genus *Aeromonas*, i.e. the high similarity between strain MTCC 3249^T and reference strains of *A. veronii* biovar sobria and veronii in *rpoD* sequencing analysis (Soler *et al.*, 2004) and the grouping of strain MTCC 3249^T with *A. veronii* reference strains in AFLP (Miñana-Galbis *et al.*, 2004). Based on new DNA–DNA hybridization data, Huys and co-workers (2005) proposed that *A. culicicola* is a later subjective synonym of *A. veronii* Hickman-Brenner *et al.* 1988. Further support to include *A. culicicola* in *A. veronii* was provided by gene sequence analysis of *rpoD* (Saavedra *et al.*, 2006) and *dnaJ* (Nhung *et al.*, 2007), and additional DNA–DNA hybridizations (Nhung *et al.*, 2007).

Analysis of *gyrB*, *rpoD* and 16S rRNA gene sequences formed the phylogenetic framework to assign representatives from a group of 48 *Aeromonas* isolates from water and skin of imported ornamental fish to the species *A. aquariorum* (Martínez-Murcia *et al.*, 2008). The taxonomic position of this new taxon was further studied in a polyphasic approach including DNA–DNA hybridizations, MALDI-TOF MS analysis, RAPD, and extensive biochemical and antibiotic susceptibility tests. Noteworthy, the original species proposal included the type strain of *A. hydrophila* subsp. *hydrophila*, but not of *A. hydrophila* subsp. *dhakensis* and subsp. *ranae*. Based on *gyrB* and *rpoD* sequence data, Martínez-Murcia and co-workers (2009) subsequently proposed to reclassify the earlier described *A. hydrophila* subsp. *dhakensis* Huys *et al.* 2002 as a member of *A. aquariorum*. According to a Note in J.P. Euzéby's List of Prokaryotic names with Standing in Nomenclature (http://www.bacterio.cict.fr/a/*aeromonas*.html), this suggestion for reclassification was not valid, probably because of a nomenclatural priority conflict.

As a follow-up to an earlier study in which a group of five strains from clinical and environmental sources was described as '*A. eucrenophila*-like' on the basis of ribotyping and AFLP data (Demarta *et al.* 2004), the phylogenetic position of this unnamed taxon was investigated into more detail using *gyrB*, *rpoD* and 16S rRNA gene sequencing (Demarta *et al.*, 2008). In conjunction with data from *gyrA*, *recA*, *dnaJ* and *dnaX*, these analyses indicated that the five isolates represented an undescribed phylogenetic line in the genus *Aeromonas* most closely related to *A. encheleia* and *A. eucrenophila*. The type strain of the latter species also showed the highest DNA reassociation values (35–45%) with the type strain of the new taxon, for which the name *A. tecta* was proposed (Demarta *et al.*, 2008). The authors reported utilization of D-cellobiose, acid production from D-cellobiose and glycerol, and indole production as the most valuable biochemical features to differentiate strains of *A. tecta* from *A. eucrenophila* and *A. encheleia*.

Multilocus sequence analysis of the housekeeping genes *rpoD*, *gyrB*, *recA* and *dnaJ* was used to determine the phylogenetic position of four isolates from moribund fish in Spain (Beaz-Hidalgo *et al.*, 2009). These analyses indicated that the unknown group was most closely related to *A. salmonicida*, *A. bestiarum* and *A. popoffii*, as evidenced by 16S rRNA gene sequencing which could not differentiate the new isolates from the aforementioned species. In line with these findings, DNA–DNA hybridizations showed 49.9–65.4% relatedness between the proposed type strain of the new taxon and *A. salmonicida*, *A. bestiarum* and *A. popoffii*. From the four housekeeping genes analysed in their study, the authors reported that *rpoD* and *gyrB* offered the best discrimination from the neighbouring species. Combined with further support from MALDI-TOF MS profiling, the four *Aeromonas* isolates were considered to represent a new species in the genus for which the name *A. piscicola* was proposed (Beaz-Hidalgo *et al.*, 2009). Hydrolysis of elastin, acid production from L-arabinose and salicin, and the inability to produce acid from lactose and use L-lactate as sole carbon source were found most useful for the biochemical differentiation of *A. piscicola* from other *Aeromonas* spp.

Supported by sequence analysis of *cpn60*, *dnaJ*, *gyrB* and *rpoD*, Miñana-Galbis and coworkers (2010a) redetermined the phylogenetic affiliation of two '*A. schubertii*-like' human isolates which had been tentatively placed in *Aeromonas* Group 501 (Hickman-Brenner *et al.*, 1988a). In combination with previously reported data on DNA–DNA reassociation values (Hickman-Brenner *et al.*, 1988b; Carnahan *et al.*, 1990a,c; Esteve *et al.*, 1995a,b; Pidiyar *et al.*, 2002; Miñana-Galbis *et al.*, 2004) and AFLP (Miñana-Galbis *et al.*, 2004, 2007), these new data indicated that *Aeromonas* Group 501 deserves the species status for which the name *A. diversa* was proposed (Miñana-Galbis *et al.*, 2010a). Phenotypically, the new taxon was readily distinguishable from other *Aeromonas* species. Differentiation from its closest neighbour *A. schubertii* could be achieved by indole and lysine decarboxylase tests and the utilization of L-lactate.

As part of a large collection of 681 bacterial isolates recovered from a karst water rivulet in northern Germany (Westerhöfer Bach) (Cousin *et al.*, 2008), *Aeromonas* isolates WB4.119^T and WB4.4-101 were subjected to a polyphasic taxonomic study (Figueras *et al.*, 2011). Both strains shared 98.6% *rpoD* gene sequence similarity, which was indicative for the fact that they are genotypically different strains. Multilocus phylogenetic analysis of concatenated sequences (totalling 3684 bp) of five housekeeping genes (*gyrB*, *rpoD*, *recA*, *dnaJ* and *gyrA*) revealed that these two strains represented a new phylogenetic line most closely affiliated *to A. molluscorum* and *A. bivalvium*. This was confirmed by DNA–DNA hybridizations which showed reassociation values below 70% between WB4.1-19^T and type strains of other *Aeromonas* species. As a result, Figueras and colleagues (2011) concluded that isolates WB4.1-19^T and WB4.4-101 should be assigned to a new species in the genus *Aeromonas* for which the name *A. rivuli* was proposed. The new species could be differentiated from the phenotypically closely related *A. molluscorum* by its ability to hydrolyse starch and no use or acid production from L-arabinose. Additional biochemical properties useful for differentiation of *A. rivuli* from other species include negative reactions for indole production, gas from D-glucose, Voges–Proskauer test, lysine decarboxylase and β-haemolysis on sheep blood agar.

Single-strain species

As is the case in many other genera, the encouragement to base bacterial species descriptions on more than one strain (Stackebrandt *et al.*, 2002) is increasingly being disregarded in descriptions of new *Aeromonas* spp. For new taxa represented by only one or a few strains, it has been proposed to create the status species proponenda (sp. pr.) as a compromise between the need for scientific description of biodiversity and data exchange and the good taxonomic practice of including a sufficient number of strains in species descriptions (Felis and Dellaglio, 2007). However, this proposal has never been put into practice. As a result, since 2010 the genus *Aeromonas* has witnessed the description of several new and officially validated species based on a single strain.

The single isolate GPTSA-6^T, recovered from the water of a warm spring in Assam (India), was found to show the highest level of 16S rRNA sequence similarity (95.04–95.13%) with the type strains of *A. sobria*, *A. molluscorum* and *A. popoffii* (Saha and Chakrabarti, 2006). On the basis of phenotypic, fatty acid profiling and phylogenetic data, but without performing DNA–DNA hybridizations, the authors concluded that strain GPTSA-6^T represented a novel species in the genus *Aeromonas* for which the name *A. sharmana* was proposed. However, soon after its description, several workers argued that *A. sharmana* does not belong to the genus *Aeromonas* (Saavedra *et al.*, 2006; Martínez-Murcia *et al.*, 2007). According to the latter authors, a 16S rRNA gene sequence similarity of 95% between strain GPTSA-6^T and its closest neighbouring species does not indicate that this isolate is a member of the genus, but on the contrary, clearly demonstrates that it does not belong to *Aeromonas*. In addition, strain GPTSA-6^T did not reduce nitrate to nitrite and showed a very weak catalase reaction, two properties which are very atypical for members of the genus *Aeromonas*. In conclusion, Martínez-Murcia and co-workers (2007) proposed to exclude strain GPTSA-6^T from *Aeromonas* and to reconsider the name '*Manjusharmella aquatica*' tentatively suggested by Saha and Chakrabarti (2006) while awaiting further taxonomic data regarding the phylogenetic position of this strain.

Previous data had indicated that strain 717^T, isolated from a water sample collected from the Muga river (Girona, Spain), showed an atypical 16S rRNA-RFLP profile (Figueras *et al.*, 2000). While subsequent 16S rRNA gene sequence analysis showed that this strain was most closely related to *A. veronii*, multilocus phylogenetic analysis of five housekeeping genes (*gyrB*, *rpoD*, *recA*, *dnaJ* and *gyrA*) suggested that isolate 717^T represented a previously unknown phylogenetic line with *A. sobria*, *A. veronii* and *A. allosaccharophila* as closest neighbouring species (Alperi *et al.*, 2010b). In DNA–DNA reassociation experiments, strain 717^T exhibited 54–66% relatedness with the type strains of the three aforementioned species. Collectively, these data suggested that this strain represented a new species for which the name *A. fluvialis* was proposed (Alperi *et al.*, 2010b). The new single-strain species was found to be biochemically different from other *Aeromonas* species due to its ability to produce acid from lactose and sucrose but not from L-arabinose or D-mannitol, and the lack of arginine dihydrolase, lysine decarboxylase and ornithine decarboxylase activity.

Phylogenetic analysis by *rpoD* gene sequencing revealed that two clinical *Aeromonas* isolates recovered from the wounds of two patients in Taiwan, A2-50^T and A2-67^T, each represented a novel and independent phylogenetic line most closely affiliated to *A. caviae* (Alperi *et al.*, 2010a). This finding was also confirmed upon extended phylogenetic analysis based on five housekeeping genes (*gyrB, rpoD, recA, dnaJ* and *gyrA*; in total encompassing 3684 bp). The mean DNA–DNA reassociation values between the two clinical isolates and the type strain of *A. caviae* ranged from 59.9% to 64.7%, indicating that strains A2-50^T and A2-67^T can be assigned to new species for which the names *A. taiwanensis* and *A. sanarellii*, respectively, were proposed (Alperi *et al.*, 2010a). The authors reported that both strains were able to produce acid from amygdalin, a relatively rare biochemical trait in the genus *Aeromonas* previously only reported in a few *A. caviae* strains (Abbott *et al.*, 2003). Other characteristics that differentiated the newly proposed single-strain species *A. taiwanensis* and *A. sanarellii* from other species included production of acid from L-arabinose, glycerol and salicin, hydrolysis of aesculin, and negative results for the Voges-Proskauer test, lysine decarboxylase activity, production of gas from glucose and acid from cellobiose and D-mannose.

An overview of all taxa currently recognized in the genus *Aeromonas* is given in Table 2.1. The phylogenetic position of all validly described *Aeromonas* species on the basis of 16S rRNA sequences is reconstructed in Fig. 2.2.

Non-validated species names

In addition to the taxa listed in Table 2.1, a number of *Aeromonas* species have been proposed in journal publications, symposia abstracts, or taxonomic information linked to public 16S rDNA sequences (Martin-Carnahan and Joseph, 2005). Using J.P. Euzéby's List of Prokaryotic names with Standing in Nomenclature (http://www.bacterio.cict.fr/) as major reference, none of these names have been officially validated as novel species and thus require further taxonomic investigation.

The name '*A. pastoria*' was first mentioned in a study of Torres and co-workers (1990). Two strains of '*A. pastoria*' were later examined in a large phenotypic study (Noterdaeme *et al.*, 1996). One of these isolates grouped into cluster 2A, which was characterized by non-fermentation of L-arabinose and alpha-methyl-D-glucoside and by assimilation of citrate. Another '*A. pastoria*' strain fell into cluster 7, which grouped strains biochemically characterized by lack of ornithine decarboxylase activity, esculin hydrolysis, assimilation and fermentation of L-arabinose, fermentation of alpha-methyl-D-glucoside, arbutin, salicin and lactose, and a positive reaction in citrate assimilation.

The name '*A. arequipensis*' was once mentioned in an abstract presented at 6th International *Aeromonas* & *Plesiomonas* Symposium (Chicago, Illinois, USA) (Matte *et al.*, 1999). The study described isolates from a lake in the Arequipa region (Peru), possibly with diarrhoeagenic potential for humans.

The name '*A. dechromatica*' was proposed by Kvasnikov and co-workers (1985) for isolates recovered from chromium-containing industrial wastewater. The organism exhibited a number of characteristics not typical for the genus *Aeromonas*, including requirement for

Table 2.1 Taxonomic composition of the genus *Aeromonas*[a]

Name	HG[b]	Type strain[c]	GenBank accession no. for 16S rRNA[c]	Reference
A. hydrophila	1			
subsp. *hydrophila*		Popoff 543 = ATCC 7966 = BCRC (formerly CCRC) 13018 = CCUG 14551 = CDC 359–60 = CIP 76.14 = DSM 30187 = HAMBI 1847 = HAMBI 1973 = IAM 12460 = JCM 1027 = LMG 2844 = LMG 12156 = LMG 13439 = NCIMB 9240 = NCTC 8049	CP000462	Stanier (1943)
subsp. *dhakensis*		P21 = LMG 19562 = CCUG 45377	AJ508765.	Huys *et al.* (2002b)
subsp. *ranae*		Au-1D12 = CCUG 46211 = LMG 19707	AM262151	Huys *et al.* (2003)
A. bestiarum	2	Popoff 218 = ATCC 51108 = CCUG 41859 = CDC 9533–76 = CIP 74.30 = DSM 13956 = LMG 13444	X60406	Ali *et al.* (1996)
A. salmonicida	3			
subsp. *salmonicida*		Smith 20 = ATCC 33658 = CAIM 346 = CIP 103209 = HAMBI 1984 = JCM 7874 = LMG 3780 = NCIMB 1102 = NCTC 12959	AY987751	Schubert (1967a,b)
subsp. *achromogenes*		Smith 6263/4/5 = ATCC 33659 = CIP 104001 = HAMBI 1990 = JCM 7875 = LMG 14900 = NCIMB 1110	X60407	Schubert (1967a,b)
subsp. *masoucida*		Schubert 1-a-1 = ATCC 27013 = CIP 103210 = IFO (now NBRC) 13784 = JCM 7873 = JCM 8191 = JCM 8234 = LMG 3782	AB027542	Kimura (1969)
subsp. *smithia*		Austin 138 = ATCC 49393 = CCM 4103 = CIP 104757 = LMG 20223	AJ009859	Austin *et al.* (1989)
subsp. *pectinolytica*		Pavan 34mel = CCUG 44929 = CIP 107036 = DSM 12609 = LMG 19569	AF134065	Pavan *et al.* (2000)
'*A. hydrophila*-like'	3	Popoff 316 = CDC 0434–84 = LMG 13451		
A. caviae	4	Popoff 309 = ATCC 15468 = CCUG 25939 = LMG 3775	X74674	Popoff (1984)
'*A. caviae*-like'	5A	Popoff 239 = ATCC 51107 = CDC 0862–83 = CIP 74.32 = LMG 13459	U88655	
A. media	5B	Austin RM = ATCC 33907 = CCM 3653 = BCRC (formerly CCRC) 14129 = CCUG 14453 = CIP 103208 = DSM 4881 = JCM 2385 = LMG 9073	X60410	Allen *et al.* (1983)
A. eucrenophila	6	Popoff 546 = ATCC 23309 = CCUG 30340 = CIP 76.17 = JCM 8238 = LMG 3774 = NCIMB 74	X60411	Schubert and Hegazi (1988)
A. sobria	7	Popoff 208 = ATCC 43979 = CCUG 14830 = CECT 4245 = CIP 74.33 = HAMBI 1974 = JCM 2139 = LMG 3783 = NCTC 12979	X60412	Popoff and Véron (1981)

Table 2.1 (Continued)

Name	HG[b]	Type strain[c]	GenBank accession no. for 16S rRNA[c]	Reference
A. veronii				
biovar *sobria*	8	Popoff 311 = ATCC 9071 = BPE 16 = Canad-212 No. 494 = CCEB 702 = CCM 1242 = CDC 360–60 = CDC-RH 36 = CIP 57.49 = CUETM 84/233 = FIRDI 375 = IAM 12333 = Kulp P2 = LMG 12158 = NCIMB 37 = NCTC 7810 = RH 252		Joseph *et al.* (1991)
biovar *veronii*	10	ATCC 35624 = CCUG 27821 = CDC 1169–83 = CECT 4257 = CIP 103438 = DSM 7386 = JCM 7375 = LMG 9075	X60414	Hickman-Brenner *et al.* (1987)
A. jandaei	9	ATCC 49568 = CCUG 29543 = CCUG 30353 = CDC 0787–80 = CECT 4228 = CIP 104171 = DSM 7311 = JCM 8316 = LMG 12221	X60413	Carnahan *et al.* (1991a)
Aeromonas HG11	11	ATCC 35941 = CCUG 30364 = CDC 1306–83 = CECT 4253 = LMG 17875 = LMG 17926 = LMG 18015 = LMG 21941 = NCIMB 13014	X60417	Hickman-Brenner *et al.* (1988a)
A. schubertii	12	ATCC 43700 = CCUG 27820 = CDC 2446–81 = CIP 103437 = DSM 4882 = JCM 7373 = LMG 9074	X60416	Hickman-Brenner *et al.* (1988b)
A. trota	13	Carnahan AH2 = ATCC 49657 = CCUG 29544 = CCUG 30367 = CECT 4255 = CIP 103677 = DSM 7312 = JCM 8315 = LMG 12223	X60415	Carnahan *et al.* (1991c)
A. allosaccharophila		Esteve E289 = ATCC 51208 = CECT 4199 = CIP 104616 = DSM 11576 = LMG 14059	S39232	Martínez-Murcia *et al.* (1992b)
A. encheleia		Esteve S181 = ATCC 51929 = CECT 4342 = CIP 104608 = DSM 11577 = LMG 16330 = NCTC 12917	AJ224309	Esteve *et al.* (1995a)
A. popoffii		Huys IK-O-a-10–3 = ATCC BAA-243 = CCUG 39350 = CIP 105493 = LMG 17541	AJ224308	Huys *et al.* (1997c)
A. simiae		IBS S6874 = CCUG 47378 = CIP 107798 = LMG 22269	AJ536821	Harf-Monteil *et al.* (2004)
A. molluscorum		Miñana-Galbis 848T = CCUG 50741 = CECT 5864 = LMG 22214	AY532690	Miñana-Galbis *et al.* (2004)
A. bivalvium		Miñana-Galbis 868E = CECT 7113 = LMG 23376	DQ504429	Miñana-Galbis *et al.* (2007)
A. aquariorum		MDC47 = CECT 7289 = DSM 18362 = LMG 24688	EU085557	Martínez-Murcia *et al.* (2008)
A. tecta		Demarta F518 = CECT 7082 = DSM 17300 = MDC 91 = LMG 24874	AJ458402	Demarta *et al.* (2008)
A. piscicola		Romalde S1.2 = CECT 7443 = LMG 24783	FM999971	Beaz-Hidalgo *et al.* (2009)
A. diversa		Hickman-Brenner 2478–85 = ATCC 43946 = CDC 2478–85 = CECT 4254 = LMG 17321	GQ365710	Miñana-Galbis *et al.* (2010a)

Table 2.1 (Continued)

Name	HG[b]	Type strain[c]	GenBank accession no. for 16S rRNA[c]	Reference
A. rivuli		Figueras WB4.1–19 = CECT 7518 = DSM 22539 = MDC 2511	FJ976900	Figueras *et al.* (2011)
A. fluvialis		717 = CECT 7401 = LMG 24681	FJ230078	Alperi *et al.* (2010b)
A. taiwanensis		A2–50 = CECT 7403 = LMG 24683	FJ230077	Alperi *et al.* (2010a)
A. sanarellii		A2–67 = CECT 7402 = LMG 24682	FJ230076	Alperi *et al.* (2010a)

[a]This table includes validly described species and unnamed HG groups awaiting formal description. The table excludes the following synonymous species: *A. enteropelogenes* Schubert *et al.* 1991 as a synonym of *A. trota* Carnahan *et al.* 1992 (Huys *et al.*, 2002a), *A. ichthiosmia* Schubert *et al.* 1991 as a synonym of *A. veronii* Hickman-Brenner *et al.* 1988 (Huys *et al.*, 2001b), *A. culicicola* Pidiyar *et al.* 2002 as a synonym of *A. veronii* Hickman-Brenner *et al.* 1988 (Huys *et al.*, 2005); also *A. sharmana* Saha and Chakrabarti 2006 is excluded as a non-*Aeromonas* species (Martínez-Murcia *et al.*, 2007).
[b]Numerical designation of DNA hybridization groups based on consensus in literature. Numbers higher than HG13 are not mentioned because of lack of such consensus.
[c]Equivalent strain numbers for the type strain and GenBank/EMBL/DDBJ accession number for the 16S rRNA gene sequence of the type strain as listed in J.P. Euzéby's List of Prokaryotic names with Standing in Nomenclature (http://www.bacterio.cict.fr/).

sodium chloride (optimal 3%), mixed flagellation, gram variable, absence of catalase activity, presence of a crystalline microcapsule.

The name '*A. guangheii*' was mentioned as taxonomic designation for isolate 95–173 recovered from eels with a 16S rRNA gene sequence submitted in GenBank (accession no. AB028881). To our knowledge, no publications mentioning this taxon are available.

Taxonomic methods

Genus level

Phenotypic methods

In preparation of phenotypic characterization, most *Aeromonas* strains can be easily cultured on standard media like tryptic soy agar or nutrient agar, without requirement for special culture conditions. To obtain optimal growth, however, the incubation temperature should be adjusted depending on whether mesophilic (range 28–30°C) or psychrophilic (range 22–25°C) *Aeromonas* strains are cultivated. As a group, aeromonads can be easily distinguished from other Gram-negative, facultatively anaerobic rods (i.e. vibrios, the *Enterobacteriaceae*, and *Plesiomonas*) by a few simple phenotypic tests (Table 2.2). Upon primary isolation, inexperienced workers may confuse aeromonads with pseudomonads when oxidase testing is used as the only differential screening criterion. Aeromonads tend to form violet-blue colour in oxidase testing whereas *Pseudomonas* spp. display a deep blue

Table 2.2 Phenotypic differentiation of *Aeromonas* from other bacterial groups

Feature	Reaction for			
	Aeromonas	*Vibrio*[a]	*Plesiomonas*	Enterobacteriaceae
Oxidase test	+	+	+	–
Resistance to				
0/129 (150 μg)	+[b]	–	–	nd
Ampicillin (10 μg)	+[c]	v	v	–

[a]Halophilic *Vibrio* spp. can be further differentiated from the other bacterial groups in Table 2.3 by their inability to grow in the absence of NaCl.
[b]0/129-susceptible (150 μg) *Aeromonas* isolates have been reported (Cahill and McRae, 1992; Sugita *et al.*, 1994b; Abbott *et al.*, 2003).
[c]*A. trota* is the sole ampicillin-susceptible species member of the genus *Aeromonas* (Carnahan *et al.*, 1991c).
+, ≥85 % positive strains; –, ≤15 % positive strains; v, variable; nd, no data found.

coloration (Schubert, 1987). Leclerc and Buttiaux (1962) noted that fermentation of lactose at 37°C is an unreliable test to differentiate aeromonads from *Enterobacteriaceae*, and further showed that the routine use of this particular test had caused false-positive counts in the determination of total coliforms in drinking water analysis. Martínez-Murcia and colleagues (2007) pointed to the value of nitrate reduction and a positive catalase test as two key features of aeromonads, and argued that weak to negative results for both characteristics provided phenotypic evidence to exclude *A. sharmana* GPTSA-6^{T} (Saha and Chakrabarti, 2006) from the genus *Aeromonas*.

Next to the production of acid from glucose, oxidase and catalase activity, Abbott and co-workers (1992) identified additional common biochemical features among members of *Aeromonas* HGs 1–15 (excluding non-motile *A. salmonicida* subspecies of HG3) including production of acid from D-trehalose, failure to utilize malonate or mucate as the sole carbon source, inability to ferment adonitol, dulcitol, erythritol, inositol, and D-xylose, and growth in nutrient broth containing 0% and 3% NaCl. Phenotypic features that were nearly always positive (98 to 99%) included motility, β-galactosidase activity, and resistance to the vibriostatic agent 0/129. Less than 2% of the strains were positive for urea hydrolysis, pectin degradation, or acid production from arabitol, D-raffinose, or D-amygdalin. In a later study, Abbott and colleagues (2003) reported that some *Aeromonas* strains may display a number of properties which are unusual or aberrant in the genus, including susceptibility to the vibriostatic agent 0/129, fermentation of *m*-inositol and D-xylose, hydrolysis of urea, and the lack of cytochrome oxidase activity. Noteworthy, also other authors had previously reported the isolation of 0/129-susceptible *Aeromonas* isolates (Cahill and McRae, 1992; Sugita *et al.*, 1994b).

In addition to conventional tests, miniaturized identification kits (API 20E and 50E, Microscan) have been used to confirm the identity of presumptive aeromonads at the genus level with changing success. In their study, Joseph and Carnahan (1994) determined that only 52% of the *Aeromonas* strains tested were correctly identified to the genus level with the API 20E system (API Systems, bioMérieux, La Balme Les Grottes, France). Here, additional tests were mandatory to obtain taxonomic information beyond the genus level. In comparison, Ogden and colleagues (1994) reported that 72% of the traditionally confirmed

Aeromonas isolates were correctly identified as a member of this genus by API 20E. However, this percentage rose to 96% when using API 20NE. As demonstrated by Abbott and associates (1998), the high risk of misidentification when using the API 20E system is even more apparent among unusual *Aeromonas* species. The latter authors reported that representatives of *A. schubertii* and *A. veronii* biovar veronii were mistakenly identified as vibrios with API 20E.

Chemotaxonomic and molecular methods

A number of chemotaxonomic and molecular methods that are usually employed for identification at the species level have also been used to obtain a rapid taxonomic confirmation of the genus identity. In most of these cases, workers are not specifically focussing on *Aeromonas* and use techniques with a broad taxonomic scope to identify as many isolates as possible among a heterogeneous strain collection. For instance, Edwards and co-workers (2001) used fatty acid methyl ester analysis to determine the chemotaxonomic diversity among a collection of planktonic heterotrophic bacteria and recognized *Aeromonas* as one of the most abundant genera.

Through the use of 16S rDNA sequence analysis, the presence of *Aeromonas* in strain collections from environmental or clinical origin can be easily verified by complete or partly sequencing of the 16S rRNA gene. Based on these sequences, probes have been developed for the *in situ* identification of aeromonads at the genus level using fluorescently labelled oligonucleotide probes (Kämpfer *et al.*, 1996; Mudaly *et al.*, 2001). Other rapid DNA-based tools for the preliminary identification of aeromonads at the genus level include RFLP-PCR analysis of the *aroA* gene (Soriano *et al.*, 1997) and PCR amplification of the intergenic *exeF–exeG* region (Karlyshev and Macintyre, 1996).

The first major study of the matrix-assisted laser desorption/ionization mass spectrometry (MALDI-MS) method in the genus *Aeromonas* revealed the presence of three common mass ions among all 17 species tested (Donohue *et al.*, 2006). The authors considered these signatures, with an average m/z of 6301, 12,160 or 12,254, and 13,450, to have potential as genus-specific biomarkers in unknown samples.

Species and HG level

For ecological studies on fermentative Gram-negatives, reliable identification of aeromonads on the genus level may be sufficient. Studies on virulence properties and antibiotic resistance at the molecular level, on the other hand, often require a correct identification of the *Aeromonas* strains under study to the species and/or hybridization group level. Literature surveys have shown that comparative studies on environmental and clinical *Aeromonas* isolates and their epidemiology highly benefit from correct identification data on the species level. An excellent review on taxonomic and identification methods for aeromonads was published by Millership (1996) in the first monograph on the genus *Aeromonas*. The current chapter expands that review with an update on methods that have emerged since then.

Phenotypic methods

A. salmonicida

The division of the genus *Aeromonas* into motile, mesophilic and non-motile, psychrophilic aeromonads (i.e. *A. salmonicida*) is not well defined because some non-motile *A. media* strains are able to grow in the temperature range of 22–37°C (Allen *et al.*, 1983) and atypical (motile, growth at 37°C) *A. salmonicida* isolates are known to exist (Austin, 1993; McIntosh and Austin, 1991). Likewise, the production of a brown soluble pigment on tryptic soy agar alone should not be considered a typical taxonomic trait for all psychrophilic aeromonads. Few studies have focussed on the development of a reliable and consistent phenotypic differentiation of the five subspecies that are currently recognized in *A. salmonicida* (Popoff, 1984; Austin *et al.*, 1989). One of the most up-to-date identification schemes was reported in the description of *A. salmonicida* subsp. *pectinolytica* (Pavan *et al.*, 2000) based on 14 biochemical and physiological properties (Table 2.3). However, it should be stressed that this table compiles data from different studies that may have used different methodologies including time and temperature parameters. The latter factors are crucial for correct reading of many tests and discrepancies have been noted between such studies (Austin *et al.*, 1989; Dalsgaard *et al.*, 1994).

Taxonomic differentiation of the five subspecies in *A. salmonicida* is further hampered by the occurrence of numerous groups of so-called atypical isolates (Belland and Trust, 1989; Austin and Adams, 1996; Lund *et al.*, 2002). There still is great controversy whether these isolates represent additional subspecies or rather belong in existing subspecies of *A. salmonicida*. Because phenotypic characterization of atypical *A. salmonicida* isolates has often lead to ambiguous taxonomic conclusions, there has been a tendency among workers to follow a polyphasic approach including molecular methods in conjunction with phenotypic characterization.

Mesophilic aeromonads

For many years, the routine identification of mesophilic *Aeromonas* isolates to the species level was based on a limited set of phenotypic tests (Popoff, 1984). These tests primarily

Table 2.3 Phenotypic characteristics for the differentiation of the five subspecies in *A. salmonicida*

	Reaction for *A. salmonicida* subsp.[a]				
Feature	*salmonicida*	*achromogenes*	*masoucida*	*smithia*	*pectinolytica*
Brown diffusible pigment	+	–	–	–	+
Growth at 35–37°C	–	–	–	–	+
Indole	–	+	+	+	+
Voges-Proskauer at 25°C	–	–	+	–	+
Lysine decarboxylase	v	v	v	v	–
KCN	–	–	–	ND	v
Fermentation of:					
D-Sorbitol	v	v	v	v	+
Mannitol	+	–	+	–	+
Sucrose	–	+	+	v	+
Gas from glucose	–	–	+	ND	+

[a]Results compiled from Popoff (1984), Austin *et al.* (1989), and Pavan *et al.* (2000).
+, >85 % positive; –, <15 % positive; v, variable (15–85%); ND, no data.

aimed at the taxonomic discrimination between the *A. hydrophila* complex (comprising HG1, HG2, mesophilic members of HG3, and *A. popoffii*), the *A. caviae* complex (comprising HG4, HG5A/B, and HG6), and the *A. sobria* complex (comprising HG7, HG8, and HG9). However, because of the increasing pathogenic relevance of specific *Aeromonas* HGs in human disease and the rising complexity of *Aeromonas* systematics, identification to the HG level became an important requirement. Initially, this need could only be fulfilled by a limited number of reference laboratories only (Altwegg *et al.*, 1990; Kuijper *et al.*, 1989a). In later studies, Carnahan and co-workers (1991b) and Abbott and colleagues (1992) recommended the use of 20 to 25 selected phenotypic features for the correct taxonomic allocation of mesophilic *Aeromonas* strains to one of the seven major clinical species (i.e. *A. hydrophila, A. caviae, A. veronii* biovars sobria and veronii, *A. jandaei, A. schubertii*, and *A. trota*), or to one of the 13 *Aeromonas* HGs, respectively. However, the latter authors conceded that it was very difficult to define reliable phenotypic markers for the rare taxa HG7 (*A. sobria*) and HG11. Nine biochemical key tests were selected by Abbott and co-workers (1992) as a primary battery to identify mesophilic aeromonads to the HG level, i.e. lysine decarboxylase, ornithine decarboxylase, arginine dihydrolase, esculin hydrolysis, gas from glucose, Voges–Proskauer, and acid production from arabinose, mannitol and sucrose. However, the use of these nine characters still did not allow taxonomic differentiation at HG level within the *A. hydrophila* complex (HGs 1 to 3) and the *A. caviae* complex (HGs 4 to 6). The fact that some species can only be separated by one or two features in these identification tables represents a significant drawback of such phenotypic schemes. In addition, the medium and temperature dependence of some key characteristics such as decarboxylase reactions (Altwegg *et al.*, 1987) may seriously affect the comparability of test results reported in literature. Likewise, Dalsgaard and associates (1998) stressed the need for standardization and harmonization of phenotypic tests in the identification of atypical *A. salmonicida.*

In an attempt to find a higher number of discriminating tests, hundreds of phenotypic characteristics were evaluated for their taxonomic value in *Aeromonas* identification. Based on spectrophotometric detection in microplates, Kämpfer and Altwegg (1992) constructed a probability matrix based on 329 characters for the phenotypic differentiation of 13 *Aeromonas* HGs. However, only three to four tests were recorded as possible HG-specific markers within each of the three major phenotypic complexes, i.e. *A. hydrophila, A. caviae,* and *A. sobria.* In their numerical analysis based on 50 phenotypic characters, Carnahan and Joseph (1993) could distinguish between HGs in the *A. caviae* complex and the *A. sobria* complex, but found that HG2 and HG3 of the *A. hydrophila* complex belonged to the same phenon. According to Hänninen (1994), a set of four biochemical key tests including DL-lactate and urocanic acid utilization and acid production from D-sorbitol and D-rhamnose was sufficient for the reliable taxonomic separation of HG1, HG2, and HG3 in *A. hydrophila.* Other studies (Esteve, 1995; Noterdaeme *et al.*, 1996) on the numerical taxonomy of fish-associated aeromonads have further contributed to the search for additional biochemical and/or physiological tests that are discriminatory at the HG level. Oakey and colleagues (1996) developed a computer-assisted identification program based on 40 biochemical and physiological tests for the discrimination of all *Aeromonas* HGs recognized at that time, and successfully challenged the system with a collection of environmental and clinical isolates. Later, Carson and associates (2001) improved the protocol of Oakey and co-workers (1996) based on 26 miniaturized tests (all incubated at 25°C) and found that 80% of the isolates were correctly allocated to their respective species using a Wilcox probability level of 95%. However, a major shortcoming in the latter system was the fact that representatives of *A. hydrophila* HG3 were not included.

Throughout the 1990s, the use of commercial miniaturized and microplate-based systems in combination with reference libraries became highly popular for phenotypic identification of bacteria. The majority of these systems were designed to facilitate routine identification of bacteria in hospitals and clinical laboratories. Millership (1996) reported that at least 14 commercial systems, most of them available in the USA only, have representatives of the major *Aeromonas* species included in their databases. However, the general experience is that the taxonomic value of these systems is often low to deal with the high phenotypic relatedness that exists between some of the *Aeromonas* taxa. In addition, it has been reported on several occasions that rarer clinical species such as *A. jandaei*, *A. schubertii* and *A. trota* may be misidentified by these systems (Abbott *et al.*, 1992, 1998; Janda and Abbott, 1998; Altwegg, 1999). Using commercial research galleries, Altwegg and Lüthy-Hottenstein (1991) investigated the value of 269 substrates included in API 50 and API ATB strips (API systems) and concluded that less than one third of these provided significant differences between *Aeromonas* HGs. A previous version of the BIOLOG microbial identification system based on spectrophotometric detection of carbon source assimilation through redox chemistry (Biolog, Inc. Hayward, CA, USA) has been used for identification of *Aeromonas* spp. (Maki *et al.*, 1998) using GN2 plates. As with other commercial systems, however, the naming of the identification entries in the BIOLOG database are in some cases not in conjunction with the accepted nomenclatural terminology of the current *Aeromonas* taxonomy (http://www.biolog.com). Vivas and associates (2000) evaluated the performance of the MicroScan WalkAway System in conjunction with the Combo Negative Type 1S panels (Dade Microscan, Inc., West Sacramento, CA, USA) for the identification of motile *Aeromonas* species. Compared to conventional tests, the authors concluded that 80% of the *A. hydrophila* group strains could be identified although no individual results of the constituting HGs were reported. In addition, low identification scores often yielded *Vibrio* species as the first choice result.

Because the number of new *Aeromonas* species has significantly increased in recent years, authors of new species descriptions attempt to determine phenotypic traits of the species nova but often compare these data with results extracted from the original taxonomic descriptions. It has been pointed out by Janda and Abbott (2010) that this practice presents a problem because test methods, test conditions (e.g. inoculation procedure, incubation period and incubation temperature) and reading method (visual versus automated) may differ significantly between studies for the same test. This leads to the situation that phenotypic markers reported to be differential between pairs or groups of species may prove not reproducible in the hands of other workers. Tables with differential characteristics have thus to be interpreted with great caution until a concerted effort is launched to define key tests and test methods under standardized conditions for biochemical identification of *Aeromonas* species (Janda and Abbott, 2010).

Chemotaxonomic methods

Chemotaxonomy refers to the classification of micro-organisms based on chemical analysis of the entire cell or a particular constituent thereof, such as fatty acids, proteins and ubiquinones. To date, almost every chemotaxonomic technique applied in the field of bacterial systematics has been evaluated for its taxonomic potential in the genus *Aeromonas*. Some of these methods were described as promising alternatives for the rather laborious methods used in phenotypic identification, whereas other techniques only seemed valuable towards the characterization of aeromonads to the strain level including serotyping, phagetyping and whole-cell protein fingerprinting.

Multilocus enzyme electrophoresis

By comparative electrophoretic profiling of esterases and dehydrogenases, Picard and Goullet (1985) demonstrated that the resulting electrophoretic data clearly supported the three species concept in *Aeromonas* (i.e. *A. hydrophila, A. caviae* and *A. sobria*). Moreover, these authors found that certain zymotypes correlated with previously established HGs. The three major *Aeromonas* species also constituted separate groups in the multilocus enzyme electrophoresis (MLEE) study of Tonolla and co-workers (1991), who concluded that clinical strains could be easily distinguished from environmental isolates in each of the species examined. A good taxonomic correlation between MLEE data and DNA–DNA hybridization results has been reported (Altwegg *et al.*, 1991), but the latter authors argued that automatization of the MLEE method was required for its routine use as an identification tool.

Fatty acid analysis

The use of gas–liquid chromatographic (GLC) analysis of cellular fatty acid methyl esters (FAMEs) to differentiate among *Aeromonas* sp. was first studied by Canonica and Pisano (1988), who reported that the ratio of the relative amounts of the fatty acids 16:0 and 17:0 was useful to separate the species *A. hydrophila, A. caviae* and *A. sobria*. In addition to these three species, Hansen and colleagues (1991) also included strains of *A. media, A. veronii* and *A. schubertii* in their FAME study. In the latter report, the FAME profiles obtained from the different species looked very similar, from which the authors concluded that fatty acid analysis had little potential for identification of mesophilic aeromonads. A far more extended FAME study, encompassing strains of *Aeromonas* HG1 to HG13, was performed by Kämpfer and co-workers (1994). Although some minor differences between several HGs could be observed, Kämpfer and co-workers concluded that a clear-cut taxonomic differentiation of all *Aeromonas* HGs was not possible. In contrast to the latter study, Huys and associates (1994) reported that GLC analysis of FAMEs (GLC-FAME) proved to be a useful chemotaxonomic marker to differentiate the majority of the phenospecies and HGs in the genus *Aeromonas* recognized at that time. It was later shown that the nutritional value of the growth medium significantly affected the cellular fatty acid composition of aeromonads and thus had consequences for the differentiation of HGs in *Aeromonas*. In this regard, Huys and colleagues (1997d) recommended the use of nutrient-rich trypticase soy agar above the nutrient-poor R2A medium previously used by Kämpfer and co-workers (1994). In the commercial TSBA50 (version 5.0) database supplied with the automated Sherlock Microbial Identification System (MIDI, Inc., Newark, DE, USA) which is used for the universal identification of aerobes, only seven validated *Aeromonas* species are included. For this reason, several workers have experienced that identification of unknown aeromonads with the TSBA database often leads to a failure in differentiating *Aeromonas* isolates at the genotypic (HG) level (Kämpfer *et al.*, 1994; Huys *et al.*, 1995).

Combined with a laboratory-based identification library, the GLC-FAME method was successfully used as an identification method in ecological studies in which ≥70% of the aquatic (Huys *et al.*, 1995; Kersters *et al.*, 1995), clinical (Kühn *et al.*, 1997), and food (Bin

Kingombe *et al.*, 1999; Neyts *et al.*, 2000) *Aeromonas* isolates under study could be reliably assigned to a known species.

Pyrolysis mass spectrometry

For use in bacterial taxonomy, pyrolysis mass spectrometry (PyMS) does not focus on one specific cellular component but gives information on the whole-cell composition. The results of Magee and colleagues (1993) showed that each of the 'biochemical' *Aeromonas* species, delineated on the basis of 48 conventional tests, corresponded to several PyMS groups. As suggested by the authors, it is possible that the PyMS grouping reflected relationships beyond the phenospecies level. Interestingly, *Aeromonas* strains with a possible pathogenic background were roughly segregated from animal and environmental isolates with PyMS.

MALDI mass spectrometry

Donohue and colleagues (2006) were the first to explore the taxonomic potential of the MALDI-MS method in the genus *Aeromonas*. Using strains of 17 species, the MALDI-MS spectra were found to yield sufficient resolving power to effectively discriminate between *Aeromonas* species. In a follow-up study (Donohue *et al.*, 2007), the strain panel was extended with the purpose to construct an identification library that was tested in two blind groups. In group 1, it was demonstrated that all unknown water isolates with a typical biochemical profile were correctly identified with the MALDI-MS database. In group 2, which harboured strains with atypical phenotypic profiles, 18 out of 27 isolates were assigned to the correct species. The latter study also reported that cultures of the same *Aeromonas* reference strain obtained from different collections yielded MALDI-MS profiles that were highly comparable.

A MALDI-TOF MS database constructed from 92 genotypically well-characterized *Aeromonas* strains achieved a taxonomic resolution at species level that was comparable to that offered by the housekeeping gene *gyrB* (Benagli *et al.*, 2012). The database was evaluated for the identification of 741 clinical and environmental *Aeromonas* isolates, of which 93% could be correctly identified at species level. The remaining 7% of unidentified isolates primarily belonged to more recently described and rarely isolated species not yet covered by the MALDI-TOF MS database. In line with these findings Lamy and co-workers (2011) showed that, compared to *rpoB* gene sequencing, MALDI-TOF MS reaches >90% species-level accuracy for the identification of taxonomic reference strains and unknown environmental and clinical isolates. In recent years, MALDI-TOF MS has been used as a taxonomic tool in support of the description of novel *Aeromonas* species such as *A. aquariorum* (Martínez-Murcia *et al.*, 2008) and *A. piscicola* (Beaz-Hidalgo *et al.*, 2009). Likely, MALDI-TOF MS is currently one of the most accurate and rapid methods for chemotaxonomic identification of aeromonads.

Genomic methods

The genomic methods used in *Aeromonas* taxonomy are either targeting the entire genome (DNA–DNA hybridizations and AFLP), the ribosomal gene operon (16S rRNA gene sequence analysis, ribotyping, 16S rRNA RFLP-PCR and oligonucleotide probes) or housekeeping genes (single or multilocus sequence analysis). For discussion of the taxonomic value of 16S rRNA gene sequencing, the reader is referred to the 'Phylogeny' section of this chapter.

Microplate DNA–DNA hybridization

Historically, DNA–DNA hybridization has always been regarded as a labour-intensive and laborious method for the classification of new bacterial taxa that is only available in a limited number of taxonomic expert labs worldwide (De Ley *et al.*, 1970; Grimont *et al.*, 1980; Huss *et al.*, 1983). Especially the need for large scale DNA preparations made this method very unpractical. Inspired by the use of microplates in biochemical fingerprinting, the latter bottleneck was overcome by the description of the fluorometric DNA–DNA hybridization method in microdilution wells (Ezaki *et al.*, 1989) which required much less DNA, allowed to determine significantly more combinations in the DNA–DNA hybridization matrix and was much faster than the classical optimal renaturation, S1 nuclease and hydroxyapatite methods.

By using the microplate hybridization method, Sugita and colleagues (1994a) were able to allocate all 65 *Aeromonas* fish isolates examined in their study to a particular species. The potential of this technique to taxonomically distinguish between *Aeromonas* HGs, however, was not discussed. In a follow-up study (Sugita *et al.*, 1995), the authors investigated 906 presumptive *Aeromonas* strains from fish origin, and found that 33% of these strains could not be identified to the species level. However, it should be noted that the authors included the type strains of only five *Aeromonas* species. Kaznowski (1995) described the application of a colorimetric DNA–DNA hybridization method in microplates for the taxonomic differentiation of aeromonads to the HG level. This study only included one strain (i.e. the type or reference strain) of each of the 14 *Aeromonas* HGs tested. In later studies, Kaznowski (1996, 1997, 1998) reported the use of microplate DNA–DNA hybridizations in conjunction with phenotypic testing for the identification of *Aeromonas* strains from human diarrhoeal stools, fish, and the environment at the genomic species level.

The remarkable lack of congruence between DNA–DNA hybridizations and 16S rRNA gene sequence similarities for several *Aeromonas* species was first described by Martínez-Murcia and colleagues (1992a). The reader is referred to 'Inter-specific phylogeny' above for further discussion of this issue.

Amplified fragment length polymorphism (AFLP) analysis

In 1996, Janssen and co-workers published their results on the evaluation of the Amplified Fragment Length Polymorphism (AFLP) method as a new tool in bacterial taxonomy which included the genus *Aeromonas* as one of the test cases. Basically, the AFLP method involves (i) the digestion of total genomic DNA by two restriction enzymes, (ii) ligation of restriction halfsite-specific adapters to all restriction fragments, and (iii) selective amplification of a subset of these fragments using two specific PCR primers. Subsequent to electrophoretic separation of the amplification products on a polyacrylamide gel, typical AFLP patterns are visualized by autoradiography or fluorescence. Huys and co-workers (1996b, 1999) demonstrated that AFLP using radionucleotide or fluorescent labelling proved to be a taxonomic tool with high discriminatory power in the genus *Aeromonas*. Numerical analysis of AFLP band patterns also provided or confirmed a number of new insights concerning the taxonomy of the species *A. eucrenophila*, *A. encheleia*, *Aeromonas* HG11, *A. ichthiosmia*, *A. enteropelogenes*, *A. allosaccharophila* and *A. culicicola* (Huys *et al.*, 1996a,b, 1997b,c, 2001b, 2002a, 2005; Miñana-Galbis *et al.*, 2004) and initiated the description of several new (sub) species including *A. popoffii* (Huys *et al.*, 1997c), *A. hydrophila* subsp. *dhakensis* (Huys *et al.*, 2002b), *A. hydrophila* subsp. *ranae* (Huys *et al.*, 2003), *A. molluscorum* (Miñana-Galbis *et al.*, 2004), *A. bivalvium* (Miñana-Galbis *et al.*, 2007), *A. tecta* (Demarta *et al.*, 2004, 2008) and *A. diversa* (Miñana-Galbis *et al.*, 2010a). In cases where AFLP grouping did not correspond to the classification based on DNA–DNA hybridization, the taxonomic value of the technique was questioned by a number of workers (Esteve, 1997; Huys *et al.*, 1997a; Martínez-Murcia, 1999). Rademaker and colleagues (2000), on the other hand, concluded from different mathematical and statistic approaches that AFLP fingerprints truly reflect the genotypic relationships among xanthomonads in high correlation with DNA–DNA hybridization.

In combination with user-generated local databases, AFLP has been employed in a range of studies to determine the taxonomic identity of aquatic (Huys *et al.*, 1996c, 2000, 2001a), sewage (Rahman *et al.*, 2007), clinical (Kühn *et al.*, 1997) and fish (Rahman *et al.*, 2002) isolates of *Aeromonas*. In addition to its taxonomic use, AFLP has also been applied as a typing tool which is not discussed here.

Ribotyping

The use of rRNA gene restriction patterns (i.e. ribotyping) as taxonomic tools for the genus *Aeromonas* has been relatively successful. The technique essentially consists of the digestion of genomic DNA by a restriction enzyme, transfer of the restriction fragments to a membrane, and hybridization with a labelled plasmid that contains the *rrnB* operon of *Escherichia coli*. Next to their use in typing studies which are not discussed here, Martinetti Lucchini and Altwegg (1992) determined that ribotyping patterns could also be correlated with taxonomic groups in *Aeromonas* when only a 567 bp restriction fragment from this operon was used as a probe. In the latter study, discrimination among 12 *Aeromonas* HGs was possible by visual comparison of banding patterns. However, Altwegg (1993) argued that the existing ribotyping methodology could be used only for research purposes but not in diagnostic laboratories. Hänninen and Siitonen (1995) successfully used the ribotyping method for the identification to the HG level of aeromonads from aquatic, food, and human clinical origin. Ribotyping was also implicated in the species descriptions of *A. bestiarum* (Ali *et al.*, 1996), *A. popoffii* (Huys *et al.*, 1997c) and *A. tecta* (Demarta *et al.*, 2004, 2008), and in a polyphasic study on the taxonomic interrelationships between *A. eucrenophila*, *A. encheleia*, and *Aeromonas* HG11 (Huys *et al.*, 1997b). Nsabimana and colleagues (2000) used the ribotyping technique in conjunction with biochemical tests to determine the genomospecies diversity of aeromonads in activated sludge.

RFLP-PCR of 16S rRNA

Based on computer analysis of published 16S rRNA gene sequences, Borrell and associates (1997) described RFLP analysis of PCR-amplified 16S rRNA genes as a promising method to discriminate among all known *Aeromonas* species. The authors reported that the use of the endonucleases *Alu*I and *Mbo*I allowed taxonomic discrimination of most species pairs except among *A. salmonicida*, *A. bestiarum* and *Aeromonas* HG11. Separation of the latter taxa required the enzyme combination *Nar*I and *Hae*II. As only one type or reference strain per *Aeromonas* taxon was included in the study of Borrell and co-workers (1997), Graf (1999) subsequently examined the validity of RFLP-PCR of 16S rRNA genes in *Aeromonas* classification using an extended collection of 62 reference strains. In the latter study, it was pointed out that *A. veronii* biovar sobria strains may produce diverse RFLP-PCR patterns leading to possible misidentifications and the further need to perform biochemical characterization.

In response, Figueras and co-workers (2000) attributed Graf's observations to an arbitrary choice instead of a computer simulated choice of the restriction enzymes, and meanwhile extended the method originally described by Borrell and colleagues (1997) with the endonucleases *Alw*NI and *Pst*I. The new 16S rRNA RFLP-PCR method was used to support the taxonomic delineation of *A. fluvialis* (Figueras *et al.*, 2000; Alperi *et al.*, 2010b).

Species-specific probes and primers

As stressed by Millership (1996), the specificity and selectivity of molecular probes as taxonomic tools in *Aeromonas* should be thoroughly investigated before adopting their widespread use. Initiated by the 16S rRNA sequencing work of Martínez-Murcia and co-workers (1992a), a number of specific primers were designed for the identification of specific *Aeromonas* species in a PCR assay. In this way, 16S rDNA-based species-specific probes have been constructed for *A. jandaei, A. schubertii, A. hydrophila, A. veronii, A. caviae* and *A. trota* (Ash *et al.*, 1993a,b; Dorsch *et al.*, 1994; Khan and Cerniglia, 1997). Demarta and colleagues (1999) described a signature region within the 16S rDNA sequence of *A. popoffii* that could be potentially used in hybridization assays for the specific detection of this species. The 16S-23S intergenic spacer regions of seven *Aeromonas* species were used by Kong and associates (1999) to develop species-specific primers for PCR-based monitoring of *Aeromonas* spp. in seawater. In other studies, DNA sequences derived from specific structural genes including the surface array protein gene *vapA* of *A. salmonicida* (Gustafson *et al.*, 1992), the extracellular lipase gene *lip* of *A. hydrophila* (Cascon *et al.*, 1996) and the haemolysin (aerolysin) gene *aerA* of *A. trota* (Husslein *et al.*, 1992; Khan *et al.*, 1999) were also successfully used in the identification of these respective species. Taxonomic probes based on randomly obtained DNA fragments have been designed for *A. salmonicida* (Hiney *et al.*, 1992; Oakey *et al.*, 1998) and *A. hydrophila* HG1 (Oakey *et al.*, 1999).

Single or multilocus sequence analysis of housekeeping genes

As alternatives for the 16S rRNA gene, *gyrB* (encoding the B subunit of DNA gyrase, a type II DNA topoisomerase) and *rpoD* (encoding the s70 factor, one of the sigma factors that confers promoter-specific transcription initiation on RNA polymerase) were the first housekeeping genes analysed in the genus *Aeromonas* (Yáñez *et al.*, 2003; Soler *et al.*, 2004). Since these first studies, one or both of these genes have been included as taxonomic support in the description of many new species including *A. aquariorum* (Martínez-Murcia *et al.*, 2008), *A. tecta* (Demarta *et al.*, 2008), *A. piscicola* (Beaz-Hidalgo *et al.*, 2009), *A. diversa* (Miñana-Galbis *et al.*, 2010a), *A. rivuli* (Figueras *et al.*, 2011), *A. fluvialis* (Alperi *et al.*, 2010b), *A. taiwanensis* (Alperi *et al.*, 2010a) and *A. sanarellii* (Alperi *et al.*, 2010a). In addition, sequence analysis of *gyrB* and *rpoD* have also been used to investigate the position of species with unclear or controversial phylogenetic affiliations (Saavedra *et al.*, 2006). In combination with the 16S rRNA and *gyrB* genes, *rpoB* (encoding the β subunit of DNA-dependent RNA polymerase) was proposed as a multilocus strategy to study taxonomic and phylogenetic relationships in the genus *Aeromonas* (Küpfer *et al.*, 2006). Partial *rpoB* gene sequence analysis was later validated as a single-locus approach for identification of environmental *Aeromonas* isolates (Lamy *et al.*, 2010). Other housekeeping genes that have been proposed for taxonomic differentiation of *Aeromonas* species include the genes *dnaJ* (encoding heat-shock protein 40; Nhung *et al.*, 2007), *recA* (encoding the RecA protein which has important functions in homologous DNA recombination, DNA damage repair and induction of the SOS response; Sepe *et al.*, 2008), *cpn60* (encoding the type I chaperonin Cpn60; Miñana-Galbis *et al.*, 2009) and *mdh* (encoding malate dehydrogenase; Farfán *et al.*, 2010).

Martínez-Murcia and co-workers (2011) pointed to the potential risk of only using a single gene to study taxonomic relationships in genera such as *Aeromonas* where recombination may be relatively frequent (Sneath, 1993; Morandi *et al.*, 2005). To compensate for these genomic rearrangement effects, a multilocus phylogenetic analysis (MLPA) scheme was proposed based on concatenated sequences of seven housekeeping gene fragments (*gyrB, rpoD, recA, dnaJ, gyrA, dnaX*, and *atpD*) totalling 4705 bp. The MLPA approach allowed phylogenetic delineation of all currently described *Aeromonas* species with uniformly high bootstrap values (Martínez-Murcia *et al.*, 2011). Likewise, a combination of sequences derived from the genes *cpn60, dnaJ, gyrB* and *rpoD* was used to verify the identity of strain CIP 57.50 (=ATCC 7965=NCTC 7812), an important taxonomic reference in the genus *Aeromonas* (Miñana-Galbis *et al.*, 2010b). In combination with phenotypic characterization, this multigene approach resulted in the reclassification of *A. hydrophila* CIP 57.50 as *A. salmonicida* CIP 57.50.

Conclusion and perspectives

Despite the fact that DNA–DNA hybridizations still form an integral part of new species descriptions, this chapter has clearly highlighted that sequence analysis of 16S rRNA and housekeeping genes currently hold the key to construct an enduring phylogenetic backbone in *Aeromonas* taxonomy. Sequencing technology has never been more accessible than in the present day time. Because of their universality and exchangeability, the resulting sequence data allow to create virtual web-based platforms which can be accessed by every taxonomist with a minimal requirement of computing technology. Already, the next step in this technological (r)evolution is awaiting us. The average amino acid identity (AAI) and the average nucleotide identity (ANI) (Konstantinidis and Tiedje, 2005; Rosselló-Mora, 2005) are two parameters resulting from pairwise genome comparisons and averaging the sequence identities of shared orthologous genes (amino acid or nucleotide, respectively) that are already being used to compare whole-genome sequences for taxonomic purposes and, ultimately, to describe new bacterial species. In an abstract presented at the 15th International Congress on Infectious Diseases (Bangkok, Thailand, 13–16 June 2012), Klinzing and co-workers reported the use of full genome sequencing to determine the phylogenetic position of one environmental and two clinical *Aeromonas* isolates. One of the isolates exhibited ANI values of conserved highly conserved genes ranging from 89.21–93.15% with other *Aeromonas* species, from which the authors concluded that it may represent a potentially new species in this genus. In the near future, more and more of such studies will reach the surface and broaden our current view on how *Aeromonas* systematics should be best approached in the (post-)genomic era.

Despite the ongoing genome-based revolution in bacterial taxonomy, it is important to realize that taxonomic interpretation of sequence data sets of ever increasing size and complexity may not be evident in routine clinical or environmental settings. To some extent, the successful introduction of MALDI-TOF MS technology in such settings may provide an acceptable user-friendly alternative to determine the species identity of new *Aeromonas* isolates. Provided that reference profiles in databases are sufficiently representative for the respective species entries and encompass most if not all validly described *Aeromonas* taxa, MALDI-TOF MS may become important to uncover potentially new *Aeromonas* species. This would especially be the case for clinical and environmental strain collections which are hardly or even not accessible to taxonomists in the field.

Despite the fact that so many tools are at the disposal of *Aeromonas* taxonomists, a number of long-lasting nomenclatural issues in the genus remain unresolved. Although the tradition to allocate new *Aeromonas* taxa to numbered HGs is no longer continued, this classification still remains relevant to indicate taxonomic groups that are still unnamed. For instance, the motile strains of HG3 that phenotypically resemble *A. hydrophila* are one of the most commonly found *Aeromonas* groups in environmental samples, yet still await a proper taxonomic designation. Other issues concern debates about suspected synonymies. Also, as extensively discussed elsewhere in this chapter, there are indications from several independently performed studies that *A. allosaccharophila* and *A. aquariorum* may be synonyms of *A. veronii* and *A. hydrophila* subsp. *dhakensis*, respectively. However, discussions on the species status of both taxa are still ongoing and will probably only find their way out in a multicentre approach.

Finally, two previous recommendations from the ad hoc committee for the re-evaluation of the species definition in bacteriology (Stackebrandt *et al.*, 2002) deserve renewed attention. Although recent practices in the genus *Aeromonas* prove differently, the description of single-strain species should be discouraged. To determine the intra-specific variation and to understand the functional and ecological behaviour of new species, it seems evident that more than one strain of that taxon is required. In the same context, one can predict that the practice to discriminate single-strain species from other species in the same genus on the basis of phenotypic profiles will increasingly suffer from a lack of both biological and diagnostic relevance. Secondly, in the case that new species descriptions are based on multiple strains, genotyping should be mandatory (Stackebrandt *et al.*, 2002). In fact, possible consequences of ignoring this guideline have been clearly illustrated with examples from the genus *Aeromonas* (Figueras *et al.*, 2006). Using ERIC-PCR fingerprinting, the latter authors showed that the species descriptions of *A. culicicola* originally based on three strains (Pidiyar *et al.*, 2002) and of *A. simiae* originally based on two strains (Harf-Monteil *et al.*, 2004) both included two isolates of the same strain type. These practices may create the false impression that new species are proposed on the basis of different strains whereas in reality they contain duplicate data from multiple isolates of the same strain.

In conclusion, it seems clear that the only way forward in *Aeromonas* taxonomy is to embrace new technological developments. At the same time, it is important to remain alert and critical when it comes to translating (often complex) biological data arising from such new methods and approaches to new phylogenetic insights and nomenclatural conclusions. Only in this way, it is safe to assume that our knowledge on the diversity of aeromonads is likely to expand in a reliable and biologically meaningful way.

Acknowledgements

The author would like to express his sincere thanks to Dr Raúl Muñoz for construction of the phylogenetic trees. This chapter is dedicated to Levon Helm (1940–2012), Adam Yauch (1964–2012), Jon Lord (1941–2012), Reg Presley (1941–2013) and Kevin Ayers (1944–2013), for continuous inspiration.

References

Abbott, S.L., Cheung, W.K.W., Kroskebystrom, S., Malekzadeh, T., and Janda, J.M. (1992). Identification of *Aeromonas* strains to the genospecies level in the clinical laboratory. J. Clin. Microbiol. *30*, 1262–1266.

Abbott, S.L., Seli, L.S., Catino, M., Hartley, M.A., and Janda, J.M. (1998). Misidentification of unusual *Aeromonas* species as members of the genus *Vibrio*: a continuing problem. J. Clin. Microbiol. *36*, 1103–1104.

Abbott, S.L., Cheung, W.K.W., and Janda, J.M. (2003). The genus *Aeromonas*: biochemical characteristics, atypical reactions, and phenotypic identification schemes. J. Clin. Microbiol. *41*, 2348–2357.

Ali, A., Carnahan, A.M., Altwegg, M., Lüthy-Hottenstein, J., and Joseph, S.W. (1996). *Aeromonas bestiarum* sp. nov. (formerly genomospecies DNA group 2 *A. hydrophila*), a new species isolated from non-human sources. Med. Microbiol. Lett. *5*, 156–165.

Allen, D.A., Austin, B., and Colwell, R.R. (1983). *Aeromonas media*, a new species isolated from river water. Int. J. Syst. Bacteriol. *33*, 599–604.

Alperi, A., Martínez-Murcia, A.J., Ko, W.C., Monera, A., Saavedra, M.J., and Figueras, M.J. (2010a). *Aeromonas taiwanensis* sp. nov. and *Aeromonas sanarellii* sp. nov., clinical species from Taiwan. Int. J. Syst. Evol. Micr. *60*, 2048–2055.

Alperi, A., Martínez-Murcia, A.J., Monera, A., Saavedra, M.J., and Figueras, M.J. (2010b). *Aeromonas fluvialis* sp nov., isolated from a Spanish river. Int. J. Syst. Evol. Microbiol. *60*, 72–77.

Altwegg, M. (1993). A polyphasic approach to the classification and identification of *Aeromonas* strains. Med. Microbiol. Lett. *2*, 200–205.

Altwegg, M. (1999). *Aeromonas*. In Manual Of Clinical Microbiology, 7th edn, P.R. Murray, E.J. Baron, M.A. Pfaller, F.C. Tenover, and R.H. Yolken, eds (ASM Press, Washington, DC), pp. 507–516.

Altwegg, M., and Lüthy-Hottenstein, J. (1991). Methods for the identification of DNA hybridization groups in the genus *Aeromonas*. Experientia *47*, 403–406.

Altwegg, M., von Graevenitz, A., and Zollingeriten, J. (1987). Medium and temperature-dependence of decarboxylase reactions in *Aeromonas* spp. Curr. Microbiol. *15*, 1–4.

Altwegg, M., Steigerwalt, A.G., Altwegg-Bissig, R., Lüthy-Hottenstein, J., and Brenner, D.J. (1990). Biochemical identification of *Aeromonas* genospecies isolated from humans. J. Clin. Microbiol. *28*, 258–264.

Altwegg, M., Reeves, M.W., Altwegg-Bissig, R., and Brenner, D.J. (1991). Multilocus enzyme analysis of the genus *Aeromonas* and its use for species identification. Zentralbl. Bakteriol. *275*, 28–45.

Ash, C., Martínez-Murcia, A.J., and Collins, M.D. (1993a). Identification of *Aeromonas schubertii* and *Aeromonas jandaei* by using a polymerase chain reaction-probe Test. FEMS Microbiol. Lett. *108*, 151–155.

Ash, C., Martínez-Murcia, A.J., and Collins, M.D. (1993b). Molecular identification of *Aeromonas sobria* by using a polymerase chain reaction-probe test. Med. Microbiol. Lett. *2*, 80–86.

Austin, B. (1993). Recovery of atypical isolates of *Aeromonas salmonicida*, which grow at 37°C, from ulcerated non-salmonids in England. J. Fish Dis. *16*, 165–168.

Austin, B., and Adams, C. (1996). Fish pathogens. In The Genus *Aeromonas*, B. Austin, M. Altwegg, P.J. Gosling, and S.W. Joseph, eds (John Wiley & Sons, Ltd., Chichester, UK), pp. 197–244.

Austin, D.A., McIntosh, D., and Austin, B. (1989). Taxonomy of fish associated *Aeromonas* spp., with the description of *Aeromonas salmonicida* subsp. *smithia* subsp. nov. Syst. Appl. Microbiol. *11*, 277–290.

Baumann, P., Baumann, L., Bang, S.S., and Woolkalis, M.J. (1980). Re-evaluation of the taxonomy of *Vibrio*, *Beneckea*, and *Photobacterium* – Abolition of the genus *Beneckea*. Curr. Microbiol. *4*, 127–132.

Beaz-Hidalgo, R., Alperi, A., Figueras, M.J., and Romalde, J.L. (2009). *Aeromonas piscicola* sp. nov., isolated from diseased fish. Syst. Appl. Microbiol. *32*, 471–479.

Belland, R.J., and Trust, T.J. (1989). *Aeromonas salmonicida* plasmids – Plasmid-directed synthesis of proteins *in vitro* and in *Escherichia coli* minicells. J. Gen. Microbiol. *135*, 513–524.

Benagli, C., Demarta, A., Caminada, A., Ziegler, D., Petrini, O., and Tonolla, M. (2012). A rapid MALDI-TOF MS identification database at genospecies level for clinical and environmental *Aeromonas* strains. PLoS One 7, e48441.

Bin Kingombe, C.I., Huys, G., Tonolla, M., Albert, M.J., Swings, J., Peduzzi, R., and Jemmi, T. (1999). PCR detection, characterization, and distribution of virulence genes in *Aeromonas* spp. Appl. Environ. Microbol. *65*, 5293–5302.

Borrell, N., Acinas, S.G., Figueras, M.J., and Martínez-Murcia, A.J. (1997). Identification of *Aeromonas* clinical isolates by restriction fragment length polymorphism of PCR-amplified 16S rRNA genes. J. Clin. Microbiol. *35*, 1671–1674.

Brown, G.R., Sutcliffe, I.C., and Cummings, S.P. (2001). Reclassification of [*Pseudomonas*] *doudoroffii* (Baumann *et al.* 1983) into the genus *Oceanomonas* gen. nov. as *Oceanomonas doudoroffii* comb. nov., and description of a phenol-degrading bacterium from estuarine water as *Oceanomonas baumannii* sp. nov. Int. J. Syst. Evol. Microbiol. *51*, 67–72.

Cahill, M.M., and Macrae, I.C. (1992). Characteristics of 0/129-sensitive motile *Aeromonas* strains isolated from freshwater on starch-ampicillin agar. Microb. Ecol. *24*, 215–226.

Canonica, F.P., and Pisano, M.A. (1988). Gas-liquid-chromatographic analysis of Fftty-acid methyl-esters of *Aeromonas hydrophila*, *Aeromonas sobria*, and *Aeromonas caviae*. J. Clin. Microbiol. *26*, 681–685.

Carnahan, A.M. (1993). *Aeromonas* taxonomy – a sea of change. Med. Microbiol. Lett. *2*, 206–211.

Carnahan, A.M., and Joseph, S.W. (1993). Systematic assessment of geographically and clinically diverse Aeromonads. Syst. Appl. Microbiol. *16*, 72–84.

Carnahan, A.M., and Altwegg, M. (1996). Taxonomy. In The Genus *Aeromonas*, B. Austin, M. Altwegg, P.J. Gosling, and S. Joseph, eds (Wiley and Sons, Chichester, UK), pp. 1–38.

Carnahan, A., Fanning, G.R., and Joseph, S.W. (1991a). *Aeromonas jandaei* (formerly genospecies DNA group 9 *A. sobria*), a new sucrose-negative species isolated from clinical specimens. J. Clin. Microbiol. *29*, 560–564.

Carnahan, A.M., Behram, S., and Joseph, S.W. (1991b). Aerokey II – a flexible key for identifying clinical *Aeromonas* species. J. Clin. Microbiol. *29*, 2843–2849.

Carnahan, A.M., Chakraborty, T., Fanning, G.R., Verma, D., Ali, A., Janda, J.M., and Joseph, S.W. (1991c). *Aeromonas trota* sp. nov., an ampicillin-susceptible species isolated from clinical specimens. J. Clin. Microbiol. *29*, 1206–1210.

Carson, J., Wagner, T., Wilson, T., and Donachie, L. (2001). Miniaturized tests for computer-assisted identification of motile *Aeromonas* species with an improved probability matrix. J. Appl. Microbiol. *90*, 190–200.

Cascon, A., Anguita, J., Hernanz, C., Sanchez, M., Fernandez, M., and Naharro, G. (1996). Identification of *Aeromonas hydrophila* hybridization group 1 by PCR assays. Appl. Environ. Microbiol. *62*, 1167–1170.

Chester, F.D. (1897). A prelimenary arrangement of the species of the genus *Bacterium*. Contribution to determinative bacteriology. Part I. 9th Annual Report of the Delaware College Acriculture Experiment Station, p. 92.

Chester, F.D. (1901). A Manual of Determinative Bacteriology (Macmillan, New York), p. 235.

Collins, M.D., Martínez-Murcia, A.J., and Cai, J. (1993). *Aeromonas enteropelogenes* and *Aeromonas ichthiosmia* are identical to *Aeromonas trota* and *Aeromonas veronii*, respectively, as revealed by small-subunit ribosomal-RNA sequence-analysis. Int. J. Syst. Bacteriol. *43*, 855–856.

Colwell, R.R., MacDonell, M.T., and De Ley, J. (1986). Proposal to recognize the family *Aeromonadaceae* fam. nov. Int. J. Syst. Bacteriol. *36*, 473–477.

Cousin, S., Brambilla, E., Yang, J., and Stackebrandt, E. (2008). Culturable aerobic bacteria from the upstream region of a karst water rivulet. Int. Microbiol. *11*, 91–100.

Dalsgaard, I., Gudmundsdottir, B.K., Helgason, S., Hoie, S., Thoresen, O.F., Wichardt, U.P., and Wiklund, T. (1998). Identification of atypical *Aeromonas salmonicida*: inter-laboratory evaluation and harmonization of methods. J. Appl. Microbiol. *84*, 999–1006.

Dalsgaard, I., Nielsen, B., and Larsen, J.L. (1994). Characterization of *Aeromonas salmonicida* subsp. *salmonicida* – a comparative study of strains of different geographic origin. J. Appl. Bacteriol. 77, 21–30.

De Ley, J. (1992). The Proteobacteria. Ribosomal RNA cistron similarities and bacterial taxonomy. In The Prokaryotes, A Handbook on the Biology of Bacteria, Ecophysiology, Isolation, Identification and Applications, 2nd edn, A. Balows, H.G. Trüper, M. Dworkin, W. Harder, and K.H. Schleifer, eds (Springer-Verlag, New York, NY), pp. 2111–2140.

De Ley, J., Cattoir, H., and Reynaert, A. (1970). The quantitative measurement of DNA hybridization from renaturation rates. Eur. J. Biochem. *12*, 133–142.

Demarta, A., Tonolla, M., Caminada, A.P., Ruggeri, N., and Peduzzi, R. (1999). Signature region within the 16S rDNA sequences of *Aeromonas popoffii*. FEMS Microbiol. Lett. *172*, 239–246.

Demarta, A., Huys, G., Tonolla, M., Swings, J., and Peduzzi, R. (2004). Polyphasic taxonomic study of '*Aeromonas eucrenophila*-like' isolates from clinical and environmental sources. Syst. Appl. Microbiol. *27*, 343–349.

Demarta, A., Kupfer, M., Riegel, P., Harf-Monteil, C., Tonolla, M., Peduzzi, R., Monera, A., Saavedra, M.J., and Martínez-Murcia, A. (2008). *Aeromonas tecta* sp. nov., isolated from clinical and environmental sources. Syst. Appl. Microbiol. *31*, 278–286.

Donohue, M.J., Smallwood, A.W., Pfaller, S., Rodgers, M., and Shoemaker, J.A. (2006). The development of a matrix-assisted laser desorption/ionization mass spectrometry-based method for the protein fingerprinting and identification of *Aeromonas* species using whole cells. J. Microbiol. Methods *65*, 380–389.

Donohue, M.J., Best, J.M., Smallwood, A.W., Kostich, M., Rodgers, M., and Shoemaker, J.A. (2007). Differentiation of *Aeromonas* isolated from drinking water distribution systems using matrix-assisted laser desorption/ionization-mass spectrometry. Anal. Chem. *79*, 1939–1946.

Dorsch, M., Ashbolt, N.J., Cox, P.T., and Goodman, A.E. (1994). Rapid identification of *Aeromonas* species using 16s rDNA targeted oligonucleotide primers – a molecular approach based on screening of environmental isolates. J. Appl. Bacteriol. *77*, 722–726.

Eardly, B.D., Wang, F.S., and van Berkum, P. (1996). Corresponding 16S rRNA gene segments in *Rhizobiaceae* and *Aeromonas* yield discordant phylogenies. Plant Soil *186*, 69–74.

Eddy, B.P. (1960). Cephalotrichous, fermentative Gram-negative bacteria: the genus *Aeromonas*. J. Appl. Bacteriol. *23*, 216–249.

Eddy, B.P. (1962). Further studies on *Aeromonas*. I. Additional strains and supplementary tests. J. Appl. Bacteriol. *25*, 137–146.

Eddy, B.P., and Carpenter, K.P. (1964). Further studies on *Aeromonas*. II. Taxonomy of *Aeromonas* + C27 strains. J. Appl. Bacteriol. *27*, 96–100.

Edwards, M.L., Lilley, A.K., Timms-Wilson, T.H., Thompson, I.P., and Cooper, I. (2001). Characterisation of the culturable heterotrophic bacterial community in a small eutrophic lake (Priest Pot). FEMS Microbiol. Ecol. *35*, 295–304.

Emmerich, R., and Wiebel, C. (1894). Über eine durch Bakterien erzeugte Seuche unter den Forellen. Arch. Hyg. *21*, 1–21.

Esteve, C. (1995). Numerical taxonomy of *Aeromonadaceae* and *Vibrionaceae* associated with reared fish and surrounding fresh and brackish water. Syst. Appl. Microbiol. *18*, 391–402.

Esteve, C. (1997). Is AFLP fingerprinting a true alternative to the DNA–DNA pairing method to assess genospecies in the genus *Aeromonas*? Int. J. Syst. Bacteriol. *47*, 245–245.

Esteve, C., Gutierrez, M.C., and Ventosa, A. (1995a). *Aeromonas encheleia* sp. nov., isolated from European eels. Int. J. Syst. Bacteriol. *45*, 462–466.

Esteve, C., Gutierrez, M.C., and Ventosa, A. (1995b). DNA relatedness among *Aeromonas allosaccharophila* strains and DNA hybridization groups of the genus *Aeromonas*. Int. J. Syst. Bacteriol. *45*, 390–391.

Esteve, C., Valera, L., Gutierrez, C., and Ventosa, A. (2003). Taxonomic study of sucrose-positive *Aeromonas jandaei*-like isolates from faeces, water and eels: emendation of *A. jandaei* Carnahan *et al.* 1992. Int. J. Syst. Evol. Microbiol. *53*, 1411–1419.

Ewing, W.H., Hugh, R., and Johnson, J.G. (1961). Studies on the *Aeromonas* group (Centers for Disease Control and Prevention, Atlanta GA) pp. 1–37.

Ezaki, T., Hashimoto, Y., and Yabuuchi, E. (1989). Fluorometric deoxyribonucleic acid deoxyribonucleic acid hybridization in microdilution wells as an alternative to membrane-filter hybridization in which radioisotopes are used to determine genetic relatedness among bacterial strains. Int. J. Syst. Bacteriol. *39*, 224–229.

Fanning, G.R., Hickman-Brenner, F.W., Farmer III, J.J., and Brenner, D.J. (1985). DNA relatedness and phenotypic analysis of the genus *Aeromonas*. Abstract at: Annual Meeting American Society for Microbiology (Las Vegas, NV, USA) C116:319.

Farfán, M., Miñana-Galbis, D., Garreta, A., Loren, J.G., and Fuste, M.C. (2010). Malate dehydrogenase: a useful phylogenetic marker for the genus *Aeromonas*. Syst. Appl. Microbiol. *33*, 427–435.

Farmer III, J.J., Hickman-Brenner, F.W., Fanning, G.R., Arduino, M.J., and Brenner, D.J. (1986). Analysis of *Aeromonas* and *Plesiomonas* by DNA–DNA hybridization and phenotype. Abstact P1 at: First International Workshop *Aeromonas* and *Plesiomonas* (Manchester, UK).

Farmer III, J.J., Arduino, M.J., and Hickman-Brenner, F.W. (1992). The genera *Aeromonas* and *Plesiomonas*. In The Prokaryotes, 2nd edn, vol. 3, A. Balows, H.G. Truper, M. Dworkin, W. Harder, and K.-H. Schleifer, eds (Springer-Verlag, New York, NY), pp. 3012–3045.

Felis, G.E., and Dellaglio, F. (2007). On species descriptions based on a single strain: proposal to introduce the status species proponenda (sp pr.). Int. J. Syst. Evol. Microbiol. *57*, 2185–2187.

Figueras, M.J., Soler, L., Chacon, M.R., Guarro, J., and Martínez-Murcia, A.J. (2000). Extended method for discrimination of *Aeromonas* spp. by 16S rDNA RFLP analysis. Int. J. Syst. Evol. Microbiol. *50*, 2069–2073.

Figueras, M.J., Alperi, A., Guarro, J., and Martínez-Murcia, A.J. (2006). Genotyping of isolates included in the description of a novel species should be mandatory. Int. J. Syst. Evol. Microbiol. *56*, 1183–1184.

Figueras, M.J., Alperi, A., Beaz-Hidalgo, R., Stackebrandt, E., Brambilla, E., Monera, A., and Martínez-Murcia, A.J. (2011). *Aeromonas rivuli* sp. nov., isolated from the upstream region of a karst water rivulet. Int. J. Syst. Evol. Microbiol. *61*, 242–248.

Fischer-Romero, C., Tindall, B.J., and Juttner, F. (1996). *Tolumonas auensis* gen. nov., sp. nov., a toluene-producing bacterium from anoxic sediments of a freshwater lake. Int. J. Syst. Bacteriol. *46*, 183–188.

Fox, G.E., Wisotzkey, J.D., and Jurtshuk, P. (1992). How close is close: 16S rRNA sequence identity may not be sufficient to guarantee species identity. Int J Syst Bacteriol *42*, 166–170.

Graf, J. (1999). Diverse restriction fragment length polymorphism patterns of the PCR-amplified 16S rRNA genes in *Aeromonas veronii* strains and possible misidentification of *Aeromonas* species. J. Clin. Microbiol. *37*, 3194–3197.

Grimont, P.A.D., Popoff, M.Y., Grimont, F., Coynault, C., and Lemelin, M. (1980). Reproducibility and correlation study of three deoxyribonucleic acid hybridization procedures. Curr. Microbiol. *4*, 325–330.

Gustafson, C.E., Thomas, C.J., and Trust, T.J. (1992). Detection of *Aeromonas salmonicida* from fish by using polymerase chain-reaction amplification of the virulence surface array protein gene. Appl. Environ. Microbol. *58*, 3816–3825.

Habs, H., and Schubert, R.H. (1962). Uber Die Biochemischen Merkmale Und Die Taxonomische Stellung Von *Pseudomonas shigelloides* (Bader). Zbl. Bakt. Parasit. *186*, 316–327.

Hänninen, M.L. (1994). Phenotypic characteristics of the three hybridization groups of *Aeromonas hydrophila* complex isolated from different sources. J. Appl. Bacteriol. *76*, 455–462.

Hänninen, M.L., and Siitonen, A. (1995). Distribution of *Aeromonas* phenospecies and genospecies among strains isolated from water, foods or from human clinical samples. Epidemiol. Infect. *115*, 39–50.

Hansen, W., Freney, J., Labbe, M., Renaud, F., Yourassowsky, E., and Fleurette, J. (1991). gas-liquid-chromatographic analysis of cellular fatty-acid methyl-esters in *Aeromonas* species. Zentralbl. Bakteriol. *275*, 1–10.

Harf-Monteil, C., Le Fleche, A., Riegel, P., Prevost, G., Bermond, D., Grimont, P.A.D., and Monteil, H. (2004). *Aeromonas simiae* sp nov., isolated from monkey faeces. Int. J. Syst. Evol. Microbiol. *54*, 481–485.

Hickman-Brenner, F.W., MacDonald, K.L., Steigerwalt, A.G., Fanning, G.R., Brenner, D.J., and Farmer, J.J. (1987). *Aeromonas veronii*, a new ornithine decarboxylase-positive species that may cause diarrhea. J. Clin. Microbiol. *25*, 900–906.

Hickman-Brenner, F.W., Fanning, G.R., Arduino, M.J., Brenner, D.J., and Farmer III, J.J. (1988a). *Aeromonas* Group 501, a new mannitol-negative species found in human clinical specimens. Abstract P-20 at: Second International Workshop on *Aeromonas* and *Plesiomonas* (Miami Beach, FL).

Hickman-Brenner, F.W., Fanning, G.R., Arduino, M.J., Brenner, D.J., and Farmer, J.J. (1988b). *Aeromonas schubertii*, a new mannitol-negative species found in human clinical specimens. J. Clin. Microbiol. *26*, 1561–1564.

Hiney, M., Dawson, M.T., Heery, D.M., Smith, P.R., Gannon, F., and Powell, R. (1992). DNA probe for *Aeromonas salmonicida*. Appl. Environ. Microbiol. *58*, 1039–1042.

Huss, V.A.R., Festl, H., and Schleifer, K.H. (1983). Studies on the spectrophotometric determination of dna hybridization from renaturation rates. Syst. Appl. Microbiol. *4*, 184–192.

Husslein, V., Chakraborty, T., Carnahan, A., and Joseph, S.W. (1992). Molecular studies on the aerolysin gene of *Aeromonas* species and discovery of a species-specific probe for *Aeromonas trota* species nova. Clin. Infect. Dis. *14*, 1061–1068.

Huys, G., Vancanneyt, M., Coopman, R., Janssen, P., Falsen, E., Altwegg, M., and Kersters, K. (1994). Cellular fatty-acid composition as a chemotaxonomic marker for the differentiation of phenospecies and hybridization groups in the genus *Aeromonas*. Int. J. Syst. Bacteriol. *44*, 651–658.

Huys, G., Kersters, I., Vancanneyt, M., Coopman, R., Janssen, P., and Kersters, K. (1995). Diversity of *Aeromonas* sp. in Flemish drinking-water production plants as determined by gas-liquid-chromatographic analysis of cellular fatty-acid methyl-esters (Fames). J. Appl. Bacteriol. *78*, 445–455.

Huys, G., Altwegg, M., Hanninen, M.L., Vancanneyt, M., Vauterin, L., Coopman, R., Torck, U., Lüthy-Hottenstein, J., Janssen, P., and Kersters, K. (1996a). Genotypic and chemotaxonomic description of two subgroups in the species *Aeromonas eucrenophila* and their affiliation to *A. encheleia* and *Aeromonas* DNA hybridization group 11. Syst. Appl. Microbiol. *19*, 616–623.

Huys, G., Coopman, R., Janssen, P., and Kersters, K. (1996b). High-resolution genotypic analysis of the genus *Aeromonas* by AFLP fingerprinting. Int. J. Syst. Bacteriol. *46*, 572–580.

Huys, G., Kersters, I., Coopman, R., Janssen, P., and Kersters, K. (1996c). Genotypic diversity among *Aeromonas* isolates recovered from drinking water production plants as revealed by AFLP(TM) analysis. Syst. Appl. Microbiol. *19*, 428–435.

Huys, G., Janssen, P., and Kersters, K. (1997a). Is AFLP fingerprinting a true alternative to the DNA–DNA pairing method to assess genospecies in the genus *Aeromonas*? Reply. Int. J. Syst. Bacteriol. *47*, 245–246.

Huys, G., Kämpfer, P., Altwegg, M., Coopman, R., Janssen, P., Gillis, M., and Kersters, K. (1997b). Inclusion of *Aeromonas* DNA hybridization group 11 in *Aeromonas encheleia* and extended descriptions of the species *Aeromonas eucrenophila* and *A. encheleia*. Int. J. Syst. Bacteriol. *47*, 1157–1164.

Huys, G., Kämpfer, P., Altwegg, M., Kersters, I., Lamb, A., Coopman, R., Lüthy-Hottenstein, J., Vancanneyt, M., Janssen, P., and Kersters, K. (1997c). *Aeromonas popoffii* sp. nov., a mesophilic bacterium isolated from drinking water production plants and reservoirs. Int. J. Syst. Bacteriol. *47*, 1165–1171.

Huys, G., Kämpfer, P., Vancanneyt, M., Coopman, R., Janssen, P., and Kersters, K. (1997d). Effect of the growth medium on the cellular fatty acid composition of aeromonads: consequences for the chemotaxonomic differentiation of DNA hybridization groups in the genus *Aeromonas*. J. Microbiol. Methods *28*, 89–97.

Huys, G., and Swings, J. (1999). Evaluation of a fluorescent amplified fragment length polymorphism (FAFLP) methodology for the genotypic discrimination of *Aeromonas* taxa. FEMS Microbiol. Lett. *177*, 83–92.

Huys, G., Rhodes, G., McGann, P., Denys, R., Pickup, R., Hiney, M., Smith, P., and Swings, J. (2000). Characterization of oxytetracycline-resistant heterotrophic bacteria originating from hospital and freshwater fishfarm environments in England and Ireland. Syst. Appl. Microbiol. *23*, 599–606.

Huys, G., Gevers, D., Temmerman, R., Cnockaert, M., Denys, R., Rhodes, G., Pickup, R., McGann, P., Hiney, M., Smith, P., *et al.* (2001a). Comparison of the antimicrobial tolerance of oxytetracycline-resistant heterotrophic bacteria isolated from hospital sewage and freshwater fishfarm water in Belgium. Syst. Appl. Microbiol. *24*, 122–130.

Huys, G., Kämpfer, P., and Swings, J. (2001b). New DNA–DNA hybridization and phenotypic data on the species *Aeromonas ichthiosmia* and *Aeromonas allosaccharophila*: *A. ichthiosmia* Schubert *et al.* 1990 is a later synonym of *A. veronii* Hickman-Brenner *et al.* 1987. Syst. Appl. Microbiol. *24*, 177–182.

Huys, G., Denys, R., and Swings, J. (2002a). DNA–DNA reassociation and phenotypic data indicate synonymy between *Aeromonas enteropelogenes* Schubert *et al.* 1990 and *Aeromonas trota* Carnahan *et al.* 1991. Int. J. Syst. Evol. Microbiol. *52*, 1969–1972.

Huys, G., Kämpfer, P., Albert, M.J., Kühn, I., Denys, R., and Swings, J. (2002b). *Aeromonas hydrophila* subsp. *dhakensis* subsp nov., isolated from children with diarrhoea in Bangladesh, and extended description of *Aeromonas hydrophila* subsp. *hydrophila* (Chester 1901) Stanier 1943 (Approved Lists 1980). Int. J. Syst. Evol. Microbiol. *52*, 705–712.

Huys, G., Pearson, M., Kämpfer, P., Denys, R., Cnockaert, M., Inglis, V., and Swings, J. (2003). *Aeromonas hydrophila* subsp. *ranae* subsp. nov., isolated from septicaemic farmed frogs in Thailand. Int. J. Syst. Evol. Microbiol. *53*, 885–891.

Huys, G., Cnockaert, M., and Swings, J. (2005). *Aeromonas culicicola* Pidiyar *et al.* 2002 is a later subjective synonym of *Aeromonas veronii* Hickman-Brenner *et al.* 1987. Syst. Appl. Microbiol. *28*, 604–609.

Janda, J.M., and Abbott, S.L. (1998). Evolving concepts regarding the genus *Aeromonas*: an expanding panorama of species, disease presentations, and unanswered questions. Clin. Infect. Dis. *27*, 332–344.

Janda, J.M., and Abbott, S.L. (2010). The genus *Aeromonas*: taxonomy, pathogenicity, and infection. Clin. Microbiol. Rev. *23*, 35–73.

Janssen, P., Coopman, R., Huys, G., Swings, J., Bleeker, M., Vos, P., Zabeau, M., and Kersters, K. (1996). Evaluation of the DNA fingerprinting method AFLP as a new tool in bacterial taxonomy. Microbiology *142*, 1881–1893.

Joseph, S.W., and Carnahan, A.M. (1994). The isolation, identification, and systematics of the motile *Aeromonas* species. Ann. Rev. Fish Dis. *4*, 315–343.

Joseph, S.W., Carnahan, A.M., Brayton, P.R., Fanning, G.R., Almazan, R., Drabick, C., Trudo, E.W., and Colwell, R.R. (1991). *Aeromonas jandaei* and *Aeromonas veronii* dual infection of a human wound following aquatic exposure. J. Clin. Microbiol. *29*, 565–569.

Kämpfer, P., and Altwegg, M. (1992). Numerical classification and identification of *Aeromonas* genospecies. J. Appl. Bacteriol. *72*, 341–351.

Kämpfer, P., Blasczyk, K., and Auling, G. (1994). Characterization of *Aeromonas* genomic species by using quinone, polyamine, and fatty acid patterns. Can. J. Microbiol. *40*, 844–850.

Kämpfer, P., Erhart, R., Beimfohr, C., Bohringer, J., Wagner, M., and Amann, R. (1996). Characterization of bacterial communities from activated sludge: culture-dependent numerical identification versus *in situ* identification using group- and genus-specific rRNA-targeted oligonucleotide probes. Microb. Ecol. *32*, 101–121.

Karlyshev, A.V., and MacIntyre, S. (1996). Study of the intergenic exeF-exeG region and its application as a simple preliminary test for *Aeromonas* spp. FEMS Microbiol. Lett. *137*, 37–44.

Kaznowski, A. (1995). A method of colorimetric DNA–DNA hybridization in microplates with covalently immobilized dna for identification of *Aeromonas* spp. Med. Microbiol. Lett. *4*, 362–369.

Kaznowski, A. (1996). Application of a colorimetric DNA–DNA hybridization method for identification of *Aeromonas* genomic species, isolated from diarrheal stools. Folia Microbiol. *41*, 505–509.

Kaznowski, A. (1997). Numerical taxonomy and DNA–DNA hybridizations of *Aeromonas* strains isolated from human diarrhoeal stool, fish and environment. Syst. Appl. Microbiol. *20*, 458–467.

Kaznowski, A. (1998). Identification of *Aeromonas* strains of different origin to the genomic species level. J. Appl. Microbiol. *84*, 423–430.

Kersters, I., Vanvooren, L., Huys, G., Janssen, P., Kersters, K., and Verstraete, W. (1995). Influence of temperature and process technology on the occurrence of *aeromonas* species and hygienic indicator organisms in drinking water production plants. Microb. Ecol. *30*, 203–218.

Khan, A.A., and Cerniglia, C.E. (1997). Rapid and sensitive method for the detection of *Aeromonas caviae* and *Aeromonas trota* by polymerase chain reaction. Lett. Appl. Microbiol. *24*, 233–239.

Khan, A.A., Nawaz, M.S., Khan, S.A., and Cerniglia, C.E. (1999). Identification of *Aeromonas trota* (hybridization group 13) by amplification of the aerolysin gene using polymerase chain reaction. Mol. Cell Probes *13*, 93–98.

Kimura, T. (1969). A new subspecies of *Aeromonas salmonicida* as an etiological agent of furunculosis on 'Sakuramasu' (*Oncorhynchus masou*) and Pink Salmon (*O. gorbuscha*) rearing for maturity. Part 1. On the morphological and physiological properties. Part 2. On the serological properties. Fish Pathol. (Tokyo). *3*, 45–52.

Kita-Tsukamoto, K., Oyaizu, H., Nanba, K., and Simidu, U. (1993). Phylogenetic relationships of marine bacteria, mainly members of thc family *Vibrionaceae*, determined on the basis of 16S rRNA sequences. Int. J. Syst. Bacteriol. *43*, 8–19.

Klinzing, D., Matias, R., Choi, S.Y., Hasan, N., Torres, B., Liles, V., Juan, M.L., Alvarez, M., Capistrano, R., Villareal, M.E.J., *et al.* (2012). Discovery of a potentially novel *Aeromonas* species by genome sequencing of isolates from a diarrhoeal outbreak in the Philippines. Abstract ISE.334 at: Fifteenth International Congress on Infectious Diseases (Bangkok, Thailand).

Kluyver, A.J., and van Niel, C.B. (1936). Prospects for a natural system of classification of bacteria. Zentralbl. Bakteriol. Parasitenk. Infektionskr. Hyg. Abt. II *94*, 369–403.

Kong, R.Y.C., Pelling, A., So, C.L., and Wu, R.S.S. (1999). Identification of oligonucleotide primers targeted at the 16S-23S rDNA intergenic spacers for genus- and species-specific detection of aeromonads. Mar. Pollut. Bull. *38*, 802–808.

Konstantinidis, K.T., and Tiedje, J.M. (2005). Genomic insights that advance the species definition for prokaryotes. Proc. Natl. Acad. Sci. U.S.A. *102*, 2567–2572.

Kühn, I., Albert, M.J., Ansaruzzaman, M., Bhuiyan, N.A., Alabi, S.A., Islam, M.S., Neogi, P.K.B., Huys, G., Janssen, P., Kersters, K., *et al.* (1997). Characterization of *Aeromonas* spp isolated from humans with diarrhea, from healthy controls, and from surface water in Bangladesh. J. Clin. Microbiol. *35*, 369–373.

Kuijper, E.J., Steigerwalt, A.G., Schoenmakers, B.S.C.I.M., Peeters, M.F., Zanen, H.C., and Brenner, D.J. (1989). Phenotypic characterization and DNA relatedness in human fecal isolates of *Aeromonas* spp. J. Clin. Microbiol. *27*, 132–138.

Küpfer, M., Kuhnert, P., Korczak, B.M., Peduzzi, R., and Demarta, A. (2006). Genetic relationships of *Aeromonas* strains inferred from 16S rRNA, *gyrB* and *rpoB* gene sequences. Int. J. Syst. Evol. Microbiol. *56*, 2743–2751.

Kvasnikov, E.I., Stepanyuk, V.V., Klyushnikova, T.M., Serpokrylov, N.S., Simonova, G.A., Kasatkina, T.P., and Panchenko, L.P. (1985). A new chromium reducing, gram-variable bacterium with mixed type of flagellation. Microbiology *54*, 69–75.

Lamy, B., Laurent, F., Verdier, I., Decousser, J.W., Lecaillon, E., Marchandin, H., Roger, F., Tigaud, S., de Montclos, H., Kodjo, A., *et al.* (2010). Accuracy of 6 commercial systems for identifying clinical *Aeromonas* isolates. Diagn. Microboil. Infect. Dis. *67*, 9–14.

Lamy, B., Kodjo, A., Laurent, F., and ColBVH Study Group (2011). Identification of *Aeromonas* isolates by matrix-assisted laser desorption ionization time-of-flight mass spectrometry. Diagn. Microbiol. Infect. Dis. *71*, 1–5.

Leclerc, H., and Buttiaux, R. (1962). Frequence des *Aeromonas* dans les eaux d'alimentation. Ann. Inst. Pasteur. *103*, 97–100.

Lin, Y.T., and Shieh, W.Y. (2006). *Zobellella denitrificans* gen. nov., sp. nov. and *Zobellella taiwanensis* sp. nov., denitrifying bacteria capable of fermentative metabolism. Int. J. Syst. Evol. Microbiol. *56*, 1209–1215.

Lund, V., Jenssen, L.M., and Wesmajervi, M.S. (2002). Assessment of genetic variability and relatedness among atypical *Aeromonas salmonicida* from marine fishes, using AFLP-fingerprinting. Dis. Aquat. Organ. *50*, 119–126.

MacInnes, J.I., Trust, T.J., and Crosa, J.H. (1979). Deoxyribonucleic-acid relationships among members of the genus *Aeromonas*. Can. J. Microbiol. *25*, 579–586.

McIntosh, D., and Austin, B. (1991). Atypical characteristics of the salmonid pathogen *Aeromonas salmonicida*. J. Gen. Microbiol. *137*, 1341–1343.

Magee, J.T., Randle, E.A., Gray, S.J., and Jackson, S.K. (1993). Pyrolysis mass spectrometry characterization and numerical taxonomy of *Aeromonas* spp. Anton. Leeuw. *64*, 315–323.

Maki, J.S., Patel, G., and Mitchell, R. (1998). Experimental pathogenicity of *Aeromonas* spp. for the zebra mussel, *Dreissena polymorpha*. Curr. Microbiol. *36*, 19–23.

Martin-Carnahan, A., and Joseph, S.W. (2005). Order XII. Aeromonadales. In Bergey's Manual of Systematic Bacteriology, 2nd edn, vol. 2, part B, D.J. Brenner, N.R. Krieg, and J.T. Staley, eds (Springer, New York, NY), pp. 556–578.

Martinetti-Lucchini, G., and Altwegg, M. (1992). rRNA gene restriction patterns as taxonomic tools for the genus *Aeromonas*. Int. J. Syst. Bacteriol. *42*, 384–389.

Martínez-Murcia, A.J. (1999). Phylogenetic positions of *Aeromonas encheleia*, *Aeromonas popoffii*, *Aeromonas* DNA hybridization Group 11 and *Aeromonas* Group 501. Int. J. Syst. Bacteriol. *49*, 1403–1408.

Martínez-Murcia, A.J., Benlloch, S., and Collins, M.D. (1992a). Phylogenetic interrelationships of members of the genera *Aeromonas* and *Plesiomonas* as determined by 16S ribosomal DNA sequencing: lack of congruence with results of DNA–DNA hybridizations. Int. J. Syst. Bacteriol. *42*, 412–421.

Martínez-Murcia, A.J., Esteve, C., Garay, E., and Collins, M.D. (1992b). *Aeromonas allosaccharophila* sp. nov., a new mesophilic member of the genus *Aeromonas*. FEMS Microbiol. Lett. *91*, 199–206.

Martínez-Murcia, A.J., Soler, L., Saavedra, M.J., Chacon, M.R., Guarro, J., Stackebrandt, E., and Figueras, M.J. (2005). Phenotypic, genotypic, and phylogenetic discrepancies to differentiate *Aeromonas salmonicida* from *Aeromonas bestiarum*. Int. Microbiol. *8*, 259–269.

Martínez-Murcia, A.J., Figueras, M.J., Saavedra, M.J., and Stackebrandt, E. (2007). The recently proposed species *Aeromonas sharmana* sp. nov., isolate GPTSA-6(T), is not a member of the genus *Aeromonas*. Int. Microbiol. *10*, 61–64.

Martínez-Murcia, A.J., Saavedra, M.J., Mota, V.R., Maier, T., Stackebrandt, E., and Cousin, S. (2008). *Aeromonas aquariorum* sp. nov., isolated from aquaria of ornamental fish. Int. J. Syst. Evol. Microbiol. *58*, 1169–1175.

Martínez-Murcia, A., Monera, A., Alperi, A., Figueras, M.J., and Saavedra, M.J. (2009). Phylogenetic evidence suggests that strains of *Aeromonas hydrophila* subsp. *dhakensis* belong to the species *Aeromonas aquariorum* sp. nov. Curr. Microbiol. *58*, 76–80.

Martínez-Murcia, A.J., Monera, A., Saavedra, M.J., Oncina, R., Lopez-Alvarez, M., Lara, E., and Figueras, M.J. (2011). Multilocus phylogenetic analysis of the genus *Aeromonas*. Syst. Appl. Microbiol. *34*, 189–199.

Matte, M.H., *et al.* (1999). Molecular epidemiology of *Aeromonas arequipensis* isolated from diarrhea, in Arequipa, Peru. Abst. P7, 6th International *Aeromonas* and *Plesiomonas* Symposium, Chicago, Illinois, p. 19.

Millership, S.E. (1996). Identification. In The Genus *Aeromonas*, Austin, B. Altwegg, M. Gosling, P.J., and Joseph, S.W. eds (John Wiley & Sons, Ltd., Chichester, UK), pp. 85–108.

Miñana-Galbis, D., Farfan, M., Loren, J.G., and Fuste, M.C. (2002). Biochemical identification and numerical taxonomy of *Aeromonas* spp. isolated from environmental and clinical samples in Spain. J. Appl. Microbiol. *93*, 420–430.

Miñana-Galbis, D., Farfan, M., Fuste, M.C., and Loren, J.G. (2004). *Aeromonas molluscorum* sp. nov., isolated from bivalve molluscs. Int. J. Syst. Evol. Microbiol. *54*, 2073–2078.

Miñana-Galbis, D., Farfan, M., Fuste, M.C., and Loren, J.G. (2007). *Aeromonas bivalvium* sp. nov., isolated from bivalve molluscs. Int. J. Syst. Evol. Microbiol. *57*, 582–587.

Miñana-Galbis, D., Urbizu-Serrano, A., Farfan, M., Fuste, M.C., and Loren, J.G. (2009). Phylogenetic analysis and identification of *Aeromonas* species based on sequencing of the *cpn60* universal target. Int. J. Syst. Evol. Microbiol. *59*, 1976–1983.

Miñana-Galbis, D., Farfan, M., Loren, J.G., and Fuste, M.C. (2010a). Proposal to assign *Aeromonas diversa* sp. nov. as a novel species designation for *Aeromonas* group 501. Syst. Appl. Microbiol. *33*, 15–19.

Miñana-Galbis, D., Farfan, M., Loren, J.G., and Fuste, M.C. (2010b). The reference strain *Aeromonas hydrophila* CIP 57.50 should be reclassified as *Aeromonas salmonicida* CIP 57.50. Int. J. Syst. Evol. Microbiol. *60*, 715–717.

Morandi, A., Zhaxybayeva, O., Gogarten, J.P., and Graf, J. (2005). Evolutionary and diagnostic implications of intragenomic heterogeneity in the 16S rRNA gene in *Aeromonas* strains. J. Bacteriol. *187*, 6561–6564.

Mudaly, D.D., Atkinson, B.W., and Bux, F. (2001). 16S rRNA *in situ* probing for the determination of the family level community structure implicated in enhanced biological nutrient removal. Water Sci. Technol. *43*, 91–98.

Neyts, K., Huys, G., Uyttendaele, M., Swings, J., and Debevere, J. (2000). Incidence and identification of mesophilic *Aeromonas* spp. from retail foods. Lett. Appl. Microbiol. *31*, 359–363.

Nhung, P.H., Hata, H., Ohkusu, K., Noda, M., Shah, M.M., Goto, K., and Ezaki, T. (2007). Use of the novel phylogenetic marker *dnaJ* and DNA–DNA hybridization to clarify interrelationships within the genus *Aeromonas*. Int. J. Syst. Evol. Microbiol. *57*, 1232–1237.

Noterdaeme, L., Bigawa, S., Steigerwalt, A.G., Brenner, D.J., and Ollevier, F. (1996). Numerical taxonomy and biochemical identification of fish associated motile *Aeromonas* spp. Syst. Appl. Microbiol. *19*, 624–633.

Nsabimana, E., Belan, A., and Bohatier, J. (2000). Analysis at the genomospecies level of microbial populations changes in activated sludge: the case of *Aeromonas*. Water Res. *34*, 1696–1704.

Oakey, H.J., Ellis, J.T., and Gibson, L.F. (1996). A biochemical protocol for the differentiation of current genomospecies of *Aeromonas*. Zentralbl. Bakteriol. *284*, 32–46.

Oakey, H.J., Ellis, J.T., and Gibson, L.F. (1998). The development of random DNA probes specific for *Aeromonas salmonicida*. J. Appl. Microbiol. *84*, 37–46.

Oakey, H.J., Gibson, L.F., and George, A.M. (1999). DNA probes specific for *Aeromonas hydrophila* (HG1). J. Appl. Microbiol. *86*, 187–193.

Ogden, I.D., Millar, I.G., Watt, A.J., and Wood, L. (1994). A comparison of three identification kits for the confirmation of *Aeromonas* spp. Lett. Appl. Microbiol. *18*, 97–99.

Pavan, M.E., Abbott, S.L., Zorzopulos, J., and Janda, J.M. (2000). *Aeromonas salmonicida* subsp. *pectinolytica* subsp. nov., a new pectinase-positive subspecies isolated from a heavily polluted river. Int. J. Syst. Evol. Microbiol. *50*, 1119–1124.

Picard, B., and Goullet, P. (1985). Comparative electrophoretic profiles of esterases, and of glutamate, lactate and malate-dehydrogenases, from *Aeromonas hydrophila*, *Aeromonas caviae* and *Aeromonas sobria*. J. Gen. Microbiol. *131*, 3385–3391.

Pidiyar, V., Kaznowski, A., Narayan, N.B., Patole, M., and Shouche, Y.S. (2002). *Aeromonas culicicola* sp. nov., from the midgut of *Culex quinquefasciatus*. Int. J. Syst. Evol. Microbiol. *52*, 1723–1728.

Pidiyar, V.J., Jangid, K., Dayananda, K.M., Kaznowski, A., Gonzalez, J.M., Patole, M.S., and Shouche, Y.S. (2003). Phylogenetic affiliation of *Aeromonas culicicola* MTCC 3249(T) based on *gyrB* gene sequence and PCR-amplicon sequence analysis of cytolytic enterotoxin gene. Syst. Appl. Microbiol. *26*, 197–202.

Popoff, M. (1984). Genus III. *Aeromonas*. Kluyver and Van Niel 1936, 398^{AL}. In Bergey's Manual of Systematic Bacteriology, 1st edn, vol.1, Krieg, N.R., and Holt, J.G. eds (Williams and Wilkins, Baltimore, MA), pp. 545–548.

Popoff, M., and Véron, M. (1976). Taxonomic study of *Aeromonas hydrophila*-*Aeromonas punctata* group. J. Gen. Microbiol. *94*, 11–22.

Popoff, M.Y., Coynault, C., Kiredjian, M., and Lemelin, M. (1981). Polynucleotide sequence relatedness among motile *Aeromonas* species. Curr. Microbiol. *5*, 109–114.

Rademaker, J.L.W., Hoste, B., Louws, F.J., Kersters, K., Swings, J., Vauterin, L., Vauterin, P., and de Bruijn, F.J. (2000). Comparison of AFLP and rep-PCR genomic fingerprinting with DNA–DNA homology studies: *Xanthomonas* as a model system. Int. J. Syst. Evol. Microbiol. *50*, 665–677.

Rahman, M., Colque-Navarro, P., Kühn, I., Huys, G., Swings, J., and Möllby, R. (2002). Identification and characterization of pathogenic *Aeromonas veronii* biovar sobria associated with epizootic ulcerative syndrome in fish in Bangladesh. Appl. Environ. Microb. *68*, 650–655.

Rahman, M., Huys, G., Rahman, M., Albert, M.J., Kühn, I., and Möllby, R. (2007). Persistence, transmission, and virulence characteristics of *Aeromonas* strains in a duckweed aquaculture-based hospital sewage water recycling plant in Bangladesh. Appl. Environ. Microb. 73, 1444–1451.

Romanenko, L.A., Schumann, P., Zhukova, N.V., Rohde, M., Mikhailov, V.V., and Stackebrandt, E. (2003). *Oceanisphaera litoralis* gen. nov., sp. nov., a novel halophilic bacterium from marine bottom sediments. Int. J. Syst. Evol. Microbiol. 53, 1885–1888.

Rosselló-Mora, R. (2005). Updating prokaryotic taxonomy. J. Bacteriol. *187*, 6255–6257.

Ruimy, R., Breittmayer, V., Elbaze, P., Lafay, B., Boussemart, O., Gauthier, M., and Christen, R. (1994). Phylogenetic analysis and assessment of the genera *Vibrio*, *Photobacterium*, *Aeromonas*, and *Plesiomonas* deduced from small-subunit rRNA sequences. Int. J. Syst. Bacteriol. *44*, 416–426.

Saavedra, M.J., Figueras, M.J., and Martínez-Murcia, A.J. (2006). Updated phylogeny of the genus *Aeromonas*. Int. J. Syst. Evol. Microbiol. *56*, 2481–2487.
Saha, P., and Chakrabarti, T. (2006). *Aeromonas sharmana* sp. nov., isolated from a warm spring. Int. J. Syst. Evol. Microbiol. *56*, 1905–1909.
Sanarelli, G. (1891). Über einen neuen Mikroorganismus des Wassers, welcher für Thiere mit veränderlicher und konstanter Temperatur pathogen ist. Zentralbl. Bakteriol. Parasitenk. *9*, 222–228.
Schubert, R.H.W. (1967a). The taxonomy and nomenclature of the genus *Aeromonas* Kluyver and van Niel 1936. I. Suggestions on the taxonomy and nomenclature of the aerogenic *Aeromonas* species. Int. J. Syst. Bacteriol. *17*, 23–37.
Schubert, R.H.W. (1967b). The taxonomy and nomenclature of the genus *Aeromonas* Kluyver and van Niel 1936. II. Suggestions on the taxonomy and nomenclature of the anaerogenic *Aeromonas* species. Int. J. Syst. Bacteriol. *17*, 273–279.
Schubert, R.H.W. (1968). The taxonomy and nomenclature of the genus *Aeromonas* Kluyver and van Niel 1936. III. Suggestions on the definition of the genus *Aeromonas* Kluyver and van Niel 1936. Int. J. Syst. Bacteriol. *18*, 1–7.
Schubert, R.H.W. (1969). Zur Taxonomie von *Aeromonas salmonicida* subsp. *achromogenes* (Smith 1963) Schubert 1967 und *Aeromonas salmonicida* subsp. *masoucida* Kimura 1969. Zentralbl. Bakteriol. I. Orig. *211*, 413–417.
Schubert, R.H.W. (1974). Genus II. *Aeromonas* Kluyver and van Niel, 1936. In Bergey's Manual of Determinative Bacteriology 8th edn, Buchanan, R.E., and Gibbons, N.E., eds (Williams & Wilkins, Baltimore, MA), pp. 345–348.
Schubert, R.H.W. (1987). Ecology of aeromonads and isolation from environmental samples. Experientia *43*, 351–354.
Schubert, R.H.W., and Hegazi, M. (1988). *Aeromonas eucrenophila* species nova *Aeromonas caviae* a later and illegitimate synonym of *Aeromonas punctata*. Zentralbl. Bakteriol. Mikrobiol. Hyg. A *268*, 34–39.
Schubert, R.H.W., Hegazi, M., and Wahlig, W. (1990a). *Aeromonas enteropelogenes* species nova. Hyg. Med. *15*, 471–472.
Schubert, R.H.W., Hegazi, M., and Wahlig, W. (1990b). *Aeromonas ichthiosmia* species nova. Hyg. Med. *15*, 477–479.
Sepe, A., Barbieri, P., Peduzzi, R., and Demarta, A. (2008). Evaluation of *recA* sequencing for the classification of *Aeromonas* strains at the genotype level. Lett. Appl. Microbiol. *46*, 439–444.
Smith, I.W. (1963). The classification of *Bacterium salmonicida*. J. Gen. Microbiol. *33*, 263–274.
Sneath, P.H.A. (1993). Evidence from *Aeromonas* for genetic crossing-over in ribosomal sequences. Int. J. Syst. Bacteriol. *43*, 626–629.
Snieszko, S.F. (1957). Genus IV. *Aeromonas* Kluyver and van Niel, 1936. In Bergey's Manual of Determinative Bacteriology 7nd edn, R.S. Breed, E.G.D. Murray, and N.R. Smith, eds (Williams & Wilkins, Baltimore, MD), p. 189.
Soler, L., Yanez, M.A., Chacon, M.R., Aguilera-Arreola, M.G., Catalan, V., Figueras, M.J., and Martínez-Murcia, A.J. (2004). Phylogenetic analysis of the genus *Aeromonas* based on two housekeeping genes. Int. J. Syst. Evol. Microbiol. *54*, 1511–1519.
Soriano, A.C., Castillo, J.A., Moral, C.H., Salazar, M.S., Marcos, J.Y., and Carrasco, G.N. (1997). RFLP-PCR analysis of the *aroA* gene as a taxonomic tool for the genus *Aeromonas*. FEMS Microbiol. Lett. *156*, 199–204.
Stackebrandt, E., Frederiksen, W., Garrity, G.M., Grimont, P.A.D., Kampfer, P., Maiden, M.C.J., Nesme, X., Rossello-Mora, R., Swings, J., Truper, H.G., *et al.* (2002). Report of the ad hoc committee for the re-evaluation of the species definition in bacteriology. Int. J. Syst. Evol. Microbiol. *52*, 1043–1047.
Staley, T.E., and Colwell, R.R. (1973). Deoxyribonucleic acid reassociation among members of genus *Vibrio*. Int. J. Syst. Bacteriol. *23*, 316–332.
Stanier, R.Y. (1943). A note on the taxonomy of *Proteus hydrophilus*. J. Bacteriol. *46*, 213.
Sugita, H., Nakamura, T., Tanaka, K., and Deguchi, Y. (1994a). Identification of *Aeromonas* species isolated from freshwater fish with the microplate hybridization method. Appl. Environ. Microb. *60*, 3036–3038.
Sugita, H., Nakamura, T., Tanaka, K., and Deguchi, Y. (1994b). Identification of O/129-sensitive *Aeromonas* by the microplate hybridization method. Fisheries Sci. *60*, 351–352.
Sugita, H., Tanaka, K., Yoshinami, M., and Deguchi, Y. (1995). Distribution of *Aeromonas* species in the intestinal tracts of river fish. Appl. Environ. Microb. *61*, 4128–4130.
Tonolla, M., Demarta, A., and Peduzzi, R. (1991). Multilocus genetic relationships between clinical and environmental *Aeromonas* strains. FEMS Microbiol. Lett. *81*, 193–200.

Torres, J.L., Shariff, M., and Law, A.T. (1990). Identification and virulence screening of *Aeromonas* spp. isolated from healthy and epizootic ulcerative syndrome (EUS)-infected fish. Paper presented at: The Second Asian Fisheries Forum (Manila, Philippines).

Véron, M. (1966). Numerical Taxonomy of Vibrios and Some Comparable Bacteria. II. Correlation between Phenotypic Similarities and DNA Base Compositions. Ann. Inst. Pasteur *111*, 671–709.

Vivas, J., Saa, A.I., Tinajas, A., Barbeyto, L., and Rodriguez, L.A. (2000). Identification of motile *Aeromonas* strains with the MicroScan WalkAway system in conjunction with the Combo Negative type 1S panels. Appl. Environ. Microb. *66*, 1764–1766.

Wayne, L.G., Brenner, D.J., Colwell, R.R., Grimont, P.A.D., Kandler, O., Krichevsky, M.I., Moore, L.H., Moore, W.E.C., Murray, R.G.E., Stackebrandt, E., *et al.* (1987). Report of the ad hoc committee on reconciliation of approaches to bacterial systematics. Int. J. Syst. Bacteriol. *37*, 463–464.

Woese, C.R. (1987). Bacterial evolution. Microbiol. Rev. *51*, 221–271.

Yáñez, M.A., Catalan, V., Apraiz, D., Figueras, M.J., and Martínez-Murcia, A.J. (2003). Phylogenetic analysis of members of the genus *Aeromonas* based on *gyrB* gene sequences. Int. J. Syst. Evol. Microbiol. *53*, 875–883.

Yarza, P., Ludwig, W., Euzéby, J., Amann, R., Schleifer, K.H., Glockner, F.O., and Rosselló-Mora, R. (2010). Update of the All-Species Living Tree Project based on 16S and 23S rRNA sequence analyses. Syst. Appl. Microbiol. *33*, 291–299.

Yarza, P., Richter, M., Peplies, J., Euzéby, J., Amann, R., Schleifer, K.H., Ludwig, W., Glockner, F.O., and Rosselló-Mora, R. (2008). The All-Species Living Tree project: a 16S rRNA-based phylogenetic tree of all sequenced type strains. Syst. Appl. Microbiol. *31*, 241–250.

Aeromonas Fish Pathogens

3

Brian Austin

Abstract

An increasing range of aeromonads, including *A. allosaccharophila*, *A. bestiarum*, *A. caviae*, *A. hydrophila*, *A. jandaei*, *A. salmonicida*, *A. schubertii*, *A. sobria* biovar sobria and *A. veronii* biovar sobria, have become associated with disease of predominantly freshwater fish in most countries. The diseases caused include fin/tail rot and generalized septicaemias often with high mortality rates. Diagnosis has progressed from phenotypic and serological approaches to molecular methods, the latter of which may be extremely sensitive and specific. Pathogenicity mechanisms include surface associated, e.g. the surface (S) layer, diffusible, e.g. proteases, and subcellular components, including lipopolysaccharide and outer membrane proteins. Disease control has progressed from use of antimicrobial compounds to include vaccines, probiotics, immunostimulants and plant products.

Introduction

Historically, fish pathogenic aeromonads have been considered to encompass motile and non-motile species, with the former labelled as *Aeromonas hydrophila* and the latter as *A. salmonicida*, which is one of the oldest known fish pathogens. Whereas identification of *A. salmonicida* is comparatively straightforward, confusion has surrounded the early work with motile aeromonads, and it is only comparatively recently that the taxonomic situation has begun to be clarified, and accurate diagnoses achieved. Thus, the early literature makes constant reference to *A. hydrophila* or its antecedents, e.g. *A. liquefaciens* and *A. punctata*, regardless of the true taxonomic position of the isolates. This situation has confused the role of motile aeromonads in fish disease although *A. hydrophila* is clearly an important cause of mortalities in aquaculture. Nevertheless, there has been a recognition that a wider range of motile aeromonads are associated with fish diseases, and the list now includes *A. allosaccharophila*, *A. bestiarum*, *A. caviae*, *A. hydrophila*, *A. jandaei*, *A. schubertii*, *A. sobria* biovar sobria and *A. veronii* biovar sobria (Table 3.1; Austin and Austin, 2007).

The diseases

Motile aeromonads

Generally, the motile aeromonads are associated with external conditions on freshwater fish, namely fin and tail rot, and generalized septicaemia in which haemorrhaging may be

Table 3.1 *Aeromonas* pathogens of fish (modified and updated from Austin and Austin, 2007)

Pathogen	Name of disease	Host range	Recovered from
Aeromonas allosaccharophila		Young eel (= elvers; *Anguilla* spp.)	Spain
Aeromonas bestiarum	–		USA
Aeromonas caviae	Septicaemia	Atlantic salmon (*Salmo salar*)	Turkey
Aeromonas hydrophila (=*Aer. liquefaciens*, *Aer. punctata*)	Haemorrhagic septicaemia, motile *aeromonas* septicaemia, redsore disease, fin rot	Many freshwater fish species	Worldwide
Aeromonas jandaei	–	Eel	Spain
Aeromonas salmonicida (subsp. *achromogenes*, *masoucida*, *salmonicida* and *smithia*) {=*Haemophilus piscium*}	Furunculosis, carp erythrodermatitis, ulcer disease	Salmonids, cyprinids, and marine species (cod, wrasse, and flatfish, e.g. dabs)	Worldwide
Aeromonas sobria	–	Pedicure fish (=*Garra rufa*), perch (*Perca fluvialitis*), gizzard shad (*Dorosoma cepedianum*), tilapia (*Oreochromis niloticus*)	China, Slovakia, Switzerland, USA
Aeromonas schubertii	Tuberculous lesions	Snakehead (*Ophiocephalus argus*)	China
Aeromonas veronii biovar sobria	Epizootic ulcerative syndrome, infectious dropsy	African catfish (*Clarias gariepinus*), rajputi (*Puntius gonionotus*), ruhu (*Labeo rohita*), catla (*Catla catla*), shole (*Channa striatus*), oscar (*Astronotus ocellatus*)	Bangladesh, India

seen extensively on internal organs and musculature (e.g. Ogara *et al.*, 1998). The most recent addition to the list of pathogens, *A. schubertii*, has been associated in China with the development of off-white nodules in the kidney of 18-month-old snakehead (*Ophiocephalus argus*) and mortalities of 45% (Liu and Li, 2012).

Aeromonas salmonicida

There are two forms of disease, namely furunculosis, which is a systemic haemorrhagic septicaemia occurring mostly in salmonids, and ulcerations, which occur in cyprinids (e.g. goldfish ulcer disease) and marine fish. Traditionally, furunculosis is recognized by the presence of lesions resembling boils, i.e. furuncles, in the musculature although this disease sign is not particularly common among affected fish (Austin and Austin, 2007).

Isolation

Primary isolation of cultures from diseased fish involves use of tryptone soya agar (TSA) or brain heart infusion agar (BHIA), which is inoculated with swabbed material, especially from diseased kidney, with incubation at 25°C for 48 hours (e.g. Candan *et al.*, 1995). On this medium, *A. salmonicida* from cases of furunculosis produces any of three colonies types termed 'rough', 'smooth' or 'G-phase' with brown diffusible pigment. Rough colonies are of ≤1 mm in diameter, have a rough appearance, and auto-agglutinate. These colonies remain intact and may be pushed across the agar surface by means of an inoculating loop. Smooth colonies are 3–4 mm in diameter after incubation at 25°C for 48 hours, do not auto-agglutinate, and have a smooth appearance. G-phase colonies are of intermediate size and appearance. Rough colonies are the most virulent, with virulence associated with the presence of a 49 kDa extracellular protein layer, the A-layer (Austin and Austin, 2007).

Supplementation of media with 10% (v/v) horse, sheep or bovine blood together with longer incubation periods of up to seven days is often needed to recover *A. salmonicida* from ulcerative conditions (Austin and Austin, 2007). Coomassie brilliant blue (CBB) agar, as developed originally by Udey (1982), is an effective medium for the recovery of *A. salmonicida* (Markwardt *et al.*, 1989), which develops dark blue colonies. Rimler–Shotts medium or peptone beef-extract glycogen agar may be used selectively for the recovery of aeromonads from pathological material (Austin and Austin, 2007).

Taxonomy

Motile aeromonads

Improvements in the understanding of the taxonomy of motile aeromonads has resulted from the widespread use of molecular techniques, including sequencing of the 16S rRNA gene. For example, *A. allosacharophila* was described after the examination of 16S rRNA sequences of two isolates from diseased elvers (Martinez-Murcia *et al.*, 1992), although phenotypic heterogeneity was recognized (Huys *et al.*, 2001). Another newcomer to fish pathology, *A. bestiarum*, was originally classified in DNA Hybridization Group (HG) 2, with isolates recovered from diseased fish (Huys *et al.*, 1996).

Aeromonas salmonicida

The traditional description of *A. salmonicida* is of non-motile, fermentative, Gram-negative rods, which produce a brown water-soluble pigment on tryptone-containing agar, which do not grow at 37°C, and which produce catalase and oxidase. The brown, diffusible pigment is a primary diagnostic feature of *A. salmonicida* (Martin-Carnahan and Joseph, 2005). However, non-pigmented and slowly pigmenting forms occur especially in fish other than salmonids (e.g. Wiklund *et al.*, 1993). The brown diffusible pigmenting forms from salmonids are classified as *A. salmonicida* subsp. *salmonicida*. All others, including the subspecies *achromogenes*, *masoucida*, *smithia* and *pectinolytica* together with the growing number of isolates that do not fit into any of the subspecies, are regarded as 'atypical'. These atypical isolates differ from the classical description of the species over a number of biochemical, physiological and genetic properties (Austin *et al.*, 1998; Dalsgaard *et al.*, 1998; Lund *et al.*, 2002). The most common differences are lack of, weak or slow pigment production (Koppang *et al.*, 2000), catalase-negativity (Kaku *et al.*, 1999), oxidase-negativity (Kaku *et al.*, 1999), nutritional fastidiousness, i.e. for blood or blood products (Austin, 1993), slow growth, i.e. ≥5 days to obtain visible colonies (Austin, 1993; Kaku *et al.*, 1999), and recovery from fish other than salmonids, i.e. cyprinids (e.g. Austin, 1993; Kaku *et al.*, 1999) and marine fish, including dab, plaice and flounder *(Platichthys flesus)* (Wiklund *et al.*, 1994; Wiklund and Dalsgaard, 1995).

Smith (1963) noted heterogeneity among isolates of *A. salmonicida,* and proposed a new species with the specific epithet *achromogenes* for non-pigmented isolates. Moreover, she questioned the relationship of *salmonicida* to the genus *Aeromonas,* and proposed that a new genus, *Necromonas,* be created with two species, i.e. *N. salmonicida* for typical isolates and *N. achromogenes* for non-pigmented strains. However, the proposal was not formally adopted by taxonomists. A second non-pigmented group became recognized as subspecies *masoucida* (Kimura, 1969). Schubert (1974) regarded these non-pigmented isolates as *A. salmonicida* subsp. *achromogenes* and *A. salmonicida* subsp. *masoucida,* respectively, in contrast to typical pigmented strains which were classified as *A. salmonicida* subsp. *salmonicida.* However, there is an ongoing issue regarding the taxonomic position of atypical isolates. The issue is that brown pigmented *A. salmonicida* subsp. *salmonicida* form a homogeneous group but not so isolates of the other subspecies and those labelled as atypical (Austin *et al.,* 1998). Clearly, atypical isolates are dissimilar from typical *A. salmonicida* (e.g. Austin *et al.,* 1998; Dalsgaard *et al.,* 1998; Høi *et al.,* 1999), but there is incongruence in the outputs from the different methods used to examine isolates. Another issue concerns the true identity of *Haemophilus piscium,* the causal agent of ulcer disease (Snieszko *et al.,* 1950). Although an original authentic strain was not deposited in a recognized culture collection, subsequent opinion places *H. piscium* in with *A. salmonicida* although its precise relationship to the existing subspecies remains to be established (Austin *et al.,* 1998).

Diagnosis

Diagnosis of fresh isolates has progressed from phenotypic, e.g. using commercial rapid identification kits, to serology, and subsequently to molecular methods (Table 3.2). Serology has advantages insofar as there is the possibility of adapting methods for field use (e.g. Eurell *et al.,* 1978). Techniques include whole cell agglutination, which is effective with smooth, non-autoagglutinating cultures of *A. salmonicida* (Rabb *et al.,* 1964); the fluorescent antibody technique (FAT; Kawahara and Kusuda; 1987), the indirect FAT (iFAT), the peroxidase–antiperoxidase enzyme immunoassay (PAP) (Sakai *et al.,* 1986), latex agglutination and co-agglutination techniques, and the enzyme linked immunosorbent assay (ELISA) (Hiney *et al.,* 1994). Molecular techniques, including terminal-restriction

Table 3.2 Approaches for the diagnosis of aeromonad fish pathogens

Category	Method	Comment
Phenotyping	Commercial kits, e.g. API 20E	Commercial kits are convenient, but have inevitably been designed for human pathogens that grow at ~37°C
Serology	A wide range of methods including whole cell agglutination, indirect or co-agglutination, direct and indirect fluorescent antibody test, peroxidase-antiperoxidase enzyme immunoassay, enzyme-linked immunosorbent assay (ELISA)	Quick, generally reliable depending on the specificity of the antiserum, some, notably ELISA, may be adapted to field use
Molecular techniques	A wide range of approaches including multiplex polymerase chain reaction and terminal-restriction fragment length polymorphism	Should be specific, sensitive and highly reliable, laboratory based

fragment length polymorphism (Nilsson and Strom, 2002) and multiplex polymerase chain reaction, the latter of which was developed for the simultaneous detection of multiple pathogens (Tapia-Cammas *et al.*, 2011), have been used for the detection of *A. hydrophila* (Chu and Lu, 2005) and *A. salmonicida* (e.g. Oakey *et al.*, 1998). Yet, the presence of aeromonads in diseased fish does not necessarily imply that they are the causal agents of the illness. Their possible presence as opportunists or contaminants needs to be considered carefully in any diagnostic work.

Ecology

Motile aeromonads, and in particular *A. hydrophila*, abound in eutrophic freshwater environments (e.g. Heuschmann-Brunner, 1978) and in close association with aquatic invertebrates (King and Shotts, 1988) where cells will be readily available to initiate infection. Chemotaxis to the mucus of freshwater fish has been described (Hazen *et al.*, 1982). However because aeromonads are commonplace in the aquatic environment their recovery from fish – particularly external surfaces – need not necessarily prove a pathogenic role even if disease signs are present.

Aeromonas salmonicida

A. salmonicida is often defined as a fish-associated organism that does not occur naturally in the aquatic environment (Popoff, 1984). However, culture-dependent and culture-independent work has pointed to the survival of cells in fresh, brackish and seawater and their sediments (e.g. McCarthy, 1980), although the various methods employed often give conflicting data. The overall outcome is that culture-independent approaches indicate the survival of intact cells for much longer periods than gleaned from culturing. Thus, Deere *et al.* (1996a) cultured *A. salmonicida* for <4 weeks, but found the DNA remained intact for > 13 weeks. In parallel, Effendi and Austin (1994) developed a laboratory microcosm and used culturing, direct counts, epifluorescence microscopy, iFAT, respiratory activity (the reduction of tetrazoliums to coloured formazans; after Effendi and Austin, 1993), and the direct viable count technique (by incorporating yeast extract and nalidixic acid; after Kogure *et al.*, 1979), and confirmed that cells of *A. salmonicida* remained after plate counts reached zero. Survival was maximal in brackish conditions, on substrates rather than in the water column. Furthermore, Wiklund (1995) observed that survival was better at 4°C than 15°C in brackish rather than sea or fresh water, and in the presence of particulates, i.e. sand. Sakai (1986) considered that the electrostatic interrelationship occurring among bacteria, humic acid and river sand explained the ability of virulent cells to survive for extended periods in river sediments. In addition, the detection of bacteriophages specific for *A. salmonicida* in river sediments was used as indirect evidence that the pathogen was present throughout the year (Dubois-Darnaudpeys, 1977).

Besides the issue of culturability in the aquatic environment, evidence has been presented that the pathogen survives in aquatic inhabitants notably sea lice (~10^4 bacterial cells/louse), marine plankton (600 cells/g of homogenized plankton) (Nese and Enger, 1993) and within the digestive tract of the ciliated protozoan, *Tetrahymena pyriformis* (King and Shotts, 1988). Furthermore, Frerichs *et al.* (1992) recovered typical cells from healthy wild wrasse (these eat ectoparasites on other fish, e.g. Atlantic salmon) captured in the open sea, although there was not any evidence of pathogenicity to Atlantic salmon smolts.

An issue that needs to be addressed is whether or not most culturing techniques are adequate for the recovery of *A. salmonicida* from environmental samples. Possibly, the difficulty for finding the pathogen in the aquatic environment reflects the lack of suitable selective recovery methods (e.g. McCarthy, 1980) and possible antagonism by other members of the microflora (e.g. Effendi and Austin, 1991). It is interesting to note that using CBB, Cipriano *et al.* (1992) recovered *A. salmonicida* from 56% of mucus samples and 6% of kidney material from salmonids. Subsequently, Cipriano *et al.* (1996a,b) reported the presence of the pathogen in gills and mucus, i.e. pointing to external carriage.

Clearly, the evidence emerging from different studies is that intact *A. salmonicida* cells may be observed for longer periods than they are cultured. The question to be resolved is whether or not these intact cells are viable, damaged or senescent, and are they pathogenic or not? Certainly, Allen-Austin *et al.* (1984) observed that culturability declined quickly when cells were inoculated into filter-sterilized river water, but microscopic evidence revealed that after an initial decrease, the number of *A. salmonicida* cells remained fairly constant and well beyond the period when colony counts had declined to zero.

The subject of dormant/non-culturable but viable (NCBV) cells of *A. salmonicida* was investigated by Morgan *et al.* (1991), who assessed survival in lake water, employing an extensive range of techniques, including cell culture, cell revival, epifluorescence microscopy, flow cytometry, membrane fatty acid analysis, plasmid maintenance and respiration. The conclusion was made that *A. salmonicida* became unculturable in sterile lake water, with flow cytometric and microscopic methods revealing the continued presence of cells. However, attempts to revive these cells to culturability by the addition of tryptone soya broth failed. Despite this, it was found that both genomic and plasmid DNA, and RNA were maintained in the cells. Subsequently by means of flow cytometry with rhodamine 123, Morgan *et al.* (1993) established a NCBV state in sterile lake water, although there was an indication that cellular properties related with viability were lost shortly after culturability disappeared in distilled water, but not so in lake water (Deere *et al.*, 1996b).

A mechanism has been suggested for the long-term survival of *A. salmonicida* in the aquatic environment, and was based on electrostatic charge differences on cells, with net negative charges on virulent, agglutinating cells, and net positive charges on avirulent non-agglutinating strains (Sakai, 1986). He suggested that the negatively charged virulent cell may persist under starvation conditions, retaining viability in river sediments. It was proposed that the decline in negatively charged cells in sediments over prolonged periods, also noted by others (e.g. Michel and Dubois-Darnaudpeys, 1980), may be attributed to the spontaneous occurrence of positively charged avirulent free-living cells which originate from virulent ones, attached to sediment particle surfaces, and subsequently detach from the sediment/sand particles. This free-living form could be considered to enter dormancy, according to Sakai (1986), because the viability of these bacteria declines due to a lack of nutrients. It was further proposed that the free-living cells represent a transitional life stage of the pathogen which would ultimately lose viability (Sakai, 1986).

An isolate of *A. salmonicida* was marked with the *xylE* gene, using the plasmid pLV1013. This isolate was culturable from normal (non-sterile) lake water for 3 weeks, after which non-culturability developed, with the NCBV cells retaining chromosomal and plasmid DNA. The NCBV state can be postponed (60 days was mentioned by Pickup *et al.*, 1996) by the addition of high levels of nutrient, especially 125 μM quantities of the amino acids arginine and methionine, to experimental microcosms. *A. salmonicida* decreased in size and became rounded, but were still culturable (Pickup *et al.*, 1996). Effendi and Austin (1991) determined that samples where the pathogen was considered to be absent (or unculturable) actually contained cells which passed through 0.22 and 0.45 µm pore size porosity filters. On this basis, these workers recorded populations of *ca.* 10^3 *A. salmonicida* cells/ml in the microcosms after corresponding colony counts had reached zero. Thus, it was suggested that the existence of specialized forms, e.g. L-forms, of *A. salmonicida*, may be a factor in the difficulties previous researchers had experienced in attempts to recover the pathogen from environmental samples. Moreover, Effendi and Austin (1995a) examined the characteristics of the so-called NCBV cells, and observed that these cells became much smaller and coccoid while retaining respiratory activity as measured by the reduction of tetrazoliums to insoluble formazans. There was not any alteration in the lipopolysaccharide (LPS) composition of the cells, but an alteration in the protein composition was recorded, with a reduction in some (15, 17, 22, 30 and 70 kDa proteins) and an increase in the 49 kDa protein. This was accompanied by a loss in DNA. That these cells were still alive was indicated by the development of large bizarre shapes following the addition of yeast extract and nalidixic acid (after Kogure *et al.*, 1979). The development of a dormant, non-culturable state of *A. salmonicida* in seawater at 4°C was apparent from the work of Ferguson *et al.* (1995), who incorporated a luciferase gene, *luxAB*, from *Vibrio fischeri* into *A. salmonicida* and followed the fate of the cells. However, intact non-culturable cells could not be resurrected.

If the pathogen exists in the aquatic environment, how is it taken up into fish? Different routes have been proposed, and include the gills, skin/mucus, lateral line, mouth, anus or a surface injury (Effendi and Austin, 1995b; Ferguson *et al.*, 1998), with mortalities leading after experimental challenge via the gill and anus (Effendi and Austin, 1995b). Evidence has been presented that shows *A. salmonicida* may translocate across the intestinal epithelia (Jutfelt *et al.*, 2006). Here, the authors exposed intestinal segments of rainbow trout for 90 minutes to isothiocyanate-labelled cells of virulent *A. salmonicida* and demonstrated translocation.

Pathogenicity

Many studies have been directed at ascertaining the pathogenic mechanisms of aeromonads (Table 3.3). However, most studies rely on single cultures, which are different to each laboratory. Therefore, it is difficult to make sweeping conclusions about species-level pathogenicity. Nevertheless, some generalisations may be made:

Surface structures

Surface structures of motile and non-motile aeromonads may be involved in autoaggregation/hydrophobicity and haemagglutination (e.g. Paula *et al.*, 1988). The surface array matrix, i.e. the S-layer, of motile aeromonads has been considered to influence the interaction between the bacterial cell and its environment (Esteve *et al.*, 2004). A major function is believed to be the provision of physical protection from lytic components, including serum proteins and bacteriophages (Dooley *et al.*, 1988). However, the majority of work with surface structures has involved *A. salmonicida* and the virulent rough cells that possess the 49 kDa (Kay *et al.*, 1981) proteinaceous extracellular layer, termed the A-layer, which is believed to be the product of a single chromosomal gene (Belland and Trust, 1985), and contributes to auto-agglutination, hydrophobicity (Van Alstine *et al.*, 1986), adherence to

Table 3.3 Overview of the pathogenicity mechanisms

Category	Component	Reported in	Comment
Surface structures	S-layer	Motile aeromonads	Role in physical protection from lysis
	A-layer (49kDa protein)	*A. salmonicida*	Role in auto-agglutination, hydrophobicity, adherence, and survival
Capsule	Capsule	*A. salmonicida*	Role in adherence, invasion, and resistance
Type I pilus system	Type I pilus system	*A. salmonicida*	Role in colonization
Type III secretion system	Type III secretion system	*A. salmonicida*	Role in virulence
Type IV pilin	Type IV pilin	*A. hydrophila*, *A. salmonicida*	Role in adherence
Extracellular products	Aerolysin, haemolysin, serine protease	*A. hydrophila*	Role in virulence
	ADP-ribosyltransferase, acetylcholinesterase, metalloprotease, phospholipase, haemolysin	*A. salmonicida*	Role in cytotoxicity, leucocytolytic and haemolytic activities
Subcellular components	Lipopolysaccharide	Motile and non-motile aeromonads	Role in endotoxicity
	Iron-regulated outer membrane proteins	*A. salmonicida*	Enables the pathogen to scavenge successfully for iron
	Siderophores	*A. salmonicida*	An iron sequestering mechanism

cells (Moki *et al.*, 1995), an insulin-binding capacity (Nisr *et al.*, 2012), protection from the action of protease (Kay and Trust, 1991) and bacteriophage (Ishiguro *et al.*, 1981), interference with the antibacterial peptides, i.e. magainin, cecropins and defensins (Henery and Secombes, 2000), and survival in macrophages (Daly *et al.*, 1996).

Capsules

Using intraperitoneal chambers in fish, it was observed that *A. salmonicida* produced capsules with virulence functions (Garduño and Kay, 1995). In particular, adherence to fish cells was slightly higher in cultures grown in conditions to promote capsule formation. Also, invasion of fish cells was more pronounced in capsulated cells (Merino *et al.*, 1996), which were partially resistant to complement (Merino *et al.*, 1997).

Type I pilus system

Using knockout mutants, it was concluded that the type I pilus system was important for colonization of *A. salmonicida* (Dacanay *et al.*, 2010).

Type III secretion system

A type III secretion system, which utilizes a 140 kbp plasmid and chromosome-encoded transmembrane injection devise incorporating membrane proteins and a needle-like structure to translocate the effector protein AexT toxin from the cytosol into the host cell, has been associated with the virulence of *A. salmonicida* subsp. *salmonicida* (Burr *et al.*, 2002; Ebanks *et al.*, 2006).

Type IV pilin

Type IV pili are important virulence determinants as adhesins. A four-gene cluster, *tapABCD*, from a virulent *A. salmonicida* isolate encoded proteins with homology to those necessary for biogenesis of type IV pili (Masada *et al.*, 2002). *TapA* encoded a protein with homology to type IV pilin subunits in common with other Gram-negative bacterial pathogens, e.g. *A. hydrophila*. A mutant *A. salmonicida* defective in *tap*A was less pathogenic to rainbow trout following i.p. injection. *Tap*B is part of the ABC-transporter family with nucleotide-binding regions; *Tap*C homologues are cytoplasmic membrane proteins that exert a role in the anchoring and/or assembly of pili; *Tap*D has homology with type IV prepilin leader peptidases (Masada *et al.*, 2002). *Tap*D was capable of restoring type IV pilin assembly and type II extracellular protein secretion and was presumed to have a similar function in *A. salmonicida* (Masada *et al.*, 2002).

Extracellular products (ECP)

A. hydrophila produces ECPs, which contains considerable enzymatic activity (Santos *et al.*, 1987), including a 64 kDa serine protease (Cascón *et al.*, 2000). It was determined that isolates with aerolysin (*AerA*), cytotoxic enterotoxin (*alt*) and serine protease genes (*ahp*) were most frequently virulent with lower lethal dose $(LD)_{50}$ doses in zebra fish than isolates with only one or two of these virulence genes (Li *et al.*, 2011). Genes considered to represent known virulence attributes included haemolysin, histone-like protein, oligoprotease A, outer membrane protein (OMP) and multidrug resistance protein. Other genes encoded synthesis of O-antigen (Zhang *et al.*, 2000).

A. salmonicida ECPs include an ADP-ribosyltransferase toxin (AexT) (Braun *et al.*, 2002), acetylcholinesterase (an ichthyotoxin with neurotoxic activity; Pérez *et al.*, 1998), several proteases namely two metalloproteases, i.e. the 37 kDa leucine aminopeptidase and the 30 kDa metalloprotease 3 (Arnesen and Eggset, 1999), P1, glycerophospholipid:cholesterol acyltransferase, extracellular metallo-caseinase AsaP1, P2 metallo-gelatinase and a 70 kDa serine caseinase (Ellis *et al.*, 1997; Gudmundsdóttir *et al.*, 1997), phospholipase, a 56 kDa haemolysin (Fyfe *et al.*, 1986), a leucocidin (Munro *et al.*, 1980; Lygren *et al.*, 1998) and 100 kDa unidentified protein (Fyfe *et al.*, 1987). When injected intraperitoneally into rainbow trout, ECP produced fatalities (Munro *et al.*, 1980). In fish cells, the ECP exhibited cytotoxic effects, and at higher concentrations was leucocytolytic and haemolytic. AsaP1 has been linked with lethal toxicity of atypical *A. salmonicida* in Atlantic salmon, with furuncles being produced by ECP with AsaP1 (Gunnlaugsdóttir and Gudmundsdóttir, 1997). A LPS-free phospholipase has been demonstrated to cause disease signs in Atlantic salmon (Huntly *et al.*, 1992).

The ability to scavenge for iron

A. salmonicida cells scavenge for iron in iron-limited conditions, with evidence that iron-regulated outer membrane proteins (IROMPs) are produced *in vivo* (Ellis *et al.*, 1997). An inducible siderophore-dependent iron-chelating system was described in typical strains and an unidentified siderophore-independent system in atypical *A. salmonicida* (Hirst and Ellis, 1996).

Disease control

An increasing range of methods are becoming available for the control of fish diseases (Table 3.4).

Genetic resistance

Evidence has accumulated that some discrete fish strains have different levels of resistance. Initially, Embody and Hayford (1925) increased resistance in brook trout to furunculosis by

Table 3.4 Methods for the control of fish diseases

Category	Specifics	Comment
Genetic resistance	Considered for control of *A. hydrophila* and *A. salmonicidaida* infections in carp and trout strains/families	Offers promise for disease control particularly if considering the developments in genetic manipulation
Management	Disinfection of eggs and equipment	An effective approach to disease control
Vaccines	Inactivated whole cells, subcellular components	Work well for motile aeromonads
	Recombinant proteins (S-layer)	Work well for *A. hydrophila*
	Live attenuated vaccines	Work well for *A. hydrophila* and *A. salmonicida*
	Inactivated whole cells with IROMPs	Effective for control of furunculosis
Immunostimulants/ dietary supplements	Vitamin C, yeast RNA, levan, β-glucans	Potential for control of *A. hydrophila* infections
	Synthetic peptides, Baypamum, lysozyme, -1,3-glucan, chitosan	Potential for use on *A. salmonicida* infections
Medicinal plants/ natural plant products	*Astragalus membranaceous*, *Lonicera japonica*, garlic (*Allium sativum*), ginger (*Zingiber officinale*), lupin (*Lupinus perennis*), mango (*Mangifera indica*), stinging nettle (*Urtica dioica*)	Effective for control of *A. hydrophila* infections
Probiotics	A wide range of bacteria, including *Bacillus pumilus*	Effective for control of *A. hydrophila* and *A. salmonicida* infection
N-acylated homoserine lactone	Application by feeding	Effective for control of *A. hydrophila* infections
Antimicrobial compounds	A wide range of compounds, including oxytetracycline	Has been a effective means of disease control, but there are concerns about residues and resistance

selective breeding. Cipriano (1982) reported varying degrees of resistance to furunculosis among 11 different strains of rainbow trout, and correlated this with the serum neutralization titre. Moreover, the resistance of four common carp strains to infection by *A. hydrophila* was documented with crosses derived from Tata and Szarvas; 15 parents being most resistant whereas fish from the wild strains Duna and Amur were the most resistant (Jeney *et al.*, 2011).

Vaccines

Inactivated whole cells, LPS, ECPs or OMPs administered by immersion, injection or orally work well with motile aeromonads (Schachte, 1978; Selvaraj *et al.*, 2009) with the host response including the production of superoxide anion by the head-kidney leucocytes (Basheera *et al.*, 2002). A rough attenuated culture produced by repeated subculturing on BHIA for 8 years conferred immune protection in rohu (*Labeo rohita*) (Swain *et al.*, 2010). A recombinant S-layer protein was immunoprotective [relative per cent survival (RPS; after Amend, 1981) = 56–87%] when administered in an adjuvant intraperitoneally (30 μg protein/fish) in common carp (Poobalane *et al.*, 2010). Zhao *et al.* (2011) used the glyceraldehyde-3-phosphate dehydrogenase (GAPDH) gene *gapA* to express *A. hydrophila* GAPDH in an attenuated *V. anguillarum* strain, which was injected intraperitoneally (10^6 [colony forming units] CFU/fish) into turbot and challenged after 4 weeks. Challenge of cytoplasm GAPDH expressing strain AV/pUC-gapA-vaccinated fish with *A. hydrophila* and *V. anguillarum* led to RPS values of 42% and 92%, respectively. Tu *et al.* (2010) used ghost cells (these were produced with a culture harbouring the lysis plasmid, pElysis, and were induced to lysis after incubation at 27–42°C) fed orally for 5 days, normal feed for 14 days and a second dose of oral vaccine for a further 5 days at 4×10^8 cells/g of fish/day to carp, and found improved RPS of 76.8% after challenge with *A. hydrophila*, compared with RPS of 58.9% in the group administered with formalin-inactivated whole cells. In fish vaccinated with ghost cells, there was greater amounts of antigen-specific antibody in serum and the intestinal mucosa (Tu *et al.*, 2010).

The development of a furunculosis vaccine started with Duff (1942), who reported success with an orally administered, chloroform-inactivated whole-cell preparation. Thereafter, work focused on a range whole cell inactivated, subcellular and attenuated live vaccines with Lund *et al.* (2003a, 2008a), confirming the need for A-layer in whole cell vaccine preparations. Other developments included success with inactivated whole vaccines based on IROMP technology (e.g. Durbin *et al.*, 1999) and live attenuated products. For example, avirulent cells, with altered A-layer, have been evaluated as a live vaccine (Thornton *et al.*, 1991). Success resulted with a live aromatic-dependent *A. salmonicida* vaccine, *aroA*, which was administered intraperitoneally at 2×10^6 to 2×10^9 live bacteria/fish, and resulted in a 253-fold increase in LD_{50} (Vaughan *et al.*, 1993) and stimulation of T-cells rather than B-cell responses in rainbow trout (Marsden *et al.*, 1996). The value of products based on ECPs was unclear with some studies indicating immunosuppression or inflammation (Mutoloki *et al.*, 2006), whereas others described immunogenicity (Kawahara *et al.*, 1990). The specificity of vaccines for protecting against challenge with atypical and typical strains has been documented, with data revealing the benefit of using autogenous products (Gudmundsdóttir and Gudmundsdóttir, 1997). Concerns about oily adjuvants has been voiced when abdominal adhesions were found in Atlantic salmon after intraperitoneal uptake (Gudmundsdóttir *et al.*, 2003). However, injection techniques are the most efficacious means of administration to fish, whereas the oral route is least promising (Midtlyng *et al.*, 1996). Ultrasound has been evaluated for the administration of vaccines specifically for the control of goldfish ulcer disease. Here, soluble A-layer protein was applied by immersion (100 µg A-protein/ml for 10 minutes) after ultrasound (1 MHz frequency of ultrasound/1 minute) pretreatment, and led to promising results in goldfish against challenge (Navot *et al.*, 2011).

There is ongoing discussion about the value of vaccines developed for use in salmonids, and containing antigens of *A. salmonicida* subsp. *salmonicida*, for application in other groups of fish, which are affected by atypical isolates. Thus, a commercial polyvalent product for salmon failed to protect turbot from experimental challenge with *A. salmonicida* subsp. *achromogenes* (Björnsdóttir *et al.*, 2005). However, Santos *et al.* (2005) was more successful with turbot, although the specific pathogen was not equated with subsp. *achromogenes*. Nevertheless, the commercial vaccine, Furovac 5, and an autogenous vaccine resulted in RPS of 72–99% when challenged 120 days after administration (Santos *et al.*, 2005).

Immunostimulants/dietary supplements

Evidence points to a role for vitamin C in stimulating the humoral and cell-mediated immune response in fish vaccinated with *A. hydrophila* vaccines (Anbarasu and Chandran, 2001). Misra *et al.* (2007) detailed a role for vitamin C in immunostimulation and protection against challenge with *A. hydrophila*. A further example was yeast RNA, which was incorporated in diets at 0.4% (w/v) and fed for 60 days leading to reduced mortalities caused by *A. hydrophila* in rohu and enhanced phagocyte respiratory burst activity (Choudhury *et al.*, 2005). Also, microbial levan incorporated into the diet at 1.25% led to protection (RPS = 60%) of Indian major carp against challenge with *A. hydrophila* (Gupta *et al.*, 2008). Finally, β-glucans enhanced the non-specific resistance to *A. hydrophila* infection (Selvaraj *et al.*, 2005; Kumari and Sahoo, 2006).

A similar approach has been used with success to control *A. salmonicida* infections with Kitao and Yoshida (1986) first reporting that synthetic peptides could enhance resistance of rainbow trout. The range of beneficial compounds quickly increased to include Baypamum (Ortega *et al.*, 1996), dimerized lysozyme which was injected into rainbow trout at a dose of 10 or 100 µg/kg and stimulated cellular and humoral mechanisms giving protection against furunculosis (Siwicki *et al.*, 1998), lysozyme (Siwicki *et al.*, 1998), β-1,3-glucan and chitosan (Anderson and Siwicki, 1994).

Plant products

Numerous plant-based products have found success, and include Chinese herbs, 0.1% (w/v) *Astragalus membranaceous* and 0.1% (w/v) *Lonicera japonica*, which were fed for 4 weeks, enhancing the non-specific immune response, i.e. phagocytic, plasma lysozyme and respiratory burst activities of tilapia and conferring resistance to *A. hydrophila* (Ardó *et al.*, 2008). Garlic (*Allium sativum*) dosed at 0.5 and 1.0 g/100 g of feed led to a stimulation in growth, feed conversion and protein efficiency, and marked reduction in mortalities (RPS = 95%) in rainbow trout after challenge with *A. hydrophila* with the mode of action reflecting the stimulation of the number of erythrocytes and leucocytes, higher haematocrit and increase in anti-protease, bactericidal, lysozyme, phagocytic and respiratory burst activities (Nya and Austin 2009a). Similar success occurred with ginger (*Zingiber officinale*), which again led to immunomodulation and protection against *A. hydrophila* (Nya and Austin, 2009b). Lupin (*Lupinus perennis*), mango (*Mangifera indica*) and stinging nettle (*Urtica dioica*) were added to feed to 1% (w/v) for 14 days, and protected rainbow trout against challenge with *A. hydrophila*, with the mode of action considered to involve immune stimulation, principally enhancement in lysozyme respiratory burst and serum bactericidal activity, increases in the number of red and white blood cells in the blood (Awad and Austin, 2010), and up-regulation of cytokine gene, i.e. IL-8, IL-1β and TGF-β expression (Awad *et al.*, 2011).

Antimicrobial compounds

A variety of inhibitory agents have been applied with varying degrees of success to the treatment of aeromonad diseases although in many countries the non-medical use of antibiotics is discouraged. Within the UK, four compounds, namely amoxicillin, oxolinic acid, oxytetracycline and potentiated sulphonamide, are licensed for fisheries use in the treatment of furunculosis. In addition, florfenicol, dosed at 10 mg/kg body weight of fish/day for 10 days, offers promise, insofar as it has already been used with some success against furunculosis in Norway (Samuelsen *et al.*, 1998). Enrofloxacin and sarofloxacin have been found to be more effective than oxolinic acid, in terms of minimum inhibitory concentration at inactivating *A. salmonicida*. Thus, the effectiveness of enrofloxacin at 10 mg/kg body weight of fish/day for 10 days has been attested by field trials with lake trout *(Salvelinus namaycush)* (Hsu *et al.*, 1995). In addition, enrofloxacin has been effective in controlling atypical *A. salmonicida* in tom cod, insofar as a single injection with 5 mg of enrofloxacin/kg of fish stopped disease (Williams *et al.*, 1997). Similarly, difloxacin, dosed at 5 or 10 days at 5 mg/kg body weight, and 10 days at 1.25 or 2.5 mg/kg body weight resulted in significantly lowered mortalities compared to controls following i.m. injection with a virulent culture of *A. salmonicida*.

Disinfection methods

Adequate husbandry practices is advocated, such as maintenance of good water quality, disinfection of fish farm equipment and utensils especially when disease outbreaks occur, and routine disinfection policies for eggs upon arrival at receiving sites (McCarthy and Roberts, 1980).

Probiotics

Bacillus pumilus fed for 14 days at 10^7 cells/g of feed controlled infection by *A. hydrophila* (Aly *et al.*, 2008). *V. alginolyticus*, previously used as probiotic in Ecuadorian shrimp hatcheries, has been effective at controlling diseases caused by *A. salmonicida* (Austin *et al.*, 1995).

N-acylated homoserine lactone (AHL)

Feeding with AHL, which had lactonase activity, from *Bacillus* A196 reduced the development of *A. hydrophila* infection in zebra fish (Cao *et al.*, 2012).

Conclusions

The range of motile aeromonad fish pathogens continues to increase with research interest focusing on molecular diagnostics, pathogenicity mechanisms and disease control principally by vaccination and the use of probiotics and plant products. In contrast, interest in ecology has declined although some key questions particularly concerning the whereabouts, nature and activity of *A. salmonicida* cells in asymptomatic fish and the state of cells in the aquatic environment remain unclear. The taxonomic status of the subspecies and atypical strains of *A. salmonicida* need clarification. Nevertheless, it is clear that the aeromonads are significant causes of predominantly freshwater fish in most countries.

References

Allen-Austin, D., Austin, B., and Colwell, R.R. (1984). Survival of *Aeromonas salmonicida* in river water. FEMS Microbiol. Lett. *21*, 143–146.

Aly, S.M., Abd-El-Rahman, John, G., and Mohammed, M.F. (2008). Characterization of some bacteria isolated from *Oreochromis niloticus* and their potential use as probiotics. Aquaculture 277, 1–6.

Amend, D.F. (1981). Potency testing of fish vaccines. Dev. Biol. Stand. *49*, 447–454.

Anbarasu, K., and Chandran, M.R. (2001). Effect of ascorbic acid on the immune response of the catfish, *Mystus gulio* (Hamilton), to different bacterins of *Aeromonas hydrophila*. Fish Shellfish Immunol. *11*, 347–355,

Anderson, D.P., and Siwicki, A.K. (1994). Duration of protection against *Aeromonas salmonicida* in brook trout immunostimulated with glucan or chitosan by injection or immersion. Progr. Fish Cult. *56*, 258–261.

Ardó, L., Yin, G., Xu, P., Váradi, L., Szigeti, G., Jeney, Z., and Jeney, G. (2008). Chinese herbs *(Astragalus membranaceus* and *Lonicera japonica)* and boron enhance the non-specific immune response of Nile tilapia *(Oreochromis niloticus)* and resistance against *Aeromonas hydrophila*. Aquaculture 275, 26–33.

Arnesen, J.A., and Eggset, G. (1999). Isolation and characterisation of two extracellular metalloproteases from *Aeromonas salmonicida* ssp. *salmonicida*. J. Fish Dis. 22, 35–43.

Austin, B. (1993). Recovery of 'atypical' isolates of *Aeromonas salmonicida*, which grow at 37°C, from ulcerated non-salmonids in England. J. Fish Dis. *16*, 165–168.

Austin, B., and Austin, D.A. (2007). Bacterial Fish Pathogens: Disease of Farmed and Wild Fish, 4th edn (Springer-Praxis, Godalming, UK).

Austin, B., Stuckey, L.F., Robertson, P.A.W., Effendi, I., and Griffith, D.R.W. (1995). A probiotic strain of *Vibrio alginolyticus* effective in reducing diseases caused by *Aeromonas salmonicida, Vibrio anguillarum* and *Vibrio ordalii*. J. Fish Dis. *18*, 93–96.

Austin, B., Austin, D.A., Dalsgaard, I., Gudmundsdóttir, B.K., Høie, S., Thornton, J.M., Larsen, J.L., O'Hici, B., and Powell, R. (1998). Characterization of atypical *Aeromonas salmonicida* by different methods. Syst. Appl. Microbiol. *21*, 50–64.

Awad, E., and Austin, B. (2010). Use of lupin, *Lupinus perennis*, mango, *Mangifera indica*, and stinging nettle, *Urtica dioica*, as feed additives to prevent *Aeromonas hydrophila* infection in rainbow trout, *Oncorhynchus mykiss* (Walbaum). J. Fish Dis. 33, 413–420.

Awad, E., Mitchell, W.J., and Austin, B. (2011). Effect of dietary supplements on cytokine gene expression in rainbow trout, *Oncorhynchus mykiss* (Walbaum). J. Fish Dis. *34*, 629–634.

Basheera, J.M., Chandran, M.R., Aruna, B.V., and Anbarasu, K. (2002). Production of superoxide anion by head-kidney leucocytes of Indian major carps immunised with bacterins of *Aeromonas hydrophila*. Fish Shellfish Immunol. *12*, 201–207.

Belland, R.J., and Trust, T.J. (1985). Synthesis, export, and assembly of *Aeromonas salmonicida* A-layer analyzed by transposon mutagenesis. J. Bacteriol. *163*, 877–881.

Björnsdóttir, B., Gudmundsdóttir, S., Bambir, S.H., and Gudmundsdóttir, B.K. (2005). Experimental infection of turbot, *Scophthalmus maximus* (L.), by *Aeromonas salmonicida* subsp. *achromogenes* and evaluation of cross protection induced by a furunculosis vaccine. J. Fish Dis. *28*, 181–188.

Braun, M., Stuber, K., Schlatter, Y., Wahli, T., Kuhnert, P., and Frey, J. (2002). Characterization of an ADP-ribosyltransferase toxin (AexT) from *Aeromonas salmonicida* subsp. *salmonicida*. J. Bacteriol. *184*, 1851–1858.

Burr, S.E., Stuber, K., Wahli, T., and Frey, J. (2002). Evidence for a type III secretion system in *Aeromonas salmonicida* subsp. *salmonicida*. J. Bacteriol. *184*, 5966–5970.

Candan, A., Küçüker, M.A., and Karatas, S. (1995). Motile aeromonad septicaemia in *Salmo salar* cultured in the Black Sea in Turkey. Bull. Eur. Assoc. Fish Pathol. *15*, 195–196.

Cao, Y., He, S., Zhou, Z., Zhang, M., Mao, W., Zhang, H., and Yao, B. (2012). Orally administered thermostable *N*-acyl homoserine lactonate from *Bacillus* sp. strain A196 attenuates *Aeromonas hydrophila* infection in zebrafish. Appl. Environ. Microbiol. *78*, 1899–1908.

Cascón, A., Fregeneda, J., Aller, M., Yugueros, J., Temprano, A., Hernanz, C., Sanchez, M., Rodríguez-Aparicio, L., and Naharro, G. (2000). Cloning, characterization, and insertional inactivation of a major extracellular serine protease gene with elastolytic activity from *Aeromonas hydrophila*. J. Fish Dis. *23*, 49–59.

Choudhury, D., Pal, A.K., Sahu, N.P., Kumar, S., Das, S.S., and Mukherjee, S.C. (2005). Dietary yeast RNA supplementation reduces mortality by *Aeromonas hydrophila* in rohu (*Labeo rohita* L.) juveniles. Fish Shellfish Immunol. *19*, 281–291.

Chu, W.-H., and Lu, C.-P. (2005). Multiplex PCR assay for the detection of pathogenic *Aeromonas hydrophila*. J. Fish Dis. *28*, 437–442.

Cipriano, R.C. (1982). Immunization of brook trout (*Salvelinus fontinalis*) against *Aeromonas salmonicida:* immunogenicity of virulent and avirulent isolates and protective ability of different antigens. Can. J. Fish. Aquat. Sci. *39*, 218–221.

Cipriano, R.C., Ford, L.A., Teska, J.D., and Hale, L.E. (1992). Detection of *Aeromonas salmonicida* in the mucus of salmonid fishes. J. Aquat. Anim. Health *4*, 14–118.

Cipriano, R.C., Bullock, G.L., and Noble, A. (1996a). Nature of *Aeromonas salmoncida* carriage on asymptomatic rainbow trout maintained in a culture system with recirculating water and fluidized sand biofilters. J. Aquat. Anim. Health *8*, 47–51.

Cipriano, R.C., Ford, L.A., Teska, J.D., Schachte, J.H., Petrie, C., Novak, B.M., and Flint, D.E. (1996b). Use of non-lethal procedures to detect and monitor *Aeromonas salmonicida* in potentially endangered or threatened populations of migrating and post-spawning salmon. Dis. Aquat. Org. *27*, 233–236.

Dacanay, A., Boyd, J.M., Fast, M.D., Knickle, L.C., and Reith, M.E. (2010). *Aeromonas salmonicida* Type I pilus system contributes to host colonization but not invasion. Dis. Aquat. Org. *88*, 199–206.

Dalsgaard, I., Gudmundsdóttir, B.K., Helgason, S., Høie, S., Thoresen, O.F., Wichardt, U.-P., and Wiklund, T. (1998). Identification of atypical *Aeromonas salmonicida*: inter-laboratory evaluation and harmonization of methods. J. Appl. Microbiol. *84*, 999–1006.

Daly, J.G., Kew, A.K., Moore, A.R., and Olivier, G. (1996). The cell surface of *Aeromonas salmonicida* determines *in vitro* survival in cultivated brook trout (*Salvelinus fontinalis*) peritoneal macrophages. Microb. Pathog. *21*, 447–461.

Deere, D., Porter, J., Pickup, R.W., and Edwards, C. (1996a). Survival of cells and DNA of *Aeromonas salmonicida* released into aquatic microcosms. J. Appl. Bacteriol. *81*, 309–318.

Deere, D., Porter, J., Pickup, R., and Edwards, C. (1996b). Direct analysis of starved *Aeromonas salmonicida*. J. Fish Dis. *19*, 459–467.

Dooley, J.S.G., McCubbin, W.D., Kay, C.M., and Trust, T.J. (1988). Isolation and biochemical characterization of the S-layer protein from a pathogenic *Aeromonas hydrophila* strain. J. Bacteriol. *170*, 2631–2638.

Dubois-Darnaudpeys, A. (1977). Epidemiologie de la furonculose des salmonides. III. Ecologie de *Aeromonas salmonicida* proposition d'un modele epidemiologique. Bull. Français Piscicult. *50*, 21–32.

Duff, D.C.B. (1942). The oral immunization of trout against *Bacterium salmonicida*. J. Immunol. *44*, 87–94.

Durbin, M., McIntosh, D., Smith, P.D., Wardle, R., and Austin, B. (1999). Immunization against furunculosis in rainbow trout with iron-regulated outer membrane protein vaccines: relative efficacy of immersion, oral and injection delivery. J. Aquat. Anim. Health *11*, 68–75.

Ebanks, R.O., Knickle, L.C., Goguen, M., Boyd, J.M., Pinto, D.M., Reith, M., and Ross, N.W. (2006). Expression of and secretion through the *Aeromonas salmonicida* type III secretion system. Microbiology *152*, 1275–1286.

Effendi, I., and Austin, B. (1991). Survival of the fish pathogen *Aeromonas salmonicida* in seawater. FEMS Microbiol. Lett. *84*, 103–106.

Effendi, I., and Austin, B. (1993). A rapid method for the determination of viability of *Aeromonas salmonicida* in seawater. Bull. Eur. Assoc. Fish Pathol. *13*, 171–173.

Effendi, I., and Austin, B. (1994). Survival of the fish pathogen *Aeromonas salmonicida* in the marine environment. J. Fish Dis. *17*, 375–385.

Effendi, I., and Austin, B. (1995a). Dormant/unculturable cells of the fish pathogen *Aeromonas salmonicida*. Microb. Ecol. *30*, 183–192.

Effendi, I., and Austin, B. (1995b). Uptake of *Aeromonas salmonicida* by Atlantic salmon (*Salmo salar* L.). Bull. Eur. Assoc. Fish Pathol. *15*, 115–118.

Ellis, A.E., doVale, A., Bowden, T.J., Thompson, K., and Hastings, T.S. (1997). *In vivo* production of A-protein, lipopolysaccharide, iron-regulated outer membrane proteins and 70-kDa serine protease by *Aeromonas salmonicida* subsp. *salmonicida*. FEMS Microbiol. Lett. *149*, 157–163.

Embody, G.C., and Hayford, C.O. (1925). The advantage of rearing brook trout fingerlings from selected breeders. Trans. Am. Fish. Soc. *55*, 135–142.

Esteve, C., Alcaide, E., Canals, R., Merino, S., Blasco, D., Figueras, M.J., and Tomás, J.M. (2004). Pathogenic *Aeromonas hydrophila* serogroup O:14 and O:81 strains with an S layer. Appl. Environ. Microbiol. *70*, 5898–5904.

Eurell, T.E., Lewis, D.H., and Grumbles, L.C. (1978). Comparison of selected diagnostic tests for detection of motile *Aeromonas* septicaemia in fish. Am. J. Vet. Res. *39*, 1384–1386.

Ferguson, Y., Glover, L.A., McGillivray, D.M., and Prosser, J.I. (1995). Survival and activity of *lux*-marked *Aeromonas salmonicida* in seawater. Appl. Environ. Microbiol. *61*, 3494–3498.

Ferguson, Y., Bricknell, I.R., Glover, L.A., MacGregor, D.M., and Prosser, J.I. (1998). Colonisation and transmission of *lux*-marked and wild type *Aeromonas salmonicida* strains in Atlantic salmon (*Salmo salar* L.). FEMS Microbiol. Ecol. *27*, 251–260.

Frerichs, G.N., Millar, S.D., and McManus, C. (1992). Atypical *Aeromonas salmonicida* isolated from health wrasse (*Ctenolabrus rupestris*). Bull. Eur. Assoc. Fish Pathol. *12*, 48–49.

Fyfe, L., Finley, A., Coleman, G., and Munro, A.L.S. (1986). A study of the pathological effect of isolated *Aeromonas salmonicida* extracellular protease on Atlantic salmon, *Salmo salar* L. J. Fish Dis. *9*, 403–409.

Fyfe, L., Coleman, G., and Munro, A.L.S. (1987). Identification of major common extracellular proteins secreted by *Aeromonas salmonicida* strains isolated from diseased fish. Appl. Environ. Microbiol. *53*, 722–726.

Garduño, R.A., and Kay, W.W. (1995). Capsulated cells of *Aeromonas salmonicida* grown *in vitro* have different functional properties than capsulated cells grown *in vivo*. Can. J. Microbiol. *41*, 941–945.

Gudmundsdóttir, B.K., and Gudmundsdóttir, S. (1997). Evaluation of cross protection by vaccines against atypical and typical furunculosis in Atlantic salmon, *Salmo salar* L. J. Fish Dis. *20*, 343–350.

Gudmundsdóttir, B.K., Jónsdóttir, H., Stenhorsdóttir, V., Magnadóttir, B., and Gudmundsdóttir, S. (1997). Survival and humoral antibody response of Atlantic salmon (*Salmo salar* L.) vaccinated against *Aeromonas salmonicida* ssp. *achromogenes*. J. Fish Dis. *20*, 351–360.

Gudmunsdóttir, S., Lange, S., Magnadóttir, B., and Gudmundsdóttir, B.K. (2003). Protection against atypical furunculosis in Atlantic halibut, *Hippoglossus hippoglossus* (L.); a comparison of a commercial furunculosis vaccine and an autogenous vaccine. J. Fish Dis. *26*, 331–338.

Gunnlaugsdóttir, B., and Gudmundsdóttir, B.K. (1997). Pathogenicity of atypical *Aeromonas salmonicida* in Atlantic salmon compared with protease production. J. Appl. Microbiol. *83*, 542–551.

Gupta, S.K., Pal, A.K., Sahu, N.P., Dalvi, R., Kumar, V., and Mukherjee, S.C. (2008). Microbial levan in the diet of *Labeo rohita* Hamilton juveniles: effect on non-specific immunity and histopathological changes after challenge with *Aeromonas hydrophila*. J. Fish Dis. *31*, 649–657.

Hazen, T.C., Esch, G.W., Dimock, R.V., and Mansfield, A. (1982). Chemotaxis of *Aeromonas hydrophila* to the surface mucus of fish. Curr. Microbiol. *7*, 371–375.

Henery, M.A., and Secombes, C.J. (2000). The A-layer influences the susceptibility of *Aeromonas salmonicida* to antibacterial peptides. Fish Shellfish Immunol. *10*, 637–642.

Heuschmann-Brunner, G. (1978). Aeromonads of the '*hydrophila-punctata* group' in freshwater fishes. Arch. Hydrobiol. *83*, 99–125.

Hiney, M.P., Kilmartin, J.J., and Smith, P.R. (1994). Detection of *Aeromonas salmonicida* in Atlantic salmon with asymptomatic furunculosis infections. Dis. Aquat. Org. *19*, 161–167.

Hirst, I.D., and Ellis, A.E. (1996). Utilization of transferrin and salmon serum as sources of iron by typical and atypical strains of *Aeromonas salmonicida*. Microbiology *142*, 1543–1550.

Høie, S., Dalsgaard, I., Aase, I.L., Heum, M., Thornton, J.M., and Powell, R. (1999). Polymerase chain reaction (PCR) based typing analysis of atypical isolates of the fish pathogen *Aeromonas salmonicida*. Syst. Appl. Microbiol. *22*, 403–411.

Hsu, H.-M., Bowser, P.R., Schachte, J.H. Jr., Scarlett, J.M., and Babish, J.G. (1995). Winter field trials of enrofloxacin for the control of *Aeromonas salmonicida* infection in salmonids. J. Wld. Aquacult. Soc. *26*, 307–314.

Huntly, P.J., Coleman, G., and Munro, A.L.S. (1992). The nature of the lethal effect on Atlantic salmon, *Salmo salar* L., of a lipopolysaccharide-free phospholipase activity isolated from the extracellular products of *Aeromonas salmonicida*. J. Fish Dis. *15*, 99–102.

Huys, G., Coopman, R., Janssen, P., and Kersters, K. (1996). High-resolution genotypic analysis of the genus *Aeromonas* by AFLP fingerprinting. Int. J. Syst. Bacteriol. *46*, 572–580.

Huys, G., Kämpfer, P., and Swings, J. (2001). New DNA–DNA hybridization and phenotypic data on the species *Aeromonas ichthiosmia* and *Aeromonas allosaccharophila*: A. *ichthiosmia* is a later synonym of *A. veronii* Hickman-Brenner *et al.* 1987. Syst. Appl. Microbiol. *24*, 177–182.

Ishiguro, E.E., Kay, W.W., Ainsworth, T., Chamberlain, J.B., Austen, R.A., Buckley, J.T., and Trust, T.J. (1981). Loss of virulence during culture of *Aeromonas salmonicida* at high temperature. J. Bacteriol. *148*, 333–340.

Jeney, G., Ardó, L., Rónyai, A., Bercsényi, M., and Jeney, Z. (2011). Resistance of genetically different common carp, *Cyprinus carpio* L., families against experimental bacterial challenge with *Aeromonas hydrophila*. J. Fish Dis. *34*, 65–70.

Jutfelt, F., Olsen, R.E., Glette, J., Ringø, E., and Sundell, K. (2006). Translocation of viable *Aeromonas salmonicida* across the intestine of rainbow trout, *Oncorhynchus mykiss* (Walbaum). J. Fish Dis. *29*, 255–262.

Kaku, Y., Yamada, Y., and Wakabayashi, H. (1999). Characterization of atypical *Aeromonas salmonicida* isolated from an epizootic ulcerative disease in carp *(Cyprinus carpio)*. Fish Pathol. *34*, 155–162.

Kawahara, E., and Kusuda, R. (1987). Direct fluorescent antibody technique for diagnosis of bacterial disease in eel. Nipp. Suis. Gakk. *53*, 395–399.

Kawahara, E., Oshima, S., and Nomura, S. (1990). Toxicity and immunogenicity of *Aeromonas salmonicida* extracellular products to salmonids. J. Fish Dis. *13*, 495–503.

Kay, W.W., and Trust, T.J. (1991). Form and functions of the regular surface array (S-layer) of *Aeromonas salmonicida*. Experientia *47*, 412–414.

Kay, W.W., Buckley, J.T., Ishiguro, E.E., Phipps, B.M., Monette, J.P.L., and Trust, T.J. (1981). Purification and disposition of a surface protein associated with virulence of *Aeromonas salmonicida*. J. Bacteriol. *147*, 1077–1084.

Kimura, T. (1969). A new subspecies of *Aeromonas salmonicida* as an etiological agent of furunculosis on 'Sakuramasu' *(Oncorhynchus masou)* and pink salmon *(O. gorbuscha)* rearing for maturity. Part 1. On the serological properties. Fish Pathol. *3*, 34–44.

King, C.H., and Shotts, E.B. (1988). Enhancement of *Edwardsiella tarda* and *Aeromonas salmonicida* through ingestion by the ciliated protozoan *Tetrahymena pyriformis*. FEMS Microbiol. Lett. *51*, 95–100.

Kitao, T., and Yoshida, Y. (1986). Effect of an immunopotentiator on *Aeromonas salmonicida* infection in rainbow trout (*Salmo gairdneri*). Vet. Immunol. Immunopathol. *12*, 287–296.

Kogure, K., Simidu, U., and Taga, N. (1979). Distribution of viable marine bacteria in neretic seawater around Japan. Can. J. Microbiol. *26*, 318–323.

Koppang, E.O., Fjølstad, M., Melgård, B., Vigerust, M., and Sørum, H. (2000). Non-pigment-producing isolates of *Aeromonas salmonicida* subspecies *salmonicida:* isolation, identification, transmission and pathogenicity in Atlantic salmon, *Salmo salar* L. J. Fish Dis. *23*, 39–48.

Kumari, J., and Sahoo, P.K. (2006). Dietary immunostimulants influence specific immune response and resistance of healthy and immunocompromised Asian catfish *Clarias batrachus* to *Aeromonas hydrophila* infection. Dis. Aquat. Org. *70*, 63–70.

Li, J., Ni, X.D., Liu, Y.J., and Lu, C.P. (2011). Detection of three virulence genes *alt*, *ahp* and *aerA* in *Aeromonas hydrophila* and their relationship with actual virulence in zebrafish. J. Appl. Microbiol. *110*, 823–830.

Liu, J.Y., and Li, A.H. (2012). First case of *Aeromonas schubertii* infection in the freshwater cultured snakehead fish, *Ophiocephalus argus* (Cantor) in China. J. Fish Dis. *35*, 335–342.

Lund, V., Jenssen, L.M., and Wesmajervi, M.S. (2002). Assessment of genetic variability and relatedness among atypical *Aeromonas salmonicida* from marine fishes, using AFLP fingerprinting. Dis. Aquat. Org. *50*, 119–126.

Lund, V., Arnesen, J.A., Coucheron, D., Modalsli, K., and Syvertsen, C. (2003). The *Aeromonas salmonicida* A-layer protein is an important protective antigen in oil-adjuvanted vaccines. Fish Shellfish Immunol. *15*, 367–372.

Lund, V., Arnesen, J.A., Mikkelsen, H., Gravningen, K., Brown, L., and Schrøder, M.B. (2008). Atypical furunculosis vaccines for Atlantic cod *(Gadus morhua);* vaccine efficacy and antibody responses, Vaccine *26*, 6791–6799.

Lygren, B., Fausa Pettersen, E., Wergeland, H.I., and Endresen, C. (1998). An extracellular low molecular weight protease with activity against gelatin and casein produced earlier during growth of *Aeromonas salmonicida* ssp. *salmonicida* than the serine protease. Bull. Eur. Assoc. Fish Pathol. *18*, 96–101.

McCarthy, D.H. (1980). Some ecological aspects of the bacterial fish pathogen – *Aeromonas salmonicida.* In Aquatic Microbiology, Symposium of the Society of Applied Bacteriology No 6 (Academic Press, London, UK), pp. 299–324.

McCarthy, D.H., and Roberts, R.J. (1980). Furunculosis of fish – the present state of our knowledge. In Advances in Aquatic Microbiology, Droop, M.A., and Jannasch, H.W., eds (Academic Press, London, UK), pp. 293–341.

Markwardt, N.M., Gocha, Y.M., and Klontz, G.W. (1989). A new application for coomassie brilliant blue agar: detection of *Aeromonas salmonicida* in clinical samples. Dis. Aquat. Org. *6*, 231–233.

Marsden, M.J., Vaughan, L.M., Foster, T.J., and Secombes, C.J. (1996). A live (Δ*aroA*) *Aeromonas salmonicida* vaccine for furunculosis preferentially stimulates T-cell responses relative to B-cell responses in rainbow trout *(Oncorhynchus mykiss)*. Infect. Immun. *64*, 3863–3869.

Martin-Carnahan, A., and Joseph, S.W. (2005). Genus I. *Aeromonas* Stanier 1943, 213[AL]. In Bergey's Manual of Systematic Bacteriology, 2nd edn, vol. 2 The Proteobacteria, part B The Gammaproteobacteria, Brenner, D.J., Krieg, N.R., and Staley, J.T. eds (Springer, New York, NY), pp. 557–578.

Martinez-Murcia, A.J., Esteve, C., Garay, E., and Collins, M.D. (1992). *Aeromonas allosaccharophila* sp. nov., a new mesophilic member of the genus *Aeromonas*. FEMS Microbiol. Lett. *91*, 199–206.

Masada, C.L., LaPatra, S.E., Morton, A.W., and Strom, M.S. (2002). An *Aeromonas salmonicida* type IV pilin is required for virulence in rainbow trout *Oncorhynchus mykiss*. Dis. Aquat. Org. 51, 13–25.

Merino, S., Aguilar, A., Rubires, X., Simon-Pujol, D., Congregado, F., and Tomás, J.M. (1996). The role of the capsular polysaccharide of *Aeromonas salmonicida* in the adherence and invasion of fish cell lines. FEMS Microbiol. Lett. *142*, 185–189.

Merino, S., Aguilar, A., Rubires, X., Abitiu, N., Regué, M., and Tomás, J.M. (1997). The role of the capsular polysaccharide of *Aeromonas hydrophila* serogroup O:34 in the adherence to and invasion of fish cell lines. Res. Microbiol. *148*, 625–631.

Michel, C., and Dubois-Darnaudpeys, A. (1980). Persistence of the virulence of *Aeromonas salmonicida* kept in river sediments. Annal. Recherch. Vét. *11*, 375–386.

Midtlyng, P.J., Reitan, L.J., Lillehaug, A., and Ramstad, A. (1996). Protection, immune responses and side effects in Atlantic salmon *(Salmo salar* L.) vaccinated against furunculosis by different routes. Fish Shellfish Immunol. *6*, 599–600.

Misra, C.K., Das, B.K., Mukherjee, S.C., and Pradhan, S.J. (2007). Effects of dietary vitamin C on immunity, growth and survival of Indian major carp *Labeo rohita*, fingerlings. Aquacult. Nutr. *13*, 35–44.

Moki, S.T., Nomura, T., and Yoshimizu, M. (1995). Effects of incubation temperature for isolation on auto-agglutination of *Aeromonas salmonicida*. Fish Pathol. *30*, 67–68.

Morgan, J.A.W., Cranwell, P.A., and Pickup, R.W. (1991). Survival of *Aeromonas salmonicida* in Lake water. Appl. Environ. Microbiol. *57*, 1777–1782.

Morgan, J.A.W., Rhodes, G., and Pickup, R.W. (1993). Survival of non culturable *Aeromonas salmonicida* in lake water. Appl. Environ. Microbiol. *59*, 874–880.

Munro, A.L.S., Hastings, T.S., Ellis, A.E., and Liversidge, J. (1980). Studies on an ichthyotoxic material produced extracellularly by the furunculosis bacterium *Aeromonas salmonicida*. In Fish Diseases – Third COPRAQ Session, Ahne, W. ed. (Springer-Verlag, Berlin, Germany), pp. 98–106.

Mutoloki, S., Brudeseth, B., Reite, O.B., and Evensen, Ø. (2006). The contribution of *Aeromonas salmonicida* extracellular products to the induction of inflammation in Atlantic salmon (*Salmo salar* L.) following vaccination with oil-based vaccines. Fish Shellfish Immunol. *20*, 1–11.

Navot, N., Sinyakov, S., and Avtalion, R.R. (2011). Application of ultrasound in vaccination against goldfish ulcer disease: a pilot study. Vaccine *29*, 1382–1389.

Nese, L., and Enger, Ø. (1993). Isolation of *Aeromonas salmonicida* from salmon lice *Lepeophtheirus salmonis* and marine plankton. Dis. Aquat. Org. *16*, 79–81.

Nilsson, W.B., and Strom, M.S. (2002). Detection and identification of bacterial pathogens of fish in kidney tissue using terminal restriction fragment length polymorphism (T-RFLP) analysis of 16S rRNA genes. Dis. Aquat. Org. *48*, 175–185.

Nisr, R.B., Moody, A.J., and Gilpin, M.L. (2012). Screening microorganisms for insulin binding reveals binding by *Burkholderia multivorans* and *Burkholderia cepacia* and novel attachment of insulin to *Aeromonas salmonicida* via the A-layer. FEMS Microbiol. Lett. doi: 10.1111/j 1574–6968.2011. 02484x

Nya, E.J., and Austin, B. (2009a). Use of garlic, *Allium sativum*, to control *Aeromonas hydrophila* infection in rainbow trout, *Oncorhynchus mykiss* (Walbaum). J. Fish Dis. *32*, 963–970.

Nya, E.J., and Austin, B. (2009b). Use of dietary ginger, *Zingiber officinale* Roscoe, as an immunostimulant to control *Aeromonas hydrophila* infections in rainbow trout, *Oncorhynchus mykiss* (Walbaum). J. Fish Dis. *32*, 971–977.

Oakey, H.J., Ellis, J.T., and Gibson, L.F. (1998). The development of random DNA probes specific for *Aeromonas salmonicida*. J. Appl. Microbiol. *84*, 37–46.

Ogara, W.O., Mbuthia, P.G., Kaburia, H.F.A., Sørum, H., Kagunya, D.K., Nduthu, D.I., and Colquhoun, D. (1998). Motile aeromonads associated with rainbow trout *(Oncorhynchus mykiss)* mortality in Kenya. Bull. Eur. Assoc. Fish Pathol. *18*, 7–9.

Ortega, C., Ruiz, I., De Blas, I., Muzquiz, J.L., Fernandez, A., and Alonso, J.L. (1996). Furunculosis control using a paraimmunization stimulant (Baypamun) in rainbow trout. Vet. Res. (Paris) *27*, 561–568.

Paula, S.J., Duffey, P.S., Abbott, S.L., Kokka, R.P., Oshio, L.S., Janda, J.M., Shimada, T., and Sakazaki, R. (1988). Surface properties of autoagglutinating mesophilic aeromonads. Infect. Immun. *56*, 2658–2665.

Peréz, M.J., Rodríguez, L.A., and Nieto, T.P. (1998). The acetylcholinesterase ichthyotoxin is a common component of the extracellular products of Vibrionaceae strains. J. Appl. Microbiol. *84*, 47–52.

Pickup, R.W., Rhodes, G., Cobban, R.J., and Clarke, K.J. (1996). The postponement of non-culturability in *Aeromonas salmonicida*. J. Fish Dis. *19*, 65–74.

Poobalane, S., Thompson, K.D., Ardó, L., Verjan, N., Han, H.-J., Jeney, G., Hirono, I., Aoki, T., and Adams, A. (2010). Production and efficacy of an *Aeromonas hydrophila* recombinant S-layer protein vaccine for fish. Vaccine *28*, 3540–3547.

Popoff, M. (1984). Genus III. *Aeromonas* Kluyver and van Niel 1936, 398[AL]. In Bergey's Manual of Systematic Bacteriology, vol. 1, Krieg, N.R., and Holt, J.G., eds (Williams & Wilkins, Baltimore, MD), pp. 545–548.

Rabb, L., Cornick, J.W., and MacDermott, L.A. (1964). A macroscopic slide agglutination test for the presumptive diagnosis of furunculosis in fish. Progr. Fish Cult. *26*, 118–120.

Sakai, D.K. (1986). Electrostatic mechanism of survival of virulent *Aeromonas salmonicida* strains in river water. Appl. Environ. Microbiol. *51*, 1343–1349.

Sakai, M., Atsuta, S., and Kobayashi, M. (1986). A streptococcal disease of cultured Jacopever, *Sebastes schlegeli*. Suisanzoshuku (Aquiculture) *34*, 171–177.

Samuelsen, O.B., Hjeltnes, B., and Glette, J. (1998). Efficacy of orally administered florfenicol in the treatment of furunculosis in Atlantic salmon. J. Aquat. Anim. Health *10*, 56–61.

Santos, Y., Toranzo, A.E., Dopazo, C.P., Nieto, T.P., and Barja, J.L. (1987). Relationship among virulence for fish, enterotoxigenicity, and phenotypic characteristics of motile *Aeromonas*. Aquaculture *67*, 29–39.

Santos, Y., García-Marquez, S., Pereira, P.G., Pazos, F., Riaza, A., Silva, R., El Morabit, A., and Ubeira, F.M. (2005). Efficacy of furunculosis vaccines in turbot, *Scophthalmus maximus* (L.): evaluation of immersion, oral and injection delivery. J. Fish Dis. *28*, 165–172.

Schachte, J.H. (1978). Immunization of channel catfish, *Ictalurus punctatus*, against two bacterial diseases. Mar. Fish. Rev. *40*, 18–19.

Schubert, R.H.W. (1974). Genus II. *Aeromonas* Kluyver and van Niel 1936, 398. In Bergey's Manual of Determinative Bacteriology, 8th edn, R.E. Buchanan, and N.E. Gibbons, eds, (Williams and Wilkins, Baltimore, MD), pp. 345–348.

Selvaraj, V., Sampath, K., and Sekar, V. (2005). Administration of yeast glucan enhances survival and some non-specific and specific immune parameters in carp (*Cyprinus carpio*) infected with *Aeromonas hydrophila*. Fish Shellfish Immunol. *19*, 293–306.

Selvaraj, V., Sampath, K., and Sekar, V. (2009). Administration of lipopolysaccharide increases specific and non-specific immune parameters and survival in carp (*Cyprinus carpio*) infected with *Aeromonas hydrophila*. Aquaculture *286*, 176–183.

Siwicki, A.K., Klein, P., Morand, M., Kiczka, W., and Studnicka, M. (1998). Immunostimulatory effects of dimerized lysozyme (KLP-602) on the nonspecific defense mechanisms and protection against furunculosis in salmonids. Vet. Immunol. Immunopathol. *61*, 369–378.

Smith, I.W. (1963). The classification of *Bacterium salmonicida*. J. Gen. Microbiol. *33*, 263–274.

Snieszko, S.F., Griffin, P., and Friddle, S.B. (1950). A new bacterium (*Haemophilus piscium* n. sp.) from ulcer disease of trout. J. Bacteriol. *59*, 699–710.

Swain, P., Behera, T., Mohapatra, D., Nanda, P.K., Nayak, S.K., Meyer, P.K., and Das, B.K. (2010). Derivation of rough attenuated variants from smooth virulent *Aeromonas hydrophila* and their immunogenicity in fish. Vaccine *28*, 4626–4631.

Tapia-Cammas, D., Yañez, A., Arancibia, G., Toranzo, A.E., and Avendaño-Herrera, R. (2011). Multiplex PCR for the detection of *Piscirickettsia salmonis, Vibrio anguillarum, Aeromonas salmonicida* and *Streptococcus phocae* in Chilean marine farms. Dis. Aquat. Org. *97*, 135–142.

Thornton, J.C., Garduño, R.A., Newman, S.G., and Kay, W.W. (1991). Surface-disorganized, attenuated mutants of *Aeromonas salmonicida* as furunculosis live vaccines. Microb. Pathog. *11*, 85–99.

Tu, F.P., Chu, W.H., Zhuang, X.Y., and Lu, C.P. (2010). Effect of oral immunization with *Aeromonas hydrophila* ghosts on protection against experimental fish infection. Lett. Appl. Microbiol. *50*, 13–17.

Udey, L.R. (1982). A differential medium for distinguishing A/r^{+} from A/r^{-} phenotypes in *Aeromonas salmonicida*. In Proceedings of the 13th Annual Conference and Workshop and 7th Eastern Fish Health Workshop, (International Association for Aquatic Animal Medicine, Baltimore, MA), p. 41.

Van Alstine, J.M., Trust, T.J., and Brooks, D.E. (1986). Differential partition of virulent *Aeromonas salmonicida* and attenuated derivatives possessing specific cell surface alterations in polymer aqueous-phase systems. Appl. Environ. Microbiol. *51*, 1309–1313.

Vaughan, L.M., Smith, P.R., and Foster, T.J. (1993). An aromatic-dependent mutant of the fish pathogen *Aeromonas salmonicida* is attenuated in fish and is effective as a live vaccine against the salmonid disease furunculosis. Infect. Immun. *61*, 2172–2181.

Wiklund, T. (1995). Survival of 'atypical' *Aeromonas salmonicida* in water and sediment microcosms of different salinities and temperatures. Dis. Aquat. Org. *21*, 137–143.

Wiklund, T., Lönnström, L., and Niiranen, H. (1993). *Aeromonas salmonicida* ssp. *salmonicida* lacking pigment production, isolated from farmed salmonids in Finland. Dis. Aquat. Org. *15*, 219–223.

Wiklund, T., Dalsgaard, I., Eerola, E., and Olivier, G. (1994). Characteristics of 'atypical' cytochrome-oxidase negative *Aeromonas salmonicida* isolated from ulcerated flounder (*Platichthys flesus* (L.). J. Appl. Bacteriol. *76*, 511–520.

Wiklund, T., and Dalsgaard, I. (1995). Atypical *Aeromonas salmonicida* associated with ulcerated flatfish species in the Baltic Sea and the North Sea. J. Aquat. Anim. Health *7*, 218–224.

Williams, P.J., Courtenay, S.C., and Vardy, C. (1997). Use of enrofloxacin to control atypical *Aeromonas salmonicida* in Atlantic tom cod. J. Aquat. Anim. Health *9*, 216–222.

Zhang, Y.L., Ong, C.T., and Leung, K.Y. (2000). Molecular analysis of genetic differences between virulent and avirulent strains of *Aeromonas hydrophila* isolated from diseased fish. Microbiology *46*, 999–1009.

Zhao, Y., Liu, Q., Wang, X., Zhou, L., Wang, Q., and Zhang, Y. (2011). Surface display of *Aeromonas hydrophila* GAPDH in attenuated *Vibrio anguillarum* to develop a noval multivalent vector vaccine. Mar. Biotechnol *13*, 963–970.

Aeromonas Infections in Humans

4

Maria José Figueras and Roxana Beaz-Hidalgo

Abstract

Even though the last review about the taxonomy, pathogenicity and infections produced by *Aeromonas* was published fairly recently, in 2010, eight new species have since then been incorporated into the genus. Interestingly during this period the species *Aeromonas dhakensis,* that resulted from the reclassification of *Aeromonas aquariorum* and *Aeromonas hydrophila* subsp. *dhakensis,* became a relevant clinical species when sequences of housekeeping genes were used for identification. This species prevailed in Malaysia above the typical dominating clinical species *Aeromonas caviae* and *Aeromonas veronii,* or showed an almost equal frequency in Australia or Taiwan. The most frequent clinical presentation remained gastroenteritis, and arguments in favour of considering *Aeromonas* a true enteropathogen are provided. Other common presentations like wound and soft tissue infections have continued to be reported and seemed to prevail in European studies, while bacteraemia affecting patients who have hepatobiliary disease or cirrhosis continued to be dominating in Asia (Thailand, Taiwan, etc.). New cases of necrotizing fasciitis, pneumonia after near drowning and infections after leech therapy have also been reported during this period. This chapter pretends to make clinicians more familiar with the infections produced by these bacteria providing tools for their correct identification.

Introduction

As indicated in 2010 by Janda and Abbott in their review, more than 100 years have passed since a bacteria presumed to be *Aeromonas* produced the red leg disease in frogs and more than half a century since the first association with human disease was made in 1951. Since then, *Aeromonas* have continually been implicated in different human and animal infections. Furthermore, we know that *Aeromonas* are ubiquitous of aquatic ecosystems, and are frequently isolated from water (Figueras and Borrego, 2010), food products (for a review see Figueras and Beaz-Hidalgo, 2014) and in association with fish disease i.e. furunculosis, septicaemia, ulcerative and haemorrhagic diseases, which causes widespread mortality damaging the economy in the aquaculture sector (for a review see Figueras *et al.*, 2011c; Beaz-Hidalgo and Figueras, 2012, 2013). Various authors have reported infections in humans mainly gastroenteritis, septicaemia and wound infections as well as other infections that affect the hepatobiliary system, the respiratory tract, bone and/or joints etc. (Janda and Abbott, 1996, 1998; Figueras, 2005; von Graevenitz, 2007; Janda and Abbott, 2010; Horneman and Ali, 2011; Parker and Shaw, 2011). Since the 1990s, Michael Janda and Sharon Abbott have regularly updated information that has accumulated on the clinical importance of *Aeromonas* (Janda and Abbott, 1996, 1998; Abbott *et al.*, 2003; Janda and Abbott, 2010 and references therein). In fact, their most recent review (Janda and Abbott, 2010) provides the most comprehensive reference of the genus. There they analyse the criteria used for defining new species in line with the current taxonomy of the genus, complementing their most recent clinical reviews with new data (Janda and Abbott, 1998; Abbott *et al.*, 2003). This chapter will further update the information presented in the last review and will provide other views on the discovery of new species and on the classically debated role of *Aeromonas* and diarrhoea. In this sense a concise summary of the existing controversial theories in relation with the association of *Aeromonas* and diarrhoea was provided in 2007 (Figueras *et al.*, 2007c) and will be re-evaluated again in this chapter in view of the new information that has become available in the literature. The data presented in the new retrospective studies that examine the characteristics of patients with *Aeromonas* bacteraemia/septicaemia from different geographical areas, i.e. France (Lamy *et al.*, 2009), French Caribbean Islands (Hochedez *et al.*, 2010), Taiwan (Tsai *et al.*, 2009; Lay *et al.*, 2010; Chuang *et al.*, 2011; Wu *et al.*, 2011, 2012a–c; Su *et al.*, 2012; Tang et at., 2014) and Japan (Morinaga *et al.*, 2011; Kimura *et al.*, 2013) will be analysed and compared with the existing studies. Furthermore, we will include the cases of sepsis (Monaghau *et al.*, 2008; Stano *et al.*, 2009: Levine *et al.*, 2010; Okumura *et al.*, 2011; Carmignani *et al.*, 2012; Coutinho *et al.*, 2012; Khalil *et al.*, 2013; Shin *et al.*, 2013) wound infections (Azzopardi *et al.*, 2011; Chao *et al.*, 2013a; Chen *et al.*, 2013), biliary tract infection (Chao *et al.*, 2013b; Fica *et al.*, 2013), urinary infection (Hua *et al.*, 2004; Ojeda-Vargas *et al.*, 2005; Al-Benwan *et al.*, 2007; Mandal *et al.*, 2010; Chao *et al.*, 2012a), empyema (Chao *et al.*, 2012b), pneumonia (Issa and Napolitano 2011; Nagata *et al.*, 2011; Chao *et al.*, 2013c), necrotizing fasciitis (Tsai *et al.*, 2009; Kao and Kao 2012; Kumar *et al.*, 2012), ecthyma gangrenosum (Avolio *et al.*, 2009; García-Lozano *et al.*, 2012), meningitis (Pampín *et al.*, 2012) and peritonitis (Chang *et al.*, 2005; Yang *et al.*, 2008; Wu *et al.*, 2009; Sahin *et al.*, 2010; Liakopoulos *et al.*, 2011; Lin *et al.*, 2013) that have been published in recent years.

The discovery of new disease associations involving serious infections has continued and *Aeromonas* have been the main protagonists following natural disasters, such as the 2001 tsunami in Thailand and hurricane Katrina in the southern states of the USA in 2005. These bacteria were the most frequent microbes found associated with wound infections in the survivors (Hiransuthikul *et al.*, 2005; Presley *et al.*, 2006). This will be presented and discussed in the chapter within the context of climate change.

Even though the last review was published fairly recently, in 2010, eight new species have since been incorporated into the genus: *Aeromonas piscicola* recovered from diseased salmon (Beaz-Hidalgo *et al.*, 2009), *Aeromonas fluvialis* (Alperi *et al.*, 2010b), *Aeromonas rivuli* (Figueras *et al.*, 2011a), *Aeromonas cavernicola* (Martínez-Murcia *et al.*, 2013) and *Aeromonas australiensis* (Aravena-Román *et al.*, 2013) from water and three new clinical species all recovered from wound infections, i.e. *Aeromonas taiwanensis* and *Aeromonas sanarellii* described by Alperi *et al.* (2010a) and *Aeromonas diversa* by Miñana-Galvis *et al.* (2010). The latter species includes two strains referred to classically as *Aeromonas* group 501 (Miñana-Galvis *et al.*, 2010). Furthermore, new information about the distribution of the species *Aeromonas aquariorum* (recently renamed as *Aeromonas dhakensis*, Beaz-Hidalgo *et al.*, 2013) has been published that fills the gap in these data underlined by Janda and Abbott (2010). For instance this species now has been found in water samples, fish and chironomid egg masses (Avena-Román *et al.*, 2011a; Figueras *et al.*, 2011d; Esteve *et al.*, 2012) and has also been recognized as an important clinical species in Australia, Taiwan, Malaysia and Japan (Avena-Román *et al.*, 2011a; Wu *et al.* 2012c,d; Puthucheary *et al.*, 2012; Chen *et al.*, 2013; Morinaga *et al.*, 2013). Other new species have also been recognized, such as '*Aeromonas lusitana*' and three others ('*Aeromonas finlandiensis*', '*Aeromonas aquatica*' and '*Aeromonas lacus*') that have been isolated from freshwater associated to cyanobacteria blooms in Finland and are in the process of publication (Martinez-Murcia *et al.*, unpublished; Beaz-Hidalgo *et al.*, unpublished). The explosion in the discovery of new species is due to new high-resolution techniques used for recognizing *Aeromonas* species, such as the sequencing of housekeeping genes like *rpoD* or *gyrB* etc. This approach is now routinely used at our and at other laboratories in order to avoid misidentification when using conventional or automatic phenotypic identification methods (Figueras *et al.*, 2009; Aravena-Roman *et al.*, 2011a,b; Senderovich *et al.*, 2012; Morinaga *et al.*, 2013); or using the 16S rRNA gene (Alperi *et al.*, 2008; Beaz-Hidalgo *et al.*, 2010). Sequences of the 16S rRNA gene can be used to assign the strains to the genus but may lead to speciation errors due to the high similarity of the sequences among certain species of the genus (Alperi *et al.*, 2008; Figueras *et al.*, 2011b). This will be discussed in light of the common use of this gene for difficult-to-identify bacteria in the clinical setting (Janda and Abbott, 2007) and in light of the new generation sequencing techniques like pyrosequencing, illumina, etc.

Aeromonas, like many other recognized pathogenic microbes, possess a great variety of virulence determinants involved in the colonization, invasion and proliferation of the bacteria leading to the development of infections. The bacteria damages host tissues and alters the immune defence system (Beaz-Hidalgo and Figueras, 2012, 2013). These virulence determinants have been reviewed recently, including the information derived from the available genomes (Tomás, 2012; Beaz-Hidalgo and Figueras, 2012, 2013), so no attention will be given to the virulence in this chapter.

As we mentioned in our previous review, the clinical importance given to *Aeromonas hydrophila* is erroneous and is due to limitations in identification methods, as several studies have verified (Figueras, 2005; Soler *et al.*, 2003b; Figueras *et al.*, 2009; Beaz-Hidalgo *et al.*, 2010; Avena-Román *et al.*, 2011a; Wu *et al.*, 2012c; Su *et al.*, 2013; Morinaga *et al.*, 2013) and this will be discussed again here. Therefore, when a particular species is mentioned in this chapter it is because we are confident that the identification of the strain is reliable, otherwise we will just refer to the strain/s as *Aeromonas*. The overall aim of this chapter is to make clinicians more familiar with these bacteria.

Exponential discovery of new species

Janda and Abbott's review (2010) criticized the fact that among the seven *Aeromonas* species published since 2002 only three (*A. molluscorum*, *A. aquariorum*, *A. tecta*) were proposed based upon the analysis of more than three strains. They commented that this was in sharp contrast to well-defined species such as *A. media*, *A. veronii*, *A. schubertii*, *A. jandaei*, *A. trota* and *A. bestiarum*, where the number of strains included in their description ranged from 7 to 15 strains (average of 10.2 strains/species). However, these authors did not consider that most of the mentioned species had been defined following the DNA–DNA hybridization (DDH) studies carried out by Popoff in the early 1980s (Martin-Carnahan and Joseph, 2005; Figueras *et al.* 2011b), that had already recognized those species as belonging to different hybridization groups (HGs), so several strains of those species were available before they were really formally named.

Until 2008, the genus had included 20 species (Janda and Abbott, 2010), but eight new species have been added since then (Table 4.1), most of them again having been described from a small number of isolates. However, as we have discussed elsewhere (Figueras *et al.*, 2011b), the discovery of a new species depends mostly on luck and on the identification methodology employed, and in our experience and based on data available at the literature,

Table 4.1 Species included in the genus *Aeromonas*

Species	Year	Source of the type strain
A. hydrophila[a,b,c,d]	1943	Tin of milk with fishy odour
A. salmonicida[a,b,c,d]	1953	Salmon
A. sobria[a,b,d]	1976	Fish
A. media[a,b,c,d]	1983	Effluent water from a fish farm
A. veronii[a,b,c,d]	1987	Sputum of a drowning victim
A. caviae[a,b,c,d]	1988	Epizootic of young guinea pig
A. eucrenophila[a,b,c,d]	1988	Freshwater fish
A. schubertii[a,d]	1988	Forehead abscess injury
A. jandaei[a,b,d]	1991	Faeces of a male with diarrhoea
A. trota[a]	1991	Human faeces
A. allosaccharophila[a,b,c,d]	1992	Diseased eel
A. encheleia[a,b,d]	1995	Healthy eel
A. bestiarum[a,b,c,d]	1996	Diseased fish
A. popoffii[a,b]	1997	Drinking water production plant
A. simiae[c]	2004	Monkey faeces
A. molluscorum[b,e]	2004	Bivalve molluscs (wedge-shells)
A. bivalvium[b,e]	2007	Bivalve molluscs (cockles)
A. tecta[a,b,d]	2008	Faeces of a child with diarrhoea
A. piscicola[b,d]	2009	Diseased salmon
A. fluvialis[b]	2010	River water
A. taiwanensis[a,b]	2010	Burn wound
A. sanarellii[a,b]	2010	Elbow wound
A. diversa[a]	2010	Leg wound
A. rivuli[b]	2011	Karst water rivulet
A. australiensis[b]	2013	Treated effluent used for irrigation
A. cavernicola[b]	2013	Water of a brook cavern
A. dhakensis[a,b,c,d]	2013	Faeces of a child with diarrhoea

The most prevalent species isolated from clinical samples are in bold.
[a–e]Indicates isolation sources for the different species: [a]clinical samples, [b]water, [c]meat, [d]fish and [e]bivalve molluscs.
The species *A. dhakensis* (Beaz-Hidalgo *et al.*, 2013) resulted of the reclassification of *A. aquariorum* (Martínez-Murcia *et al.*, 2008) and *A. hydrophila* subsp. *dhakensis* (Huys *et al.*, 2002).

only a few strains of a new taxa are found in practice. Something else may occur when a re-evaluation of a genus or a species is made using many previously known strains and new molecular tools with higher discrimination capacity. The descriptions of the eight new *Aeromonas* species mentioned above followed the requirements laid down for defining new bacterial species reviewed in a previous paper (Figueras *et al.*, 2011b). Those requirements include a phenotypic characterization to define diagnostic properties for the differentiation of the new species from the existing ones, a phylogenetic analysis based on the 16S rRNA gene sequences (> 1300 bp) and, if the similarity of the sequence with the closely existing species is above 97%, DDH studies should be performed with those species (Figueras *et al.*, 2011b). A phylogenetic analysis of at least five housekeeping genes and genotyping of the isolates are also among the recommended requirements (Figueras *et al.*, 2006, 2011b). In fact, the 16S rRNA gene similarity threshold (>97%) used in other genera to determine what are the most closely related species to perform DDH experiments is not useful in *Aeromonas* because most of the species show similarities above this threshold and also above the newly proposed threshold values 98.7–99% (Figueras *et al.*, 2011b). However, the analysis of the sequences of housekeeping genes (i.e. *gyrB*, *gyrA*, *rpoD*, *rpoB*, *recA*, *dnaJ*, *dnaX*, *atpD*, *cpn60*, etc.) have proven to have a higher capacity to discriminate the *Aeromonas* species (Soler *et al.*, 2004; Lamy *et al.*, 2010a; Figueras *et al.*, 2011a,b and references therein). In fact, the concatenated sequences of seven of the latter genes (*gyrB*, *rpoD*, *recA*, *dnaJ*, *gyrA*, *dnaX*, and *atpD*) were used to carry out a phylogenetic analysis, the so-called Multilocus Sequence Analysis (MLSA) or, more correctly in our view, the multilocus phylogenetic analysis (MLPA) to establish relationships among the species of the genus (Martínez-Murcia *et al.*, 2011). This approach, as commented proved to be a more discriminative tool than the phylogeny derived from the 16S rRNA gene and showed a good congruence with the taxonomy of the genus based on the DDH results (Martínez-Murcia *et al.*, 2011). Therefore, it has been proposed as an alternative technique to the DDH for defining new species (Figueras *et al.*, 2011b; Martínez-Murcia *et al.*, 2011). Another MLPA that used six different genes (*dnaK*, *gltA*, *radA*, *rpoB*, *tsf* and *zipA*) was developed by Roger *et al.* (2012b), these authors reached similar conclusions to those provided by Martínez-Murcia *et al.* (2011). Another approach based on the information derived from the sequences of several genes is the multilocus sequence typing (MLST) used to characterize the epidemiological relationships among isolates of a bacterial species (Figueras *et al.*, 2011b, and references therein). The MLST uses the sequences of internal fragments of multiple housekeeping genes (approximately 450–500 bp of seven genes) and for each gene the sequences of different strains are compared and each unique sequence is assigned a specific number (allele). The strains that show the same alleles (equal sequences of all genes) belong to the same sequence type (ST), which is also identified by a specific number. The first open-access MLST scheme for *Aeromonas* was developed in 2011 by Martino *et al.* (2011) (http://pubmlst.org/*aeromonas*), using six genes (*gyrB*, *groL*, *gltA*, *metG*, *ppsA*, and *recA*), and included mostly environmental and fish isolates. In the study mentioned above by Roger *et al.* (2012b) the same type of analysis was carried up but using five different genes (*dnaK*, *radA*, *rpoB*, *tsf* and *zipA*). They found 175 different STs among the 191 isolates studied from human, animal and environmental sources, although many ST (93.7%, 164/175) were only recognized once; these results indicate a high level of population diversity in the genus *Aeromonas* (Roger *et al.*, 2012b). However, these STs are not available in a database as the one proposed by Martino *et al.* (2011).

The hybridization group (HG) terminology introduced originally by Popoff to define the species and that appeared still in the latest edition of *Bergey's Manual* (Martin-Carnahan and Joseph, 2005), and in recent studies (Ghenghesh *et al.*, 2008; Pablos *et al.*, 2011; Igbinosa *et al.*, 2012) it is considered totally outdated. We agree with Janda and Abbott (2010) that this terminology should be abandoned.

Impact of the molecular identification methods on the discovery of new species

The introduction of molecular identification methods has been an important driver for the discovery of new *Aeromonas* species. Apart from the above-mentioned DNA–DNA studies by Popoff, which recognized many species, another big step forward occurred around 10 years later when Martínez-Murcia *et al.* (1992) published the first phylogeny of the genus *Aeromonas* based on the analysis of the 16S rRNA gene. That study recognized that the 16S rRNA gene was highly conserved, showing a high similarity (96.7–100%) for some species for instance *A. hydrophila* (ATCC 7966^{T}) exhibited only three nucleotide differences with the type strain of *A. media* (ATCC 33907^{T}) and the type strain of *A. trota* (ATCC 49657^{T}) was also almost identical to the one of *A. caviae* (NCIMB 13016^{T}). Nevertheless, the 16S rRNA gene is useful for the identification of more distant species and enabled the first clinical case attributed to *A. popoffii* to be demonstrated (Hua *et al.*, 2004). Despite the analysis of this gene was considered a reliable and straightforward tool for the identification of *Aeromonas* spp. for a certain time (Figueras, 2005) soon other important limitations were revealed, such as the presence of mutations or microheterogeneities in the different copies of the 16S rRNA gene that might interfere with identification (Martínez-Murcia *et al.*, 2005; Al-Benwan *et al.*, 2007; Alperi *et al.*, 2008; Roger *et al.*, 2012a). Also, the sequences of short fragments of the 16S rRNA gene (200–500 nt) have been used to identify many bacteria or microbial communities, but they are unreliable for identifying *Aeromonas* to the species level (Janda and Abbott, 2007; Figueras *et al.*, 2011b). For example, strains identified as *A. sobria* on the basis of a short fragment of the 16S rRNA gene in fact belonged to a new species *A. rivuli* (Figueras *et al.*, 2011a). The introduction of pyrosequencing techniques has made sequencing cheaper and therefore more popular, however, researchers should be aware of the mentioned limitation.

Many molecular methods (different from sequencing) have been proposed for identifying *Aeromonas* (Borrell *et al.*, 1997; Figueras *et al.*, 2000b, 2007b, and references therein). It was recommended that caution should be taken with molecular methods because some *Aeromonas* contaminated reagents have been detected which may interfere in the results (Liu *et al.*, 2012). A molecular genus probe was developed in our laboratory by Chacón *et al.* (2002) that targets the glycerophospholipid-cholesterol acyltransferase gene (*gcat*), which despite being considered an important virulence factor, it is present in all *Aeromonas* species (Chacón *et al.* 2002; Park *et al.*, 2003; Soler *et al.*, 2003b; Beaz-Hidalgo *et al.*, 2010). This genus probe was designed because confusion with other genera occurred in the identification of 10.8% to 24% of the isolates (Borrell *et al.*, 1997; Soler *et al.*, 2003b; Beaz-Hidalgo *et al.*, 2010). The presence of this gene has been used in several studies to verify that the isolates belonged to the genus *Aeromonas* (Beaz-Hidalgo *et al.*, 2010; Puthucheary *et al.*, 2012). For a review of all molecular methods defined for the identification of *A. salmonicida*, see Beaz-Hidalgo and Figueras (2012), and for those used for food, see Figueras and Beaz-Hidalgo (2014).

A specific protocol based on restriction fragment length polymorphism (RFLP), designed from the *Aeromonas* 16S rRNA diagnostic regions, generated species-specific banding patterns that enable the identification of the 14 accepted species up to the year 2000 (Borrell *et al.*, 1997; Figueras *et al.*, 2000a,b). The 16S rRNA-RFLP method includes an initial digestion of the 16S rRNA amplified gene (1503 bp) with two restriction enzymes (*Alu*I + *Mbo*I) and identification is made by comparing the obtained banding patterns with those defined as specific for 10 species or as a common pattern for *A. salmonicida*/*A. bestiarum*/*A. encheleia*/*A. popoffii* and also now *A. piscicola*. Separation of these species requires further digestions with other enzymes (Figueras *et al.*, 2000b). The method has proven to be useful to us and to others for identifying *Aeromonas* strains to the species level (Castro-Escarpulli *et al.*, 2003; Ørmen *et al.*, 2005; Aguilera-Arreola *et al.*, 2005, 2007; Beaz-Hidalgo *et al.*, 2010; Lamy *et al.*, 2010a; Puthuchearay *et al.*, 2012, among others). However, the RFLP identification method should be applied following the order of digestions in the described protocol (Borrell *et al.*, 1997) to avoid mistake (Figueras *et al.*, 2007b). Using this method we discovered new RFLP patterns that either corresponded to new species or were due to the existence of microheterogeneities or mutations mentioned above (Martínez-Murcia *et al.*, 2005; Alperi *et al.*, 2008; Figueras *et al.*, 2011b; Figueras and Beaz-Hidalgo, 2014). In fact, we have discovered that those mutations appeared in 8.1% of the strains and were significantly more prevalent in clinical than in environmental strains of the species *A. caviae* and *A. veronii*, although we do not yet know the significance of this finding (Alperi *et al.*, 2008). Using PCR temporal temperature gel electrophoresis (PCR-TTGE) Roger *et al.* (2012a) found microheterogeneities in 41.5% (81/195) of the *Aeromonas* strains. In agreement with our results they were more frequent in human than in environmental strains.

The addition of 11 new species in the genus since 2000 limits the usefulness of the commented 16S rRNA-RFLP method because the same patterns are obtained for different species, e.g. *A. dhakensis* (previously *A. aquariorum*) and *A. caviae*, both common in clinical samples; or other species implicated in fish disease, e.g. *A. salmonicida*, *A. bestiarum* and *A. piscicola* (Beaz-Hidalgo *et al.*, 2009, 2010, 2013). Nevertheless, new species-specific RFLP patterns have been obtained (using the *Alu*I + *Mbo*I) for the recently described species *A. tecta*, *A. molluscorum*, *A. simiae* and *A. rivuli* (Figueras and Beaz-Hidalgo, 2014). Owing to the limitation of this method, we explored and demonstrated that other genes that codify essential proteins for the survival of the bacteria, such as *gyrB* (β-subunit DNA gyrase) or *rpoD* (σ70 factor of the RNA polymerase), have a higher resolution than the 16S rRNA gene and are therefore more useful for separating *Aeromonas* species (Yañez *et al.*, 2003; Soler *et al.*, 2004). The use of the sequences of one single gene (*gyrB* and/or *rpoD*) have proven to be an excellent tool for identifying species and for recognizing potentially new species (Yañez *et al.*, 2003; Soler *et al.*, 2004; Martínez-Murcia *et al.*, 2008; Alperi *et al.*, 2010a,b; Beaz-Hidalgo *et al.*, 2009; Figueras *et al.*, 2011a) and has allowed wrongly described taxa like *A. hydrophila* subsp. *dhakensis* (now known as *A. dhakensis*), among others, to finally be correctly identified (Figueras *et al.*, 2009; Beaz-Hidalgo *et al.*, 2010; Figueras *et al.*, 2011b; Beaz-Hidalgo *et al.*, 2013). Furthermore this approach has also helped to recognize strains belonging to rare *Aeromonas* species, such as *A. simiae* (Fontes *et al.*, 2010, 2011), *A. allosaccharophila* (Saavedra *et al.*, 2007; Aravena-Román *et al.*, 2011a) and *A. bivalvium* (Nagar *et al.*, 2013) or new isolates of the species *A. aquariorum* now known as *A. dhakensis* (Figueras *et al.*, 2009; Aravena-Román *et al.*, 2011a; Puthucheary *et al.*, 2012; Sedláček *et al.*, 2012; Wu *et al.*, 2012c,d; Morinaga *et al.*, 2013; Chen *et al.*, 2013; Shin *et al.*, 2013), *A. tecta*

(Beaz-Hidalgo *et al.*, 2010), *A. taiwanensis* (Senderovich *et al.*, 2012; Figueira *et al.*, 2011; Beaz-Hidalgo *et al.*, 2012; Chen *et al.*, 2013) and *A. sanarellii* (Figueira *et al.*, 2011; Beaz-Hidalgo *et al.*, 2012; Chen *et al.*, 2013). The 16S rRNA-RFLP method in combination with sequencing the *rpoD* gene for verification of doubtful results has been an excellent approach in several studies (Figueras *et al.*, 2009; Beaz-Hidalgo *et al.*, 2010; Puthucheary *et al.*, 2012; Pérez-Valdespino *et al.*, 2013).

The use of the sequences of housekeeping genes have indeed broadened the diversity of *Aeromonas* species and evidenced that the terminology referring to a 'complex of species' still used in some recent studies (Wang *et al.*, 2009; Hochedez *et al.*, 2010) should be avoided because it is imprecise and ignores the true incidence of the individual species. In our view the most correct approach is that laboratories unable to properly identify strains by sequencing one housekeeping gene (*rpoD, rpoB, gyrB*, etc.) should refer to their isolates as *Aeromonas* sp. Then they should send their strains to reference laboratories for a proper identification before they publish their reports. In our view, this is the only way to avoid the classical bias that considers *A. hydrophila* the most frequent and important clinical species as commented previously (Figueras, 2005; Su *et al.*, 2013). In fact, *A. hydrophila* is still the most named species in the titles of the references included in this chapter and we hope this will change in the near future.

Prevalent clinical species using reliable identification methods

Until recently, only three species were clearly predominating in clinical samples, i.e. *A. caviae, A. veronii* and *A. hydrophila*, which accounted for up to 92% of the clinical strains (Figueras, 2005; Figueras and Beaz-Hidalgo, 2014). Additional species found in clinical samples are *A. media, A. jandaei, A. bestiarum, A. salmonicida, A. sobria, A. eucrenophila, A. schubertii, A. encheleia, A. allosaccharophila*, and *A. trota* (Figueras *et al.*, 2000a). However, considering recent data now we have to add *A. popoffii, A. dhakensis* (=*A. aquariorum*), *A. tecta, A. diversa, A. taiwanensis* and *A. sanarellii* (Hua *et al.*, 2004; Demarta *et al.*, 2008; Martínez-Murcia *et al.*, 2008; Miñana-Galbis *et al.*, 2010; Alperi *et al.*, 2010a; Beaz-Hidalgo *et al.*, 2013). Table 4.1 lists all the species of the genus, indicating if they have been isolated or not from clinical samples and Table 4.2 shows the distribution by origin of more than 800 clinical isolates identified with molecular methods.

It is important to insist, as we have previously stated (Figueras *et al.*, 2000a, 2005), that *A. veronii* bv. *sobria* is in fact one of the main species of clinical importance and not *A. sobria*, as still being frequently mentioned erroneously in several studies (Taneja *et al.*, 2004; Tsai *et al.*, 2009; Wang *et al.*, 2009, among others). In fact, *A. sobria sensu stricto* has seldom been isolated from clinical samples (Figueras *et al.*, 2000a), rather it is a typical environmental species frequently recovered from diseased fish (Beaz-Hidalgo *et al.*, 2010). Nevertheless, *A. veronii* has also been associated with diseased catfish using molecular identification methods (Nawaz *et al.*, 2006).

Since the discovery of the new clinical species *A. dhakensis* (=*A. aquariorum*), which was previously misidentified using biochemical methods as *A. caviae, A. hydrophila* or *A. veronii* (Martínez-Murcia *et al.*, 2008; Figueras *et al.*, 2009; Aravena-Román *et al.*, 2011a; Putchucheray *et al.*, 2011; Roger *et al.*, 2012b; Wu *et al.*, 2012c; Beaz-Hidalgo *et al.*, 2013; Morinaga *et al.*, 2013; Chen *et al.*, 2013; Pérez-Valdespino *et al.*, 2013), the prevalence of the three classical dominating species has changed (Tables 4.2–4.4). Globally, ca. 96% of the clinical strains belong now to four species: *A. caviae* (30.0%), *A. dhakensis* (25.7%), *A.*

Table 4.2 Distribution of *Aeromonas* spp. by origin of isolation in studies that identified the strains using molecular methods[a]

Species	Faeces	Wound	Blood	Respiratory tract[b]	Urine	Peritoneal dialysate	Bile	Misc.[c]	Total
A. caviae	123	36	69	3	6	4	2	6	249
A. veronii	89	40	46	9			2	3	189
A. dhakensis[d]	54	82	36	3	3	5	2	3	188
A. hydrophila	47	65	22	5	2	4	1	2	148
A. media	8	3	5		1				17
A. trota	1	2					1		4
A. taiwanensis	1	3							4
A. salmonicida[e]		2						1	3
A. jandaei	1		1		1				3
A. sanarellii		3							3
A. allosaccharophila	1		1						2
A. tecta	2								2
A. diversa		2							2
A. schubertii		1							1
A. bestiarum			1						1
A. popoffii					1				1
Total	327	239	181	20	14	13	8	15	817

[a]Data derived from Huys *et al.* (2002), Hua *et al.* (2004), Al Benwan *et al.* (2007), Bossi-Küpfer *et al.* (2007), Demarta *et al.* (2008), Figueras *et al.* (2009), Tena *et al.* (2009), Alperi and Figueras (2010), Miñana-Galbis *et al.* (2010), Aravena-Román *et al.* (2011a), Esteve *et al.* (2012), Puthucheary *et al.* (2012), Roger *et al.* (2012b), Sedláček *et al.* (2012), Senderovich *et al.* (2012), Wu *et al.* (2012b, c), Morinaga *et al.* (2013), Chen *et al.* (2013), Shin *et al.* (2013), Perez-Valdespino *et al.* (2013) and from the results of the 76 faecal isolates referred in the text (unpublished).
[b]10 strains from sputum.
[c]Miscellaneous specimens include *A. caviae* ($n=6$) from: vagina (3), appendix (1), ascites (1) and abscess (1); *A. veronii* ($n=3$) from: ascites (2) and ear fluid (1); *A. dhakensis* ($n=3$) from: ascites (1), joint fluid (1) and bone chip (1); *A. hydrophila* ($n=2$) from: ulcer liquid (1) and abscess (1) and *A. salmonicida* from abscess (1).
[d]In many of the studies *A. aquariorum* was the name used for *A. dhakensis*.
[e]An additional case associated with a peritoneal effluent has been described by Yang *et al.* (2008) but the identification method was not specified.

veronii (22%) and *A. hydrophila* (18%) as derived from the data of recent studies (Table 4.3). For instance the genetic re-identification of 141 *Aeromonas* clinical strains recovered from human extraintestinal infections in Taiwan ($n=138$) and from faeces of patients with diarrhoea in Spain ($n=3$) revealed that *A. dhakensis* (=*A. aquariorum*) was the second most prevalent species (17.7%; 25/141) after *A. caviae* (51.8%; 73/141) while *A. hydrophila* was relegated to 3rd place (12.1%; 17/141) closely followed by *A. veronii* (11.3%; 16/141) as shown in Table 4.3 (Figueras *et al.*, 2009). In this study two strains isolated from infected wounds of two hospitalized patients were recognized as new species, which have been described and named elsewhere as *A. sanarellii*, and *A. taiwanensis* (Alperi *et al.*, 2010a). Two new isolates of each of these species from Taiwan were recently identified from wound infections by Chen *et al.* (2013).

Table 4.3 Distribution of *Aeromonas* spp. obtained from clinical sources in different studies that identified the strains using molecular methods

Species	No. (%) of strains in each study								Total
	Figueras *et al.* (2009)[a]	Aravena-Román *et al.* (2011a)[b]	Puthucheary *et al.* (2012)[c]	Roger *et al.* (2012b)[d]	Senderovich *et al.* (2012)[e]	Wu *et al.* (2012b)[f]	Chen *et al.* (2013)[g]	Pérez-Valdespino *et al.* (2013)[e]	
A. caviae	73 (51.8)	34 (23.8)	18 (19.1)	24 (23.3)	11 (64.7)	15 (31.3)	11 (13.7)	10 (35.7)	196 (30.0)
A. dhakensis[h]	25 (17.7)	34 (23.8)	47 (50.0)	6 (5.8)		16 (33.3)	37 (46.3)	3 (10.7)	168 (25.7)
A. veronii	16 (11.3)	36 (25.2)	12 (12.8)	46 (44.6)	5 (29.4)	12 (25.0)	14 (17.5)	3 (10.7)	144 (22.0)
A. hydrophila	17 (12.1)	33 (23.1)	16 (17.0)	23 (22.3)		5 (10.4)	13 (16.2)	11 (39.3)	118 (18.0)
A. media	5 (3.5)	2 (1.4)		1 (1.0)			1 (1.3)	1 (3.6)	10 (1.5)
A. trota	3 (2.1)		1 (1.1)						4 (0.6)
A. taiwanensis	1 (0.7)				1 (5.9)		2 (2.5)		4 (0.6)
A. sanarellii	1 (0.7)						2 (2.5)		3 (0.4)
A. salmonicida		1 (0.7)		1(1.0)					2 (0.3)
A. allosacharophila		1 (0.7)		1(1.0)					2 (0.3)
A. schubertii		1 (0.7)							1 (0.2)
A. bestiarum		1 (0.7)							1 (0.2)
A. jandaei				1 (1.0)					1 (0.2)
Total	141	143	94	103	17	48	80	28	654

[a–d]Origin of the strains: [a]blood (41.1%, 58/141), wound (25.5%, 36/141), stool (3.6%, 5/141), others (14.2%, 20/141), unknown (15.6%, 22/141); [b]wound (37.8%, 54/143), stool (23.1% 33/143), blood (23.1%, 33/143), others (16.3%, 23/143); [c]stool (51.1%, 48/94), wound (23.4%, 22/94), blood (8.5%, 8/94), others (17.0%, 16/94); [d]wound (36.0%, 37/103), stool (32.0%, 33/103), blood (16.5%, 17/103), others (15.5%, 16/103).
[e–g]Studies including only: [e]stool, [f]blood and [g]wound isolates.
[h]In many studies *A. aquariorum* was the name used for the species *A. dhakensis*.

After Figueras *et al.* (2009) evidenced the clinical relevance of this species, soon other authors recognized it from other geographical regions. For instance, Aravena-Román *et al.* (2011a) in Australia, using the sequences of the *rpoD* and *gyrB* genes to identify the clinical strains, found *A. dhakensis* with a similar prevalence (23.8%) as *A. veronii*, *A. caviae*, and *A. hydrophila* which incidence ranged between 23.1% and 25.2% (Table 4.3). In fact, *A. dhakensis* was the most frequently isolated species in the water samples investigated in that study (54.5%) followed by far by *A. veronii* (27.2%) and *A. hydrophila* (9%) as the most prevalent species (Aravena-Román *et al.*, 2011a). In a study carried out by Puthucheary *et al.* (2012) in Malaysia, also using *rpoD* sequences, 50% of 94 clinical strains previously identified as *A. hydrophila* with the API20E system, were in fact *A. dhakensis* (=*A. aquariorum*), being this species again by far the most prevalent (Table 4.3). The same was found when seven blood *A. hydrophila* strains (identified with the Vitek2 and 55 Phoenix100 systems) recovered from Japanese patients with septicaemia (Morinaga *et al.*, 2011) were re-identified in our laboratory (sequencing the *rpoD* gene) because 57.1% of the strains were *A. dhakensis* (=*A. aquariorum*) (Morinaga *et al.*, 2013). In fact, only two of the seven strains (28.6%) had been correctly phenotypically identified when the results were compared with those of the original study (Morinaga *et al.*, 2011). Similarly Chen *et al.* (2013) found this species to be the most prevalent 46.3% among 80 strains from wound infections in Taiwan.

A case of septicaemia and enterocolitis in a 93-year-old woman with cirrhosis associated with two isolates of *A. dhakensis* (=*A. aquariorum*) was identified by Wu *et al.* (2012c) using the *rpoD* and *rpoB* sequences. These isolates misidentified phenotypically (Vitek system) as *A. hydrophila* were obtained from blood and faeces and showed to be the same strain on the basis of their identical *rpoD* and *rpoB* sequences and identical pulse-field gel electrophoresis (PFGE) profiles. The authors indicate that this finding suggested that oral ingestion was probably the route of infection and highlighted the importance of improving food safety in order to reduce the risk of *Aeromonas* bacteraemia (Wu *et al.*, 2012c). A similar case, in 77-year-old male patient with liver cirrhosis occurred in Korea, the isolate was identified by the (Vitek2 system) as *A. hydrophila* but corresponded to *A. dhakensis* (=*A. aquariorum*) after sequencing the same genes, because 16S RNA sequences provided inconclusive identification (Shin *et al.*, 2013). The patient frequently consumed alcohol with sliced raw fish, and it was suggested a possible translocation of the bacteria from the intestine due to the pathological changes associated with the liver cirrhosis, dysfunction of host immune cells, and bacterial virulence factors. Furthermore, Shin *et al.* (2013) indicated that these adverse host environments may have contributed to the failure of i.v. ceftriaxone treatment, although the strain was shown to be sensitive to this antibiotic in the *in vitro* susceptible test. Phenotypic misidentifications of *Aeromonas* species using the Vitek and other miniaturized methods have been reported in other studies (Park *et al.*, 2003; Soler *et al.*, 2003b; Lamy *et al.*, 2010b). Another case of clonal relationship similar to the one reported by Wu *et al.* (2012c) was found among three clinical extraintestinal strains of *A. dhakensis* (=*A. aquariorum*) from Taiwan recognized by their identical *gyrB* and *rpoD* sequences and enterobacterial repetitive intergenic consensus (ERIC-PCR) genotype (Figueras *et al.*, 2009). The latter method has proven to have a higher or equal resolution to other genotyping techniques, like PFGE or amplified fragment length polymorphism (AFLP), etc. (Martínez-Murcia *et al.*, 2000; Soler *et al.*, 2003a). The mentioned examples highlight that the use of housekeeping gene sequences might also help to recognize potentially clonally related strains if their sequences are identical (Figueras *et al.*, 2011c).

As we have suspected and suggested *A. dhakensis* (=*A. aquariorum*) is a clinically relevant species distributed worldwide (Figueras *et al.*, 2009) as it is now confirmed in the recently published studies where the species was found from many geographical regions and many types of clinical samples (Tables 4.3 and 4.4). However, this species seems to be more predominant in tropical climates such us in Taiwan, Malaysia and Australia, although this requires further confirmation. For instance, among the 76 clinical isolates recovered mainly from faeces in Spain identified by sequencing the *rpoD* gene in our laboratory, the dominant species were: *A. caviae* (47, 61.8%), *A. veronii* (23, 30.3%), *A. hydrophila* (3, 4.0%), *A. dhakensis* (2, 2.6%) and *A. media* (1, 1.3%) (unpublished results). Tables 4.2 and 4.4 illustrate the different origins of *A. dhakensis* strains along with other *Aeromonas* species encountered

Table 4.4 Distribution by source of isolation of the *Aeromonas dhakensis* strains recovered in different studies that use samples from several origins and molecular identification techniques

	No. (%) of strains in each study					
	Figueras *et al.* (2009)	Aravena-Román *et al.* (2011a)	Puthucheary *et al.* (2012)	Chen *et al.* (2013)[a]	Total	Total considering other studies
Country	Taiwan[b]	Australia	Malaysia	Taiwan		
Clinical	25 (100)	34 (100)	47 (100)	37 (100)	143	175[c]
Blood	11 (44.0)	3 (8.8)			14	36
Wound	6 (24.0)	22 (64.7)	13 (27.7)	37 (100)	78	81
Faeces	3 (12.0)	4 (11.8)	27 (57.5)		34	41
Bilis	2 (8.0)				2	2
Ascitis	1 (4.0)				1	1
Peritoneal dialysate			5 (10.6)		5	5
Joint fluid	1 (4.0)				1	1
Sputum	1(4.0)	1 (2.9)	1[d] (2.1)		3	3
Urine		2 (5.9)	1 (2.1)		3	3
Others		2[e] (5.9)			2	2
Environmental		26 (100)			26	67
Water		24 (92.3)			24	27[f]
Fish		2 (7.7)			2	28[g]
Others						12[h]

[a]Only wound infections.
[b]Except the three strains from faeces that were isolated in Spain.
[c]Additional strains from: Esteve *et al.* (2012) $n=3$; Roger *et al.* (2012) $n=4$; Sedláček *et al.* (2012) $n=1$; Wu *et al.* (2012b) $n=16$; Morinaga *et al.* (2013) $n=4$; Pérez-Valdespino *et al.* (2013) $n=3$ and Shin *et al.* (2013) $n=1$.
[d]Tracheal secretion.
[e]One strain from bone chip and another from unknown origin.
[f]Three strains were recovered from a river and a water of a cooling system pond in Spain (Esteve *et al.*, 2012).
[g]Additional strains from: Esteve *et al.* (2012) $n=3$; Yi *et al.* (2013) $n=22$ and Soto-Rodriguez *et al.* (2013) $n=1$.
[h]Strains from: chironomid egg masses from Israel (Figueras *et al.*, 2011d) $n=11$ and 1 from pig carcass from Portugal (Fontes *et al.*, 2011).

in those studies that used molecular methods, mainly housekeeping gene sequences, for identification. Most of the speculation in the clinical literature about the predominance of *A. sobria*, *A. hydrophila* or *A. caviae* and about their virulence, antibiotic resistance, and clinical characteristics has to be regarded as probably unreliable if the identification of the isolates was done with phenotypic methods. The conclusions of the recent study by Chuang and co-workers (2011), which indicated that mortality and severity of *A. caviae* bacteraemias are lower than those produced by *A. veronii* bv. *sobria* or *A. hydrophila*, should be verified after a genetic identification of the strains. Furthermore, the statement that *A. caviae* is the most frequent species to cause bacteraemia in Japan (Kimura *et al.*, 2013), or *A. hydrophila* in the French Caribbean islands or in Taiwan (Hochedez *et al.*, 2010; Tang *et al.*, 2014) are doubtful as identification in these studies were again based on phenotypic methods. Considering this, a correction was made in the latter study replacing '*A. hydrophila* complex' for *Aeromonas* species (Hochedez *et al.*, 2011). Incorrect identification has a negative impact on the knowledge and understanding of the epidemiology of *Aeromonas* infections because true interspecies variation in the invasive or enterotoxigenic capacities, antimicrobial susceptibility and geographical distribution is masked. Furthermore confusion between *A. hydrophila* and *A. dhakensis* may have clinical consequences considering that the latter species is more virulent and exhibited more cytotoxicity in human normal skin fibroblast cells, as measured by the released leucocyte lactate dehydrogenase levels (29.6 ± 1.2% vs. 20.6 ± 0.6%, $P < 0.0001$) as described by Chen *et al.* (2013). Furthermore, 81% of the isolates of *A. dhakensis* recovered from blood exhibited either cefotaxime resistance and this was not found in other clinical strains of *A. hydrophila, A. caviae, A. veronii* or *A. trota* (Wu *et al.*, 2013). In this study it was indicated that a cefotaxime monotherapy for severe *A. dhakensis* infections should be used cautiously. It is now clearly recognized that proper identification requires the use of molecular methods. The more reliable methods in our opinion are the sequences of the mentioned housekeeping genes (Soler *et al.*, 2004; Figueras, 2005; Lamy *et al.*, 2009; Janda and Abbott, 2010; Beaz-Hidalgo *et al.*, 2010; Figueras *et al.*, 2011b–d; Aravena-Román *et al.*, 2011a; Beaz-Hidalgo and Figueras, 2012; Puthucheary *et al.*, 2012; Morinaga *et al.*, 2013; Figueras and Beaz-Hidalgo, 2014).

An additional confusing problem is the use of old names that are still appearing in new studies (Janda and Abbott, 2010). In our view, this is due to the labels associated to the 16S rRNA sequences deposited at the GenBank (Figueras *et al.*, 2011b). For instance, the name *Aeromonas enteropelogenes*, used instead of its correct name *A. trota*, appeared in the study of Wu *et al.* (2013) and in a study that described a strain of this species (ATCC 49803) as the only known example of an *Aeromonas* species that produces only one beta-lactamase belonging to molecular class C (De Luca *et al.*, 2010). Other old names are *Aeromonas punctata* and *A. hydrophila* subsp. *anaerogenes* instead of *A. caviae*, and *Aeromonas culicicola* or *Aeromonas ichthiosma* instead of *A. veronii*. These names correspond to species that have been synonymized or that are pending judicial decisions for priority of the name and are never used by specialists of this genus (Martin-Carnahan and Joseph, 2005; Figueras *et al.*, 2011b; Janda and Abbott, 2010).

Aeromonas dhakensis previously known as *A. aquariorum* or *Aeromonas hydrophila* subsp. *dhakensis*

As commented before *A. dhakensis* has very quickly become an important clinical species recovered from different countries with a varying frequency (Tables 4.2–4.4). It was originally discovered in 2008 by Martínez-Murcia *et al.* from aquarium water and ornamental fish from Portugal and named *A. aquariorum*. However, 1 year after its description it was recognized that three isolates of *A. hydrophila* subsp. *dhakensis*, not studied before, grouped with the *A. aquariorum* strains in a phylogenetic tree constructed with the concatenated sequences of the *rpoD* and *gyrB* genes and it was suggested that they represented the same taxon (Martínez-Murcia *et al.*, 2009). In fact, almost at the same time we discovered that it was the second most prevalent clinical species recovered from extra intestinal infections obtained from Taiwan, and we also found this species among strains of our collection recovered from faeces (Figueras *et al.*, 2009). Faeces were in fact the origin of *A. hydrophila* subsp. *dhakensis* defined on the basis of 10 strains isolated from children with diarrhoea in Dhaka, Bangladesh, by Huys *et al.* (2002). The formal reclassification of the latter subspecies as *A. aquariorum* and of both as *A. dhakensis* has been done after performing a phenotypic and genetic characterization of several strains of the closely related species (Beaz-Hidalgo *et al.*, 2013). Furthermore, and considering the priority of the subspecies epithet '*dhakensis*', the correct name for this species was considered *A. dhakensis* to replace the name *A. aquariorum* originally proposed by Martínez-Murcia *et al.* in 2008 (Beaz-Hidalgo *et al.*, 2013). As commented before, this species has been described as a prevalent clinical species in Taiwan (17.7%), Australia (23.8%) and Malaysia (50%) as summarized in Table 4.3, and has been mainly implicated in gastroenteritis, soft tissue infections, bacteraemia and in infections associated to peritoneal dialysates as show in Tables 4.2 and 4.4. Sedláček *et al.* (2012) reported a case of acute profuse diarrhoea of 10 days' duration in a 25-year old man after returning from Egypt attributed to *A. dhakensis*. These authors concluded that this species occurs in the Czech Republic (central Europe) as an imported illness. However, so far this species has not been reported in Egypt, but this is not strange because no molecular identification studies have been performed in this country. A case of fatal sepsis was reported by Shin *et al.* (2013).

The genome of a clinical strain of *A. dhakensis* (AAK1) has recently been published (Wu *et al.*, 2012d) and it was found to possess a blaAQU-1 chromosomal gene that encodes a class C β-lactamase (Wu *et al.*, 2013). Esteve *et al.* (2012) reported this species from river water and in water from a pond of a cooling system (17.7%), in human faeces (6.25%) and in diseased freshwater European eels (5%) from Valencia (Spain). In farmed eels from Korea *A. dhakensis* showed to be the prevailing species associated with aeromoniasis (Yi *et al.*, 2013). Other environmental strains have been isolated from aquarium water and ornamental fish in Portugal (Martinez-Murcia *et al.*, 2008) and from water and fish samples in Australia (Aravena-Román *et al.*, 2011a) as commented, but also from egg masses of the non-biting midge *Chironomus* sp. in Israel (Figueras *et al.*, 2011d). Furthermore, different strains of *A. dhakensis* (=*A. aquariorum*) showed a multiresistant behaviour that has been explained by the fact that they carry genes involved with the β-lactam and sulphonamide resistance (Puah *et al.*, 2013). As expected, not only most of the 47 *Aeromonas* strains examined in the latter study showed resistance to amoxicillin (91%) but also to amoxicillin-clavulanic acid (98%), and were less resistant to gentamicin (13%), trimethoprim–sulfamethoxazole (11%), kanamycin (6%) and all of them were sensitive to ciprofloxacin or amikacin.

In conclusion, we can say that the doubt manifested by Janda and Abbott (2010) about the few data available on the distribution in the environment of *A. aquariorum* (now *A. dhakensis*) has been dissipated by many recent reports (Figueras *et al.*, 2009, 2011d; Fontes *et al.*, 2011; Aravena-Román *et al.*, 2011a; Esteve *et al.*, 2012; Sedláček *et al.*, 2012).

Aeromonas taiwanensis and *Aeromonas sanarellii*

The species *A. taiwanensis* and, *A. sanarellii* were described on the basis of one strain each recovered from wounds of hospitalized patients in Taiwan in 2010 (Alperi *et al.*, 2010a). The strain of *A. taiwanensis* (A2–50^T) was isolated together with strains of *Enterobacter cloacae* and *Enterococcus* sp. on the tenth day of hospitalization from a burn wound of a 40 year-old male with burns in 52% of his body and having *Aeromonas* and *Enterococcus* bacteraemia. The strain of *A. sanarellii* (A2–67^T) was isolated from a 70-year-old woman with non-insulin-dependent diabetes mellitus with trauma and abrasion wounds over her right elbow where *Aeromonas, Pseudomonas aeruginosa* and *Bacillus* species were isolated. Since then, four environmental isolates of *A. sanarellii* and one of *A. taiwanensis* have been recorded from waste water in Portugal (Figueira *et al.*, 2011), and also in our laboratory the former species in waste water, tomatoes and parsley irrigated with reclaimed waters (unpublished results). Additional clinical strains of both species have been found by Chen *et al.* (2013) while *A. taiwanensis* was isolated from the faeces of a patient with diarrhoea in Israel (Senderovich *et al.*, 2012). In addition, we discovered eight isolates of *A. sanarellii* (6.2%) and three of *A. taiwanensis* (2.3%) among the 129 characterized *Aeromonas* isolates from chironomid egg masses found at a waste stabilization pond in northern Israel (Beaz-Hidalgo *et al.*, 2012). This finding complements previous information in relation to the potential role of chironomid egg masses as a natural reservoir for *Aeromonas* spp. (Figueras *et al.*, 2011d). In fact the isolation of these two species increases the diversity of *Aeromonas* spp. present in this specific ecological habitat, because so far only the species *A. dhakensis, A. caviae, A. veronii* and *A. hydrophila* have previously been found (Senderovich *et al.*, 2008; Figueras *et al.*, 2011d). The *A. taiwanensis* strain obtained from faeces was recognized from a patient living in the same area from where the chironomid strains were obtained (Senderovich *et al.*, 2012). However, the *rpoD* sequences of the strain were different indicating that they were not epidemiologically related. Beaz-Hidalgo *et al.* (2012) suggested that waste water might be the dispersion route of *A. taiwanensis* into natural water environments, such as those inhabited by chironomids, from where they might contaminate drinking water and therefore humans. However, this has not been demonstrated. Again a common feature of all the studies that recognized these new species was also the use of housekeeping gene sequences for their identification.

Species that show atypical behaviours within the genus

The most recently described species within the genus *Aeromonas* have been *A. cavernicola* and *A. australiensis*, both recovered from freshwater, the former from the water of a brook in a cavern in the Czech Republic and the latter from a treated effluent used for irrigation in the wheat-belt region in Western Australia (Aravena-Román *et al.*, 2013; Martínez-Murcia *et al.*, 2013). An interesting characteristic of these two species is that both show sensitivity to the vibriostatic agent O/129 (2,4-diamino-6,7-diisopropypteridine; 150 mg/disc) while the majority of *Aeromonas* species are resistant, which has always been considered a distinctive feature of the genus. Abbott *et al.* (2003) found this sensitivity only in two strains of *A. eucrenophila* and in one strain of *A. veronii*. The inhibition zone generated around a disk containing the vibriostatic agent O/129 (150 µg) was 8.5 mm in diameter for *A. cavernicola* and ~10.8 mm for *A. australiensis* (Aravena-Román *et al.*, 2013; Martínez-Murcia *et al.*, 2013). The species *A. cavernicola* showed other atypical reactions since it is one of the two species in the genus, together with *A. fluvialis*, to be simultaneously negative for the lysine decarboxylase (LDH), arginine dihydrolase (ADH) and ornithine decarboxylase (ODC) (Aravena-Román *et al.*, 2013; Martínez-Murcia *et al.*, 2013).

Another atypical behaviour observed has been the resistance to cephalotin shown by certain genetically identified isolates of *A. veronii* (Sánchez-Céspedes *et al.*, 2009) when the susceptibility to the latter antibiotic is one of the specific differential characteristics established for this species (Abbott *et al.*, 2003).

The type strain of the species *A. taiwanensis* (A2–67^T = CECT 7402^T = LMG 24682^T) also displayed some peculiarities, i.e. two colony types of the same size (4–5 mm in diameter). One was opaque, circular and beige in colour, while the other was translucent on TSA after 48 hours at 30°C. These two morphologies were also observed on blood agar. The two colony types produced the same biochemical responses and showed identical ERIC-PCR profiles (Alperi *et al.*, 2010a). Different colony morphologies were also found in a strain of *A. veronii* bv. *sobria* involved in a case of cholangitis (Sánchez-Céspedes *et al.*, 2009).

Clinical infections and disease presentation

The major clinical manifestations produced by *Aeromonas* involve cases of diarrhoea, wound and soft tissue infections followed by bacteraemia (Figueras, 2005; Janda and Abbott, 2010). The human population suffering from *Aeromonas* diarrhoea are principally young children and immunocompromised individuals, the latter being the ones most often affected by bacteraemia or other infections. In contraposition wound and soft tissue infections mostly occur in healthy people after accidents or trauma, despite being able to be acquired in a hospital following surgery or after the application of leech therapy.

A retrospective study carried out by Wu *et al.* (2007) over 2 years, at a 900-bed hospital in Tainan (Taiwan) revealed that the spectrum of diseases from which a total of 116 *Aeromonas* isolates were obtained corresponded to primary bacteraemia (40%) followed by soft tissue (27%) and hepatobiliary tract (15%) infections. It was observed that the global mortality rate due to *Aeromonas* infections was 26% and that the majority of the patients with soft tissue infections (71%) had previously suffered from water- and/or soil-related injuries. In that study, as in previous ones, the authors discuss the higher incidence of extraintestinal infections in Taiwan in comparison with other Western countries, from where there are relatively few reports and always include a smaller number of cases. In fact, over a 15-year period in Spain, Tena *et al.* (2007) reported 38 extraintestinal infections attributed to *Aeromonas*. Eighteen (40%) corresponded to abdominal infections (10 cases, 55 % related with the hepatobiliary system), 11 (28%) to skin and soft tissue infections, three (7%) to urinary tract infections and three (7%) to episodes of primary bacteraemia with a global mortality of 16.2%. As in the previous study of Wu *et al.* (2007) most of the patients (91%) that showed wound infections had traumatic accidents. As commented wound infections involve healthy people; for instance among the 11 patients with soft tissue infections, 7 (64%) were initially healthy individuals in the Tena *et al.* (2007) study and 17 (85%) in the one of Wu *et al.* (2007). Wound infections were the predominant presentation in studies conducted in France in 2008 and 2009 and represented between 44% and 50% of the total cases (Fraisse *et al.*, 2008; Lamy *et al.*, 2009). Also, a higher proportion of isolates were recovered from wounds (38%), followed with an equal prevalence of 23% from stools and blood in a study carried out in Australia (Aravena-Román *et al.*, 2011a). However, in a study in Malaysia by Puthucheary *et al.* (2012), acute gastroenteritis was the prevailing presentation (55.3%)

among the 94 patients followed by soft tissue infection (23.4%), infections associated with peritoneal dialysis (13.8%) and primary bacteraemia (4.25%). In the latter study 72% of the patients were adults (age range between 36 to 49 years), and the rest corresponded to children (age range between 1 month and 4 years).

Ghenghesh *et al.* (2008) extensively reviewed *Aeromonas* clinical presentations in developing countries comparing studies from countries on three continents, Africa, Asia and Latin America (i.e. Bangladesh, Brazil, China, Cuba, Egypt, India, Iran, Libya, Nigeria, Venezuela and Vietnam) and highlighted the reported significant association between *Aeromonas* sp. and diarrhoea in children. The other common presentations were soft-tissue infections and aspiration pneumonia following contact with water and after floods.

Aeromonas-associated gastroenteritis

As commented above, enteric infections by *Aeromonas* are common and have been described worldwide (Janda and Abbott, 1998; Figueras, 2005; von Graveniz, 2007; Ghenghesh *et al.*, 2008; Janda and Abbott, 2010; Surek *et al.*, 2010). However, these infections are considered to be probably underestimated due to the fact that this microbe is not routinely searched for in the diarrhoeal stools and in many occasions it is accidentally recognized on routine enteric isolation media [MacConkey, xylose lysine dextrose (XLD) agar, Hektoen Enteric (HE) agar, Salmonella–Shigella (SS) agar, cefsulodin–Irgasa–novobiocin (CIN) agar, etc.] used to detect other enteropathogens (Figueras, 2005; Gröbne *et al.*, 2007; Janda and Abbott, 2010). In fact some authors propose that sheep blood agar should be more frequently used for stool microbial diagnosis, at least when the specimen is negative for the common bacterial pathogens causing gastroenteritis, in order to enhance the recognition of *Aeromonas* (Gröbne *et al.*, 2007). Exiting reports are probably biased towards cases with chronic symptoms, detection in children, immunocompromised and elderly people or patients who are hospitalized and develop gastrointestinal symptoms (Figueras, 2005). Cases have previously been reviewed (Figueras, 2005; Janda and Abbott, 2010) and recently new reports have been documented. For instance, isolates multiresistant, specially to cephalosporins, have been found to predominate in Egyptian children under two years of age suffering from *Aeromonas* diarrhoea (Mansur *et al.*, 2012). The first case of inflammatory diarrhoea due to *Aeromonas* in association with Behçet disease occurred in a 30-year-old Spanish woman while receiving immunosuppressive therapy (Lete *et al.*, 2013). Behçet disease is a vasculitis of unknown cause that manifests as aphthous ulcers and intestinal involvement with erosions similar to those found in the inflammatory bowel disease. Additionally, a severe case of enterocolitis due to *A. dhakensis* (previously *A. aquariorum*) occurred in a patient with cirrhosis (Wu *et al.*, 2012c) and cases of *Aeromonas* diarrhoea associated with *Vibrio* species have been reported (Hasan *et al.*, 2012; Mendes-Marques *et al.*, 2012; Ottaviani *et al.*, 2013). Ghenghesh *et al.* (2008) reported that in the developing world, isolation of *Aeromonas* spp. ranges between 1% and 88% in patients with diarrhoea, and 0–45% in asymptomatic patients.

The most common *Aeromonas* gastroenteritis presentation is the secretory (watery) enteritis, which represented 75–89% of all cases in which this microbe was the only associated pathogen (Janda and Abbott, 2010 and references therein). Normally *Aeromonas*-associated diarrhoea, such as diarrhoea caused by viruses and other bacteria, is a self-limiting process, despite there are cases when the clinical course is severe (cholera-like and dysentery-like) and protracted (Figueras, 2005; Janda and Abbott, 2010). Secondary diarrhoea complications like those associated to the haemolytic uraemic syndrome (HUS), although rarely, can be induced by *Aeromonas* (Figueras *et al.*, 2007a). A HUS case occurred in a 40-year-old Spanish woman produced by a strain of *A. veronii* bv. *sobria* that possessed the genes of the Type III secretion system which deliver toxins directly into the cytoplasm of the host cells (Chacón *et al.*, 2004; Figueras *et al.*, 2007a). The latter represented the second HUS case described in adults of those that appeared in the reviewed literature (Figueras *et al.*, 2007a and references therein). Among the cases reviewed some lacked information to ascertain if they were indeed HUS, while others were well documented (Figueras *et al.*, 2007a; Janda and Abbott, 2010). Probably the clearest evidence that *Aeromonas* can be responsible of the HUS is the fact that the Shiga toxin genes (*stx1* and *stx2*) considered important virulence factors in the development of HUS by *E. coli* O157:H7, have been detected and sequenced (Alperi and Figueras, 2010). The *stx1* gene was detected in 23.7% of the *Aeromonas* strains tested (19/80) and was predominantly encountered in *A. caviae* (42.1% 8/19), *A. hydrophila* (31.6% 6/19) and *A. veronii* (10.5% 2/19), which are the species commonly associated with gastroenteritis and even with a dysentery-like syndrome and HUS (Figueras, 2005; Figueras *et al.*, 2007a; Alperi and Figueras, 2010). In addition, 10.5% (2/19) of the *Aeromonas* strains that possessed the *stx1* gene produced the Stx1 toxin (Alperi and Figueras, 2010). Only one *A. caviae* strain from urine possessed both genes *stx1* and *stx2* simultaneously and the similarity of the sequences with the corresponding regions of these genes from *E. coli* O157:H7 was above 99.5% (Alperi and Figueras, 2010 and references therein).

According to Janda and Abbott (2010) the role of *Aeromonas* as etiologic agents of bacterial diarrhoea is still problematic because there are studies that argue in favour of this role while others indicate totally the opposite. For instance Chu *et al.* (2006) screened by PCR the presence of three haemolysin genes (*ahh1*, *asa1* and *aerA*) in 253 *Aeromonas* strains associated with gastroenteritis in outpatients from public clinics in Hong Kong and reported no statistically significant association between the haemolysin genotypes, the amount of growth (on the primary culture) and the presentation of diarrhoeal symptoms defined as bloody, mucous or watery stool specimens. They conclude there was a lack of association between presentation of diarrhoeal symptoms and faecal isolation of *Aeromonas.* However, in relation to the Chu *et al.* (2006) study we published a letter to the editor stating that these authors derived a highly unsubstantiated conclusion from very limited data on a topic that has been under discussion and heavily researched by specialists in this genus for many years worldwide (Figueras *et al.*, 2007c). Our arguments in favour of considering *Aeromonas* an enteropathogen are summarized in Table 4.5. In fact the latter table includes the classical arguments, underlined by Janda and Abbott (2010), used against considering *Aeromonas* an enteropathogen i.e. the lack of outbreaks, the non-fulfilment of the Koch's postulates on the basis of the results obtained in the human voluntary challenging study, the lack of animal model etc. However, in our view for each con a pro can be found (Figueras *et al.*, 2007c) and this is what is also reflected in Table 4.5. For instance, recently a murine model was used to evaluate *in vitro* and *in vivo* bactericidal and immunomodulating activities of hesperidin (HES) and ellagic acid (EA) against *Aeromonas* which are compounds from medicinal plants and a dilatation of the central lacteal of the villi indicated diarrhoea, and necrosis in the intestinal glands and atrophy of some villi were noticed (Abuelsaad *et al.*, 2013). These alterations are indications of the ability of *Aeromonas hydrophila* (ATCC 7966T) to cause fluid accumulation and destruction of intestinal mucosa resembling the activity of toxigenic *Vibrio cholerae.* So we can say that a murine model exits.

The asymptomatic or subclinical infections revealed by the existence of *Aeromonas*

Table 4.5 Arguments for and against considering *Aeromonas* a true enteropathogen

Arguments against	Arguments in favour
There is isolation of the bacteria in an equal proportion in symptomatic and asymptomatic patients	Despite some studies not showing statistical differences others do (Ghenghesh *et al.*, 2008). 2–20% of the diarrhoeal cases from children around the world have been induced solely by species of *Aeromonas* but only 0–2% of carriers had no symptoms (Figueras *et al.*, 2007a, and references therein). Asymptomatic carriers of *Vibrio cholerae* are also common (Grimes, 2006), and this is not a reason to doubt its enteropathogenicity
Only 2 of 57 volunteers challenged had diarrhoea	In many studies, *Aeromonas* has been isolated from faeces in cases of enteric disease and has disappeared when symptoms were in remission
	A strain with the same pulsotype[a] was isolated from the faeces and from blood of a patient that presented enterocolitis and septicaemia (Wu *et al.*, 2012)
Koch's classic postulates have not been fulfilled	Koch's postulates were fulfilled by the accidental ingestion of *Aeromonas* by a healthy laboratory worker, who developed severe diarrhoea that lasted for two days, and from whom the same strain that was ingested was isolated (Carnahan *et al.*, 1991)
	Other pathogens, including *V. cholerae*, do not completely fulfil all the postulates (Grimes, 2006)
	Serological evidence of infection has been provided in some cases, which supports its true enteropathogenicity (see review by Janda and Abbott, 1998; Crivelli *et al.*, 2001)
There is no animal model to reproduce diarrhoeal disease	*Campylobacter* also lacks an animal model that reproduces diarrhoeal disease but this is not a reason to doubt its enteropathogenicity. A murine model was recently described (Abuelsaad *et al.*, 2013)
Considering the ubiquitous nature of these organisms (in food and water) they should have caused many outbreaks of diarrhoeal disease but those have not been found	Many cases of *Aeromonas*-associated diarrhoea are unrecognized because a microbiological study of the faeces was not carried out, and this could be the reason why so few outbreaks have been identified. However, the same strain (genotype) producing the diarrhoea has been found in water and food (Altwegg *et al.*, 1991; Pablos *et al.*, 2010; Khajanchi *et al.*, 2010)

[a]Genotype by pulsed-field gel electrophoresis (PFGE).

carriers are also commonly found in association with other microbes like *Salmonella, Vibrio cholerae* or *Campylobacter*, etc. (Grimes, 2006) and this does not alter the enteropathogenic role attributed to the latter microbes. Other objective proof is the demonstrated specific secretory immunoglobulin A (sIgA) response at the intestinal mucosa against the extracellular products that appeared in the faeces of patients with *Aeromonas* diarrhoea (Crivelli *et al.*, 2001). Serological responses have also been described in other studies previously reviewed by Janda and Abbott (1998). In addition despite few outbreaks exists, an epidemiological relationship has been found between the source of infection and the clinical isolates because the same *Aeromonas* clone that caused diarrhoea was isolated from drinking water (Khajanchi *et al.*, 2010; Pablos *et al.*, 2010), from the consumed shrimp cocktail (Altwegg *et al.*, 1991) and from the household environment (Demarta *et al.*, 2000 and references therein). The genetic relatedness of *Aeromonas* isolates obtained from HIV/AIDS patients suffering from gastroenteritis and those recovered from their household drinking-water was demonstrated also in a study performed in South Africa by Ramalivhana *et al.* (2010). A recent *Aeromonas* outbreak that produced in over 200 college students' acute diarrhoea occurred in Guizhou (China) as reported by Zhang *et al.* (2012) who concluded that the most probable cause was the salad ingredients washed with water from a contaminated tank. In another study the same isolate of *A. dhakensis* (=*A. aquariorum*) recovered from a patient's faeces with enterocolitis was found in his blood, reinforcing the believe that the origin of many cases of bacteraemia is the intestine (Wu *et al.*, 2012c). For instance in a review study of 30 cases of acute suppurative cholangitis produced by *Aeromonas*, they provided evidence for the ascent of the microbe from the intestine to the biliary tract secondary to the instrument used during exploration (Chan *et al.*, 2000).

Furthermore, Koch's postulates were fulfilled in the case of a healthy laboratory worker that ingested accidentally an *Aeromonas* and that developed a severe diarrhoea that lasted for 2 days, and from whom the same strain that was ingested was isolated (Carnahan *et al.*, 1991). On the basis of the arguments described above and shown in Table 4.5, it is clear that *Aeromonas* can produce diarrhoea if the right host and circumstances occur in a similar way as when it produces important extra intestinal infections (Figueras, 2005; Figueras *et al.*, 2007c).

Diarrhoea in children

In the studies that evaluate the microbes involved in acute childhood diarrhoea they indicate that *Aeromonas* incidence ranges from 2.3% Taiwan up to 13% in Nigeria (Esser *et al.*, 2000; Juan *et al.*, 2000; Nzeako and Okafor, 2002; Figueras, 2005, and references therein). In a case–control study performed in Cuba by Bravo *et al.* (2012) *Aeromonas* showed a significantly higher isolation rate in the cases with diarrhoea (166, 7.15%) than in the control group (35, 1.76%). In a similar type of study carried out in a rural area in Egypt *Aeromonas* was isolated from 1.4% of stool samples from children with diarrhoea and from 0.5% of stools from asymptomatic children (Mansour *et al.*, 2012). In the latter study it was proven that the probability of having diarrhoea was higher in symptomatic patients who had *Aeromonas* as the sole pathogen with an incidence rate (IR) of 0.07 episodes/child/year. In comparison the IR for enterotoxigenic *Escherichia coli* (ETEC) was 1.26, while for *Campylobacter* was 0.47, for *Shigella* 0.06 and for *Salmonella* 0.02 episodes/child/year (Mansour *et al.*, 2012). Pablos *et al.* (2010) in León (Spain) found a frequency of *Aeromonas* diarrhoea of 4% (32 positive patients of the 800 investigated), mainly associated with paediatric patients (22/32, 68.8%). In a recent Global Enteric Multicenter Study to identify the aetiology and population-based burden of paediatric diarrhoeal disease in sub-Saharan Africa and South Asia, *Aeromonas* was considered a leading pathogen in Pakistan and Bangladesh, with the peak of adjusted population attributable fraction at age 24–59 months (Kotloff *et al.*, 2013). The analyses performed support that *Aeromonas* causes diarrhoeal disease in young children. This again supports the role of these bacteria in diarrhoeal diseases (Figueras *et al.*, 2007c).

The relative incidence in relation to other enteropathogenic bacteria varies in different studies. For instance *Aeromonas* normally occupies the third place after *Campylobacter*, if the latter was included in the investigation with *Salmonella* having the highest incidence (Esser *et al.*, 2000). However, in other studies where *Campylobacter* was not investigated *Aeromonas* prevailed over *Salmonella* and *Shigella* like in a study from Nigeria (Nzeako and Okafor, 2002), or occupied the second place after *Shigella* in Iran (Soltan Dallal and Moezardalan, 2004). The isolation rate has also been comparable to that of *Shigella* and higher than *Salmonella* in a rural area of Egypt (Mansour *et al.*, 2012) or just higher than *Salmonella* in Cuba (Bravo *et al.*, 2012). Furthermore, in many studies *Aeromonas* has been found to be more prevalent than the enteropathogenic *E. coli* (Esser *et al.*, 2000; Nzeako and Okafor, 2002; Soltan Dallal and Moezardalan, 2004; Bravo *et al.*, 2012). In fact in a recent prospective matched case control study that investigated diarrhoeal disease in children it was found that *Aeromonas* was the leading pathogen in Pakistan and Bangladesh (Kotloff *et al.*, 2013).

In general, the common clinical picture caused by *Aeromonas* in children is the acute onset of watery diarrhoea with slightly elevated body temperature, being vomiting less common than in diarrhoea caused by other pathogens (Figueras *et al.*, 2007c).

Diarrhoea in adults

A rate of 2% isolation of *Aeromonas* in adult patients with diarrhoea has been found in several studies conducted in different parts of the world (Figueras, 2005 and reference therein). For instance, in a case control study in Sweden, the 2% *Aeromonas* prevalence in the patients with diarrhoea was similar to that found for enteroaggregative *E. coli, Giardia, Cryptosporidium* and astroviruses, and furthermore *Aeromonas* was not isolated in the control group (Svenungsson *et al.*, 2000). In Israel Senderovich *et al.* (2012) also found an isolation rate of ca. 2%, and recovered in 5 months 17 cases of *Aeromonas* diarrhoea mainly from adult patients (15/17, 88.2%). Interestingly, in a Swedish study *Aeromonas* prevailed in outpatients who had travelled abroad within the previous 2 weeks (Svenungsson *et al.*, 2000). In a study performed in Barcelona by our group, the incidence of *Aeromonas* traveller's diarrhoea was also 2% i.e. 18 of the 863 studied patients (Vila *et al.*, 2003). In the latter study the frequency of *Aeromonas* spp. was similar in patients returning from Africa (1.72%), Latin and South America (1.83%) and Asia (2.32%) and approximately 50% of the patients showed fever and abdominal cramps. The persistence of the symptoms for more than 14 days made the use of antimicrobial treatment necessary. Third-generation cephalosporins were active against *Aeromonas* spp., while resistance to chloramphenicol, tetracycline or trimethoprim–sulfamethoxazole ranged from 23% to 45%. These levels of resistance were suspected to be associated to the extensive use of these antimicrobial agents in developing countries (Vila *et al.*, 2003). A case of travellers' diarrhoea in a patient returning from Ecuador to Italy was recently described by Ottaviani *et al.* (2013). An incubation period of 1 or 2 days has been associated with travellers' diarrhoea (Janda and Abbott, 2010). *Aeromonas* was isolated from drinking water and stools in some patients who were diagnosed of traveller's diarrhoea among the tourists that visited specific hotels in the costal area of Tenerife in the Canary Islands in Spain (Martin Delgado *et al.*, 2001). Failures in the drinking water system produced a faecal contamination of the water as reveal by the presence of bacteria indicators of faecal pollution, high levels of organic matter and the presence of *Aeromonas* (Martin Delgado *et al.*, 2001). Once the water distribution system was repaired the cases of diarrhoea among the tourist population when back to normal.

Wound or burn infection after trauma or injury

After gastroenteritis, wounds are the second most common source of *Aeromonas* clinical isolation and occur in healthy persons when a traumatic injury becomes environmentally contaminated with water or soil (Figueras, 2005; Janda and Abbott, 2010). In fact, this was the predominant origin of the clinical isolates included in recent studies performed in France and Australia (Lamy *et al.*, 2009; Avena-Roman *et al.*, 2011a). Furthermore, patients who had suffered an injury, i.e. burns, traumas, are candidates for developing an *Aeromonas* bacteraemia (Figueras, 2005; Janda and Abbott, 2010). Near-drowning accidents can generate pneumonia cases and those are described under the heading 'Pneumonia and other respiratory infections' (Bossi-Küpfer *et al.*, 2007; Mukhopadhyay *et al.*, 2008; Lamy *et al.*, 2009; Janda and Abbott, 2010; Chao *et al.*, 2013a). *Aeromonas* may also infect burns, because normally the lesions are immersed in water that is the typical habitat for these bacteria (Janda and Abbott, 2010). Yasti *et al.* (2009) described a case of a major burn that was extinguished with tap water, and this was considered to be the source of *Aeromonas*. A review of infections in burned patients performed by Azzopardi *et al.* (2011) reveals that *Aeromonas* infections constituted 76% of the identified publications, showing a mortality of 10.7%. In 96.2% of the cases, contact with ambient untreated water and/or rolling in dirt to extinguish flames was recognized as the main risk factor associated with the infection and should be considered a suggestive alert of a possible *Aeromonas* infection. In a recent case–control study in which Sherry *et al.* (2013) compared the microbes involved in infected burns linked to wildfire with those of routine patients (control group), it was observed that the wildfire victims were more likely to culture *Enterobacteriaceae* (5/18 vs. 2/36) and *Aeromonas* spp. (3/18 vs. 1/36) than the control group. In a review performed from soft tissue infections caused by marine bacteria, *Aeromonas* was included among the most relevant microbes (Finkelstein and Oren, 2011). Necrotizing soft-tissue infections and sepsis caused by *Vibrio vulnificus* a typical marine microbe were compared with those caused by *Aeromonas* spp. by Tsai *et al.* (2007). They found that the patients who died with an *Aeromonas* infection had significantly lower white blood-cell counts ($P=0.03$) with significantly fewer numbers of segmented white blood cells than those who died in the *V. vulnificus* group ($P=0.01$). In fact they concluded that a contact history of the patients with seawater or raw seafood could alert of *V. vulnificus* while contact with fresh or brackish water, soil, or wood could alert of an *Aeromonas* infection, sharing both microbes as characteristic the fast onset of cellulites. Chao *et al.* (2013a) reviewed retrospectively 129 cases of *Aeromonas* skin and soft-tissue infection that were attended in a 900-bed hospital in Tainan (Taiwan) during a 3-year period and found that 78% of the patients had trauma related wound infections and 30% had a history of exposure to water. The majority of the patients (79%) had polymicrobial infection and 3.1% had simultaneously *Aeromonas* bacteraemia. Chen *et al.* (2013) performed a study in the same area comparing the characteristics of *A. dhakensis* and *A. hydrophila* isolates recovered from 80 wound infections. They found *A. dhakensis* to be the prevailing species (43.6%) and found that more isolates of this species than those of *A. hydrophila* showed a significant association with water environments (32.4% vs. 0%, $P>0.05$). On the basis of their susceptibility results they discourage the use of ceftriaxone and cefuroxime in the treatment of severe wound infections.

Injury maybe also the source of entry for *Aeromonas* septic arthritis, as occurred 2 days after a freshwater injury in a girl in UK as described by Elwitigala *et al.* (2005), that also reviewed seven additional cases described at the literature. Another case of septic arthritis affected a 51-year-old man who had spent approximately 1 hour driving a personal watercraft on Okaloosa Bay near the Gulf of Mexico (Danaher and Mueller, 2011). Eight days later, he developed septic arthritis that required surgical drainage and levofloxacin treatment before it cured after 21 days. According to Danaher and Mueller (2011) this study represents the ninth case of *Aeromonas* septic arthritis described in the English literature. A previous case of septic arthritis and bacteraemia with no source of entry, no history of trauma or risk factor due to *A. veronii* (verified by sequencing the 16S rRNA gene) occurred in 81-year-old healthy man that was admitted at UK hospital with a 1-week history of progressive pain and immobility of the right shoulder joint (Roberts *et al.*, 2006).

Another example of wound infection after a trauma is that of a 40-year-old female who was hit by a water tanker, causing a lacerated wound in the leg, and that presented fever and pain 3 days later (Behera *et al.*, 2011). *Aeromonas* was recovered from the purulent aspirate from the previously sutured wound and the initial empirical treatment was changed from amoxicillin–clavulanic acid and amikacin to piperacillin with tazobactam and netilmicin for 2 weeks, and the patient recovered. Trauma was also the origin of a brain abscess infection that originated in a laceration of the right frontal scalp following an assault in a 30-year-old male and that motivated an emergency craniectomy with drainage of the abscess and debridement of the wound (Mahabeer *et al.*, 2014). A pus swab collected from the abscess cavity revealed on blood agar, large beta-haemolytic colonies resembling *Aeromonas* spp. The Vitek 2 system identified the organism as '*Aeromonas hydrophila* complex' with 98% probability. The patient died despite being treated with ceftriaxone, to which the organism was susceptible, and aggressive surgical treatment. The post-mortem examination confirmed the presence of pus over the right parietal area and a midline brainstem haemorrhage. This indicated that the infection progressed significantly over a longer period than the 2 days reported that the patient had not sought medical attention. Furthermore, a skull fracture, which was most likely not detected at the clinic initially, may have facilitated the development of the brain abscess by allowing the infection to extend rapidly to the brain. Previous studies on central nervous system *Aeromonas* infections were discussed by Mahabeer *et al.* (2014).

Burns or wound infections produced by *Aeromonas* mimic those of streptococcal or staphylococcal soft-tissue infections, although the standard empirical treatment for the latter is not effective for the former and therefore this is an important concern highlighted by different authors (Figueras, 2005; Elwitigala *et al.*, 2005). For instance, no improvement was observed in a 21-year-old male patient who showed a cutaneous abscess and fever despite the abscess was drained and empirically treated with oxacillin–clindamycin (Coutinho *et al.*, 2012). The lesion started as a local small erythema in the axillary region after the patient had bathed in a freshwater lake in Brazil that progressed to an abscess of ca. 15 cm in diameter. When *Aeromonas* was recovered from the abscess aspirate and blood, the treatment was changed to ciprofloxacin–cefepime on the basis of the antimicrobial sensitivity of the isolate and the fever disappeared and the patient was cured (Coutinho *et al.*, 2012). Another example is a case of *Aeromonas* meningitis from which a strain resistant to amoxicillin–clavulanic acid was recovered (Pampin *et al.*, 2012). This involved a 35-year-old patient who had a motorbike accident and had fallen into a pond and suffered a severe cranial trauma and died 15 days later.

An interesting outbreak of wound infection associated with a mud football game that affected 26 people was described in Australia (Vally *et al.*, 2004). All patients had scratches and abrasions that became infected when exposed to the mud irrigated with river water. Fever was reported in 35% of the cases and two patients developed complications that required surgical intervention. Since *Aeromonas* was isolated from the river water that was used to irrigate the football field, this was considered the contamination source.

Wound infection by *Aeromonas* can be severe, evolving to necrotizing fasciitis or ecthyma gangrenosum. Abuhammour *et al.* (2006) reported a necrotizing fasciitis in a 14-year-old boy with no underlying systemic diseases or immunodeficiency. However, two fatal cases of necrotizing fasciitis associated with *Aeromonas* were reported by Tsai *et al.* (2009) in 2 diabetic patients. Kumar *et al.* (2012) described a new case in a 71-year-old male patient, with history of foot trauma due to foreign body that occurred 10 days previously, showing a wound over the right foot discharging pus, along with fever and subcutaneous swelling of the lower right extremity extending up to the knee joint.

Ecthyma gangrenosum is a cutaneous necrotic manifestation that is usually fatal and the erythematous lesions can progress to necrosis of the soft tissue within 24 hours (Avolio *et al.*, 2009). A case of ecthyma gangrenosum occurred in a 53-year-old woman with type I diabetes who was under chemotherapy for a uterine cancer (García-Lozano *et al.*, 2012). The patient developed fever and vesiculobullous lesions over the left breast that was also disseminated in the trunk and extremities from which *Aeromonas* was isolated (García-Lozano *et al.*, 2012). Another case occurred in a previously healthy 63-year-old diabetic man who arrived at the hospital with fever and complaining of an ulcer in the lower left leg that was painful and swelling, and it was empirically treated with amoxicillin–clavulanic acid (Avolio *et al.*, 2009). The next day the lesion worsened with necrotic-haemorrhagic bullae and was diagnosed as a possible pseudomonal ecthyma gangrenosum. Blood cultures were negative but the bloody exudate collected by aspiration from the necrotic bullae was positive for only one resistant ampicillin isolate identified as *Aeromonas,* and hyperbaric oxygen chamber therapy was applied. Two months later he was healthy and waiting for admission to another hospital for skin grafting surgery. The patient was questioned and revealed that a few days prior to admission he had worked in a well near his house without taking precautions.

Monaghan *et al.* (2008) reported a case in a patient that also showed a sepsis caused after an open femur fracture, and Kao and Kao (2012) another one after a penetrating injury. Those cases normally evolve from bullae to skin necrosis and require an aggressive and extensive surgical debridement of the wound and sometimes amputation (Figueras, 2005; Janda and Abbott, 2010). In fact, an apparent superficial cellulitis that fails to respond to standard therapy must raise suspicion of a more extensive underlying subcutaneous infection. For instance Shak *et al.* (2011) reported an aminoglycoside-resistant *Aeromonas* strain recovered from a wound exudate of a 35-year-old man who showed a leg laceration after a traumatic fall into freshwater. One day after the accident the patient had fever and due to a rapidly advancing cellulitis, was operated twice for surgical debridement of the wound (Shak *et al.*, 2011). Koth *et al.* (2012) described a case of an infected wound with *Aeromonas* co-isolated with *Enterobacter asburiae* after a car accident, the patient required repeated extensive surgical debridement. Recently, Grim *et al.* (2013) using an intramuscular mouse model and functional genomics demonstrated the highly infectious nature of some strains of different *Aeromonas* spp. showing that they possessed virulence signatures and the ability to cause necrotizing fasciitis.

Aeromonas species were the most frequently isolated (22.6% of all isolates) bacteria from southern Thailand tsunami survivors (Hiransuthikul *et al.*, 2005). In this case, exposure to contaminated water was again the source of infection. A similar situation was recorded after the hurricane Katrina in New Orleans (Presley *et al.*, 2006). This should be borne in mind in the future following this type of disasters. In fact, flooding, together with droughts, are considered to be the main impacts of climate change as reviewed by Figueras and Borrego (2010 and references therein) and there is evidence that both low rainfall and heavy rain precede many drinking water outbreaks. In addition, as indicated in the latter review, an increase in temperature threatens water quality with regard to waterborne diseases especially cholera in Asia and South America. *Aeromonas* have been co-isolated with *Vibrio cholerae* in some outbreaks including the one that occurred in Haiti (Hasan *et al.*, 2012; Mendes-Marquez *et al.*, 2012). Both microbes can be commonly misidentified as we have indicated previously (Chacón *et al.*, 2002) and as occurred in a case of acute severe diarrhoea in a 46-year-old woman who presented dehydration and acute renal failure (Chelimilla *et al.*, 2013). The initial stool study was suggestive of *Aeromonas* but in fact the bacteria was a *Vibrio cholerae* resistant to trimethoprim/sulfamethoxazole that was the initial empiric selected treatment (Chelimilla *et al.*, 2013).

It is expected that in temperate zones, climate change will reduce the number of rainy days but increase the average volume of each rainfall event (Borrego and Figueras, 2010).

Nosocomial infection related to surgical site infections and leech therapy

As commented in a previous study surgical site infections (SSIs) are the most common nosocomial infections in surgical patients and are associated with substantial morbidity and mortality, and with high medical costs (Tena *et al.*, 2009 and references therein). The aetiological agents vary with the procedure, particularly with the anatomical site and despite extremely rare they can be caused by *Aeromonas*. Eight cases that occurred at a 400-bed hospital (University Hospital of Guadalajara), between January 1990 and December 2007, and one case from a 200-bed hospital from another area in Spain (Royo Villanova Hospital of Zaragoza) have been reported by Tena *et al.* (2009) along with a review of 15 existing cases. These cases were linked to appendectomies, cholecystectomy, and colectomy and almost all reported SSI cases developed in persons with pre-existing gastrointestinal or biliary disease and were mainly polymicrobial with a fatality rate of 5%. In that study when clinical manifestations of SSIs due to *Aeromonas* were compared with those produced by other microorganisms the most important difference was the fast onset of the infection, i.e. seven of the nine patients (77.7%) suffered the infection during the first or second postoperative day with a mean onset of SSI of 2.2 days (Tena *et al.*, 2009).

Leech therapy is used in plastic and reconstructive surgery to relive venous congestion and to improve the microvascularization of flaps, grafts or replants (Graf, 2000; Sartor *et al.*, 2002; Steer *et al.*, 2005; Figueras, 2005; Whitaker *et al.*, 2011). The application of this treatment for surgical unsalvageable venous obstruction of skin flaps and replants have shown a success rates of up to 80–83% (Sartor *et al.*, 2002; Steer *et al.*, 2005). However, acute and delayed infections by *Aeromonas* may compromise the success of the procedure which may decrease to less than 30% (Sartor *et al.*, 2002; Figueras, 2005; Schnabl *et al.*, 2010; Whitaker *et al.*, 2011). Two periods with the highest risk of *Aeromonas* infection in leech therapy have been described, the first is the acute onset, occurring within the first couple of days and the second is with delayed onset of 10 days or more after leech application (Schnabl *et al.*, 2010; Mumcuoglu *et al.*, 2014, and references therein). The hypothesis is that the acute infection results from the bite of the leeches contaminating the wound while the delayed infections may result from *Aeromonas* colonization of necrotic tissue with later invasion of surrounding tissue being the ischaemic muscle especially susceptible.

Some authors consider that the potential risks of infections associated with leech therapy is generally underestimated and not sufficiently discussed in the modern literature (Schnabl *et al.*, 2010). Leeches are known to have a symbiotic relationship with *Aeromonas* as they provide specific enzymes needed by the leeches to digest blood (Graf, 2000; Figueras,

2005). It has been demonstrated by 16S rRNA sequencing that not *A. hydrophila* as usually thought, but *A. veronii* bv. *sobria* (often incorrectly referred as *A. sobria*) is the digestive tract symbiont of the medicinal leech *Hirudo medicalis,* although, under experimental conditions strains of *A. hydrophila* and *A. veronii* isolated from faeces could colonize the leech in a similar extent than the symbiotic strain (Graf, 2000). Sequencing of the *gyrB* gene showed that in addition to *A. veronii* bv *sobria, A. media* may also be involved (Mumcuoglu *et al.,* 2009). The incidence of cases of *Aeromonas* infection after leech therapy ranges from 7% to 20% (Janda and Abbott, 1998, 2010; Sartor *et al.,* 2002; Figueras, 2005; Whitaker *et al.,* 2010).

To prevent these nosocomial infections various strategies have been proposed including (i) the prophylactic use of broad-spectrum antibiotics (Sartor *et al.,* 2002; Ouderkirk *et al.,* 2004; Figueras, 2005; Mumcuoglu *et al.,* 2009, 2014; Levine *et al.,* 2010) and (ii) paying attention to the quality of water in which the leeches are maintained. The latter was discovered on the basis of a 5 year-prospective study that investigated this nosocomial infection at different French hospitals and in which they discovered the importance of paying attention to the quality of water in which the leeches were maintained (Sartor *et al.,* 2002). They noticed that the unit of the hospital with a poorly cleaned own aquarium that was filed with tap water from the hospital (contaminated with *Aeromonas* in 33% of the samples) had a higher estimated incidence (4.1%) of *Aeromonas* infection after leech therapy than other hospital units that did not store leeches (2.4%). In the 14 identified cases of *Aeromonas* nosocomial infections, those occurred 2–11 days after the leeches were first applied (Sartor *et al.,* 2002). At the time of leech application two patients received amoxicillin-clavulanic acid, but this treatment was ineffective for the *Aeromonas* infection and a total of nine additional surgical procedures were necessary to control the infection. Three patients developed the infection after discharge from the hospital, and two of them required re-hospitalization (Sartor *et al.,* 2002). An *Aeromonas* meningitis occurred as a complication of leech therapy in a 40-year-old man who underwent a surgical procedure with post application of the leech therapy in close proximity to the CNS (Ouderkirk *et al.,* 2004). *Aeromonas* was recovered from the wound specimens, from the broth where leeches were maintained and from the leeches themselves (Ouderkirk *et al.,* 2004). Other cases of meningitis not associated with leech therapy have been associated with cranial trauma or injury (Pampín *et al.,* 2012). Two cases of *A. veronii* bv. *sobria* septicaemia occurred after mammary reconstructions in a 55- and 56-year-old woman 6 and 7 days post operation and leech therapy (Maetz *et al.,* 2012).

In relation to the prophylactic strategy it has been suggested that although third-generation cephalosporins, carbapenem and aztroeonam are active for *Aeromonas,* the quinolones should be considered for prophylaxis and/or therapy, particularly since they are available for both oral and parenteral administration. Ampicillin-sulbactam prophylaxis failed in a case of *Aeromonas* sepsis following leech therapy after surgery for a replantation of the thumb (Levine *et al.,* 2010). Blood and wound samples that were positive for *A. veronii* bv. *sobria* were resistant to ampicillin–sulbactam. The antibiotic was changed to ciprofloxacin on the basis of the sensitivity profile of the organism. Cultures from the leech bathwater confirmed it to be the source of *Aeromonas.* Another study involved five male patients who suffered from a trauma and developed a local infection of the injured extremity after leech treatment that could be associated with the fact that none of the patients were receiving adequate antibiotics when leeches were applied (Schnabl *et al.,* 2010). The infections were cured with fluoroquinolone antibiotics but the reconstructive procedures in all five cases were unsuccessful.

It has been demonstrated that the concentration of *Aeromonas* inside the digestive tracts of leeches largely decreases when the patient is under antibiotic therapy, and despite the therapy decreases the leech efficacy, this antibioprophylaxysh has been proposed as another preventative strategy (Schnabl *et al.*, 2010; Maetz *et al.*, 2012). The use of fluoroquinolone during the whole period of treatment has been proposed (Maetz *et al.*, 2011). However, infections can be developed despite the patient being on what was considered an appropriate antibiotic, when the specific *Aeromonas* strains are resistant to the antibiotic used as occurred in patients infected with a ciprofloxacin resistant strains (Pattel *et al.*, 2013; Sartor *et al.*, 2013; Giltner *et al.*, 2013). Sartor *et al.* (2013) reported that in Marseille Public Hospitals, the contamination of leeches by a ciprofloxacin-resistant *Aeromonas* isolate led to an increase in infection rate at surgical sites from 0% of 23 patients treated in 2011 to 28.5% (two of seven patients treated) in 2012. This occurred despite oral ciprofloxacin prophylaxis (1 g per day) was given prior the leech therapy to the two patients from which an *Aeromonas* ciprofloxacin resistant isolate was recovered from the infected flaps (Sartor *et al.*, 2013). A ciprofloxacin-resistant *Aeromonas* also grew from the water of the two leech aquariums but no growth was obtained from the bottled mineral water used to fill the aquaria. A new batch of leeches that were tested on arrival was also positive for a ciprofloxacin resistant *Aeromonas*. This lead the authors to speculate that the presence of fluoroquinolones in the poultry blood used to feed medicinal leeches may contribute to the selection of ciprofloxacin-resistant *Aeromonas* spp. in their digestive tract (Sartor *et al.*, 2013). They suggested that further investigation of the presence of quinolone residues in poultry blood should be done to test this hypothesis. To avoid this problem the ciprofloxacin used as prophylaxis was replaced with cotrimoxazole, and no other infections have occurred (Sartor *et al.*, 2013). It is a pity that in that study they did not genotype the *Aeromonas* isolates to find out if they were clonally related, furthermore the identification method used to characterize the *Aeromonas* isolates was not mentioned.

Another approach proposed has been feeding leeches artificially with antibiotics, Mumcuoglu *et al.* (2010) demonstrated that ciprofloxacin reduced the number of leech-associated aeromonads to undetectable levels for extended periods suggesting the possibility of using ciprofloxacin-treated leeches instead of chemoprophylaxis in patients undergoing leech therapy.

A recent study has shown a differential gene expression of *A. veronii* during symbiosis by comparing the transcriptome in culture and inside the leech (*Hirudo verbena*) gut and revealed that symbionts are experiencing environmental stress that make the microbe enter a stationary phase survival state in which it may persist for months within the host (Bomar and Graf, 2012).

Bacteraemia/septicaemia

As in our previous review (Figueras, 2005) and in the one performed by Janda and Abbott (2010), the terms bacteraemia and septicaemia will be used interchangeably in this chapter as this is the approach currently used at the literature. Several new cases of *Aeromonas* bacteraemia/septicaemia have been published from different geographical areas i.e. India (Mukhopadhyay *et al.*, 2008), France (Lamy *et al.*, 2009), French Caribbean Islands (Hochedez *et al.*, 2010), Spain (Tena *et al.*, 2007; Pampin *et al.*, 2012), Taiwan (Tsai *et al.*, 2006, 2007, 2009; Wu *et al.*, 2007, 2011, 2012a,b,c; Lay *et al.*, 2010; Chuang *et al.*, 2011; Su *et al.*, 2012; Tang *et al.*, 2014) and Japan (Morinaga *et al.*, 2011; Shizuma *et al.*, 2011; Kimura *et al.*, 2013).

Janda and Abbott (2010) classified the *Aeromonas* bacteraemia in four groups depending upon the type of population affected. The main group that accounts for more than 80% of the cases is the one that occurs in patients with underlying disease or in immunocompromised persons, followed by those that had suffered trauma or injury, healthy persons and finally, patients who underwent reconstructive surgery and to whom leech therapy has been applied.

In our pervious review in 2005 we recorded that according to Janda and Abbott (1998) the most common underlying disease were malignancy (40–50%), hepatobiliary disease (15–30%) and diabetes (3–5%) and showed that the new data that we have evaluated (studies published between 2000 and 2005) indicate an increasing incidence (36–54%) in patients with hepatobiliary disease as shown in Table 4.6 (Ko *et al.*, 2000; Lau *et al.*, 2000; Campo *et al.*, 2001; Llopis el al., 2004; Figueras, 2005). However, we were already indicating that the latter tendency along with the characteristics of the *Aeromonas* bacteraemia observed, i.e. preferentially monomicrobial, community-acquired and affecting mainly male patients had to be verified in the future (Figueras, 2005). In this sense the recent review by Janda and Abbott (2010) confirmed this tendency. Also the new bacteraemia surveys published after the latter review (see Table 4.6) ratify that the predominant affected population are men with hepatobiliary diseases (Hochedez *et al.*, 2010; Lay *et al.*, 2010: Chuang *et al.*, 2011). However, a recent study performed in Taiwan by Tang *et al.* (2014) showed that almost 50% of the bacteraemia cases (47.3%) were healthcare-associated infections, and patients within this group have more cancer (62.8%), liver cirrhosis (37.2%) than patients with community-acquired bacteraemia. A review of *Aeromonas* infections of the hepatobiliary system was carried out by Clark and Chenoweth (2003) and has been extensively commented on before (Figueras, 2005). Among the recent studies, Kimura *et al.* (2013) and Tang *et al.* (2013) are the only ones that show that malignancy prevailed over hepatobiliary disease (58% vs. 36% and 44% vs. 40%) as shown in Table 4.6. However, the study by Kimura *et al.* (2013) that evaluated retrospectively the cases of *Aeromonas* bacteraemia from 1994 to 2010 in two hospitals, one from Tokyo and another from Kangawa, among the 21 (58%) patients who had solid tumours, 14 (67%) had hepatic, biliary or pancreatic malignancies. In another study in Nigasaki (Japan), six of the seven patients were females (median age of 76), 85% showed malignant diseases, none had liver cirrhosis as an underlying condition, and two patients (28.6%) died of *Aeromonas* septicaemia (Morinaga *et al.*, 2011). The characteristics of the bacteraemia in patients with liver cirrhosis produced by *Aeromonas* were compared with those caused by enterobacteria in another study in Japan (Shizuma *et al.*, 2011). Despite no significant differences in the short-term, mortality rate, renal dysfunction and percentage of shock between these groups of patients being observed, the severity of the liver dysfunction was greater in patients with *Aeromonas* bacteraemia (Shizuma *et al.*, 2011). Liver disease, diabetes and malignancy predominated (62.5%) among the eight cases of extraintestinal infections reported in a 2-month period in south Karnataka coastal region (India) of which 37.5% showed bacteraemia (Mukhopadhyay *et al.*, 2008). Tsai *et al.* (2006) revised the characteristics of 41 adult patients with hematologic malignancies (Table 4.6), and reported that 29 (64.4%) of the 31 patients with nosocomial *Aeromonas* bacteraemia had received recent antineoplastic chemotherapy. The predominating symptoms were fever (88.9%), septic shock (40%), and altered consciousness (26.7%). Fatality occurred in 35.6% of the patients within 14 days of onset of bacteraemia and the only independent prognostic factor associated with their death was the alteration of consciousness. In the study by Tang

Table 4.6 Characteristics of *Aeromonas*-associated bacteraemia in different surveys

Reference	Country	No patients	Age (mean)	M/F ratio	Monomicrobial infection (%)	Underlying disease (%)			Mortality (%)
						Malignancy	Hepato-biliary disease	Diabetes	
Previous reviews[a]		144	54	1.8	70	42	38	NS	31
Ko *et al.* (2000)	Taiwan	104	54	3.9	100[b]	21	54	NS	32[c]
Lau *et al.* (2000)	Taiwan	42	62	2	64	33	50	NS	70
Campo *et al.* (2001)	Spain	12[d]	50	0.7	75	58	0	NS	25
Llopis *et al.* (2004)	Spain	73[e]	61	2.5	72	33	36	NS	36
Tsai *et al.* (2006)	Taiwan	41	53	2	76	100[f]	–	–	36/14 days
Lamy *et al.* (2009)	France	20	72	1.2	70	40	35	10	35
Lay *et al.* (2010)	Taiwan	216	60	2	71	40	49	30	32
Hochedez *et al.* (2010)	Caribbean Islands	37	55	2.4	62	27	32	11	27
Chuang *et al.* (2011)	Taiwan	154	58	2.6	100	26	58	28	36
Kimura *et al.* (2013)	Japan	36[g]	63	2	55	58	36	25	19/30 days
Tang *et al.* (2014)	Taiwan	91[h]	64	2.4	71	44	40	29	23

[a]Mean values of previously reviewed studies by Janda and Abbott (1996) evaluated in Figueras (2005).
[b]Study including only monomicrobial bacteraemias, 1/3 of patients died 72 hours after arrival at the hospital.
[c]Spontaneous peritonitis associated with 50% fatality and 67% with necrotizing fasciitis.
[d]6% of cases July-August.
[e]72% of the cases occurred in summer or autumn.
[f]Haematological malignancies was the most predominant: acute myelogenous leukaemia (37.8%), myelodysplastic syndrome (26.7%) and non-Hodgkin lymphoma (17.8%).
[g]The majority of cases in August–September.
[h]30% required UCI admission and mechanical ventilation; 47.3% (43/90) were classified as healthcare-associated infections and in this group 62.8% (27/43) had active cancer vs. 27.1% (13/48) ($P = 0.001$) in the community-acquired group (48/90).
NS, not specified.

et al. (2014) the in-hospital day mortality was significantly associated only with underlying cancer ($P < 0.001$), and initial shock ($P < 0.001$), and it was higher in patients who were in the intensive care unit (33.3%) or that showed acute respiratory failure (28.6%).

Mortality associated with *Aeromonas* bacteraemia varies depending on the survey but the predominating rates were around 30% (Table 4.6). However, higher rates, i.e. 70% were found in a study in Taiwan (Lau *et al.*, 2000). In a Japanese study, 62.5% the patients died shortly after the blood cultures became positive (Nakasome *et al.*, 2001) and also in a Spanish study 56% of the patients died within the first 48 hours (Llopis *et al.* 2004).

Incidence of *Aeromonas* bacteraemia has been established at a 0.3% by Llopis *et al.* (2004) on the basis of 75 episodes recorded from 1997 up to 2002 in a Spanish hospital, where the number of episodes per year ranged from 0 to 7. These authors indicated that the incidence remained quite constant throughout the years. Interestingly a 0.3% incidence was also reported by Hochedez *et al.* (2010) from the Caribbean islands of Martinique and Guadalupe. However, in a review of 12 cases recorded over 6 years in another Spanish study it was only 0.12% (Campo *et al.*, 2001). Nakasome *et al.* (2001) in Okinawa (Japan) reported a higher incidence (3.3%, 14 cases in 7 months), which is in contrast with the more recent report from Nigasaki (Japan) where only 7 cases were found between 2000 and 2010, and the incidence of *Aeromonas* bacteraemia was estimated to be 0.07 cases per 1000 admissions (Morinaga *et al.*, 2011). The estimated incidence of *Aeromonas* septicaemia in England and Wales and the United States was set at 1.5 per million population (Janda and Abbott, 2010), although it was indicated that this was probably a minimum estimate considering that many cases either go undetected or are not reported. Wu *et al.* (2012a) established that in a 3-year study (2008 to 2010) the cases of bacteraemia produced by *Aeromonas*, *Vibrio* and *Salmonella* in South Taiwan were 76, 38 and 103 cases/million population respectively. They also indicated that in Taiwan there are a substantially higher number of cases of *Aeromonas* bacteraemia than in western countries with no clear seasonal variation, providing for these 3 possible explanations for the observed geographical variation. The first is related to the fact that 10–15% of the general population in Taiwan are hepatitis B carriers (i.e. around 2.5 to 3 million in total) and it is clear that patients with chronic liver disease are highly susceptible to invasive *Aeromonas* infections. The second is the fact that *Aeromonas* are highly abundant in the environment, i.e. 88% of seafood from retailed markets is positive for these bacteria and third there is the warm climate all year-round.

What is clear from the new reviewed cases is that the seasonality patterns and the underlying diseases seem to depend on the geographical areas and their climate (Table 4.6). Regarding that, liver disease including cirrhosis are underlying diseases predominant in tropical climates where the isolation of *Aeromonas* does not show a seasonality pattern and where certain endemic conditions exist, i.e. Taiwan and Caribbean French Islands (Ko *et al.*, 2000; Lau *et al.*, 2000; Hochedez *et al.*, 2010; Chuang *et al.*, 2011).

Hochedez *et al.* (2010) in the French Caribbean islands, reported that among the 37 cases, 89% had underlying diseases, hepatobiliary and pancreatic disease being the most frequent condition followed by recent trauma and malignancies. Interestingly, one case was related to severe strongyloidiasis, which is a life-threatening disseminated infection, this being the first case of *Aeromonas* bacteraemia with this presentation.

Although it seems that the overall prevalence of *Aeromonas* bacteraemia in different parts of the world is starting to be elucidated, it still remains relatively unclear because most studies are retrospective and there is a tendency to publish only unique and/or interesting cases.

Predisposing conditions for bacteraemia

Morinaga *et al.* (2011) in Japan reported that haematological malignancy or cholangitis, and recent ingestion of raw fish or shellfish were considered important characteristics linked to the development of *Aeromonas* bacteraemia. The congestion of bile or pancreatic fluid was considered a risk factor in bacteraemia caused by *Aeromonas* (Kimura *et al.*, 2013). The APACHE II (Acute Physiology and Chronic Health Evaluation II) score, which is used to determine disease severity was the most important risk factor associated to bacteraemia mortality (Chuang *et al.*, 2011). Older age, occurrence of multiorgan failure, and impaired renal function were the underlying conditions that favour death during hospitalization (Hochedez *et al.*, 2010).

Llopis *et al.* (2004) observed that in Spain *Aeromonas* bacteraemia is associated with very elderly healthy patients. A similar situation was reported by Lamy *et al.* (2009) in France because very elderly patients with ages above 80 years were predominating among the bacteraemia patients, 30% (6/20) of whom had no comorbidities. These authors also indicate that considering the increasing life span of the Western populations this type of infections may become more frequent. However, one interesting finding of the study of Lamy *et al.* (2009) was that the fatality rate was much lower in the group of elderly patients with no comorbidity (17%, 1/6) than in the group of younger patients with comorbidities (43%, 6/14).

Aeromonas infection might be a complication associated to secondary effects of specific treatments (Stano *et al.*, 2009; Okumura *et al.*, 2011; Papadakis *et al.*, 2012; Turner *et al.*, 2012). Regarding that, *Aeromonas* was isolated from blood and the bile of a 72-year-old Japanese patient under treatment for rheumatoid arthritis with tocilizumab, a drug that includes infection as the most common adverse event (Okumura *et al.*, 2011). A fatal case of *Aeromonas* sepsis was associated to the treatment with corticosteroids in a patient that had HIV infection and hepatitis C (Stano *et al.*, 2009) while another patient also suffered a fulminant infection when treated for lymphoblastic leukaemia (Papadakis *et al.*, 2012). However, another patient with acute lymphoblastic leukaemia that showed *Aeromonas* bacteraemia responded to ciprofloxacin once the empiric ineffective selected treatment was changed (Fatima *et al.*, 2013). A rare case associated with a catheter sepsis in a patient with chronic kidney disease that received also steroids and dialysis, responded well to the antibiotic treatment and to the catheter removal as described by Khalil *et al.* (2013). In addition these authors reviewed all cases of *Aeromonas* infections associated to chronic kidney disease patients. Another case was described after a transrectal prostate biopsy within a study that evaluated bacteraemia complications after the procedure (Carmignani *et al.*, 2012). The isolated strain showed an intermediate sensitivity to ciprofloxacin, and the patient developed signs of urosepsis 6 days after the biopsy following suspension of antibiotic prophylactic therapy. Three cases of *Aeromonas* bacteraemia, one of them fatal, were described in pregnant women who carried a non-viable foetus, and it was suggested that sticks used to induce abortion may have been the source of infection (Turner *et al.*, 2012). Other cases in pregnant woman have been found by Hochedez *et al.* (2010).

Pneumonia and other respiratory tract infections

Aeromonas pneumonia is a rare presentation but cases have been reported from different regions (Kao *et al.*, 2003; Mukhopadhyay *et al.*, 2003; Issa and Napolitano, 2011; Nagata *et al.*, 2011; Chao *et al.*, 2013c). The largest-scale study on *Aeromonas* pneumonia included 84 patients with possible, probable, and definitive cases during an 8-year period at a hospital in southern Taiwan (Chao *et al.*, 2013c).

Aeromonas pneumonia tends to occur in immunosuppressed patients. For instance, 70% of the patients in Chao *et al.* (2013c) study had this condition while among the 20 cases reviewed by Nagata *et al.* (2011), 11 (55%) occurred in patients with alcohol abuse, renal failure, liver cirrhosis, bronchiectasis, lymphoma or cerebrovascular accident. There are other cases not mentioned by those authors, such as one involving a 13-year-old girl with multiple sclerosis who showed a community-acquired pneumonia produced by a multidrug-resistant *Aeromonas* strain recovered from bronchoalveolar lavage and sensitive only to meropenem (Mukhopadhyay *et al.*, 2003). Clinical symptoms improved after treatment with carbapenems. In other cases *Aeromonas* have produced fatal fulminant pneumonias that have been recognized in autopsy (Murata *et al.*, 2001; Kao *et al.*, 2003; Nagata *et al.*, 2011). For instance, in Taiwan a case of pneumonia that was fatal 4 hours after admission occurred in a previously healthy girl who, during the previous 24 hours, had a high fever, coughing, vomiting and blood-tinged sputum (Kao *et al.*, 2003). The authors commented that there was no history of contact with water. There was a similar case of the rapid death of a 75-year-old woman who had suddenly become ill with pneumonia 2 weeks after an operation for colon cancer. Despite intensive support measures, she died 6 hours after admission and 12 hours after her first symptom. Autopsy showed necrotizing pneumonia and *Aeromonas* was isolated from the blood taken for culture at admission and from the lung at autopsy (Nagata *et al.*, 2011).

Near-drowning is one of the typical accidents associated to *Aeromonas* pulmonary infections including pneumonia (Bossi-Küpfer *et al.*, 2007; Mukhopadhyay *et al.*, 2008; Lamy *et al.*, 2009; Janda and Abbott, 2010; Chao *et al.*, 2013c). Of the five cases of respiratory tract infection reported in a study by Lamy *et al.* (2009), three were near-drowning victims. In the study of Chao *et al.* (2013c), 6 of the 84 patients presented pneumonias associated with drowning while in the one of Nagata *et al.* (2011), near-drowning accidents occurred in 40% of the cases. Huys *et al.* (2012) developed a protocol for the detection of *Aeromonas* in tissues that could be used post-mortem in forensic studies that aim to determine if death occurred by drowning, considering that the recovery of the bacteria of this genus could be a potential marker. The method enabled detection of *Aeromonas* in a lung swab, blood and the bone marrow of drowned bodies ($n = 3$), but was negative in these three matrices for all controls ($n = 90$), indicating the usefulness of the test. A case of *Aeromonas* pneumonia not associated with near drowning but with a trauma was described by Issa and Napolitano (2011), who reported the case in a patient who had a motor vehicle crash and sustained pelvic fractures and an injury to the spleen. The patient required temporary venovenous extracorporeal membrane oxygenation (ECMO) for acute lung failure and recovered.

Bossi-Küpfer *et al.* (2007) reported a case of acute tracheobronchitis following a near-downing episode in a river. During a bronchoscopy examination, a 19-year-old man had a reddish-yellow bronchial secretion from which *Aeromonas* was isolated as the predominant microorganism and was identified as *A. veronii* using *gyrB* gene sequences. These authors in agreement with others (Soler *et al.*, 2003b; Aravena-Román *et al.*, 2011b; Su *et al.*, 2013) manifested that the identification of most *Aeromonas* species by conventional biochemical tests is often inaccurate.

The conclusions drawn from the study by Chao *et al.* (2013c) in Taiwan indicate that *Aeromonas* pneumonia linked to drowning have high morbidity and mortality rates, especially among patients with underlying liver cirrhosis. The latter, together with shock as the initial presentation, were considered significant risk factors for death.

Spontaneous bacterial peritonitis (SBP) and empyema

As described in our previous review, *Aeromonas* have been linked to SBP, which is a frequent complication that occurs in 15% of patients with cirrhosis (Figueras, 2005). For instance, in a Spanish study patients with cirrhosis showed a higher incidence of *Aeromonas* peritonitis (20%) than of *E. coli* (7%) or *Pseudomonas aeruginosa* (1.9%) as reported by Llopis *et al.* (2004). In a matched, case–control study of SBP in cirrhosis patients in Korea by Choi *et al.* (2008), the incidence of SBP due to *Aeromonas* was significant higher (63% vs. 25%) in the warm season (July-September). There were also more episodes of diarrhoea (26% vs. 6%) when compared with the control group with SBP due to other bacteria. The in-hospital mortality rate associated with *Aeromonas* was 23% and septic shock was the only independent prognostic factor. A 50% fatality was associated with *Aeromonas* SBP in a study by Ko *et al.* (2000) in Taiwan while in a new retrospective review of cirrhosis patients there is a reported cause mortality rate of 56% (Wu *et al.*, 2009). Yang *et al.* (2008) described a case of peritonitis in a 68-year-old diabetic woman who had eaten fish. This was attributed to *A. salmonicida*, a species typically associated with diseased fish (Beaz-Hidalgo and Figueras, 2012, 2013) but sporadically isolated from infected wounds (Aravena-Román *et al.*, 2011a; Roger *et al.*, 2012b) or other clinical samples (Figueras *et al.*, 2000; Yang *et al.*, 2008). Lin *et al.* (2013) reviewed the characteristics of 50 patients with *Aeromonas* peritonitis recorded over 2004–2011 and found an overall in-hospital mortality rate of 20%. Only nine cases (18%) corresponded to spontaneous bacterial peritonitis, and were more likely to be immunocompromised ($P = 0.0013$) and more frequently had an initial presentation of shock ($P = 0.0129$), abnormal liver function ($P < 0.05$) and concomitant bacteraemia ($P = 0.0024$) than the patients with secondary peritonitis.

Peritonitis is a major cause of morbidity in continuous ambulatory peritoneal dialysis (CAPD) patients. Isolation of *Aeromonas* from peritoneal dialysates is not uncommon and has been reported by Puthucheary *et al.* (2012) from Malaysia (Tables 4.2 and 4.4). Sahin *et al.* (2010) reported *Aeromonas* peritonitis in a 44-year-old female patient from Turkey who was on CAPD and had accidentally dropped her peritoneal dialysis catheter into the toilet. The patient was treated successfully with intravenous ceftazidime for 2 days followed by intraperitoneal administration for 12 days, and she completely recovered after 2 weeks of antibiotic treatment (Sahin *et al.*, 2010). Another case of peritonitis was reported from Greece in a CAPD patient with systemic lupus erythematosus that was also resolved after treatment with ceftazidime, but this time in combination with vancomycin. The Aeromonas injection was thought to have been acquired during gardening and involved in addition *Streptococcus viridans* (Liakopoulos *et al.*, 2011).

In contrast to SBP, which is quite a common complication in cirrhosis patients, spontaneous bacterial empyema (SBE) is rarer, but *Aeromonas* has also been implicated in such cases (Figueras, 2005; Chao *et al.*, 2012b). Six cases of *Aeromonas* empyema have been reported by Chao *et al.* (2012b), all of which occurred in patients who had underlying diseases (five with liver cirrhosis and one with diabetes).

Urinary tract infections

Urinary tract infections by *Aeromonas* are rare although several well-documented cases have been reported and several species have been molecularly identified, including *A. dhakensis* (Tables 4.2 and 4.4). Interesting to mention is an urinary tract infection produced by *A. popoffii* (Hua *et al.*, 2004), a species so far reported in freshwater and seawater, and that shows an important virulence potential (Soler *et al.*, 2002). In Spain, *Aeromonas* was isolated alone or in combination with other pathogens from the urine of seven patients (Ojeda-Vargas *et al.*,2005) and other strains recovered from urine have been reported in other studies (Llopis *et al.*, 2004; Tena *et al.*, 2007). Al-Benwan *et al.* (2007) reported a case of cystitis produced by *A. caviae*, which affected a 39-year-old healthy man. In that study, previous cases of urinary infection were also reviewed and, contrary to their case, all of them involved patients with an underlying disease, such as one having congenital spina bifida, a newborn baby with bladder and bilateral renal dilatation, and a 69-year-old diabetic patient who also presented leucocyturia and bacteruria (Al-Benwan *et al.*, 2007). It is interesting to note in that study that the API 20E profile identified the strain as *V. fluvialis*/*A. caviae*, but *gyrB* sequencing confirmed it as *A. caviae*, and 16S rRNA gene sequencing provided inconclusive results. Hua *et al.* (2004), however, recognised *A. popoffii* using the 16S rRNA gene sequencing approach, although as we have commented this gene may have limitations for recognising other species, as occurred in the Al-Benwan *et al.* (2007) study. These authors support the *gyrB* gene as an alternative molecular marker for the 16S rRNA gene.

Mandal *et al.* (2010) reported seven cases of urinary tract infection from India, and the recurrence of the isolation and the clearance after antibiotic treatment clearly demonstrated that *Aeromonas* was responsible for the infection. Chao *et al.* (2012a) have recently reviewed 19 cases of genitourinary tract infections from Taiwan. They found that more than half of the patients were > 65 years of age and that 57.9% were women, with only one case in 5-year-old child. Only three cases were nosocomial and 47.4% of the patients were immunocompromised, the underlying diseases being, in order of frequency, cancer, cirrhosis and diabetes (Chao *et al.*, 2012a).

References

Abbott, S.L., Cheung, W.K.W., and Janda, J.M. (2003). The genus *Aeromonas*: biochemical characteristics, atypical reactions, and phenotypic identification schemes. J. Clin. Microbiol. *41*, 2348–2357.

Abuelsaad, A.S., Mohamed, I., Allam, G., and Al-Solumani, A.A. (2013). Antimicrobial and immunomodulating activities of hesperidin and ellagic acid against diarrheic *Aeromonas hydrophila* in a murine model. Life Sci. *93*, 714–722.

Abuhammour, W., Hasan, R.A., and Rogers, D. (2006). Necrotizing fasciitis caused by *Aeromonas hydrophilia* in an immunocompetent child. Pediatr. Emerg. Care 22, 48–51.

Aguilera-Arreola., M.G., Hernandez-Rodriguez, C., Zuniga, G., Figueras, M.J., and Castro-Escarpulli, G. (2005). *Aeromonas hydrophila* clinical and environmental ecotypes as revealed by genetic diversity and virulence genes. FEMS Microbiol. Lett. *242*, 231–240.

Aguilera-Arreola, M.G., Hernández-Rodríguez, C., Zúñiga, G., Figueras, M.J., Garduño, R.A., and Castro-Escarpulli G. (2007). Virulence potential and genetic diversity of *Aeromonas caviae*, *Aeromonas veronii*, and *Aeromonas hydrophila* clinical isolates from Mexico and Spain: a comparative study. Can. J. Microbiol. *53*, 877–887.

Al-Benwan, K., Abbott, S., Janda, J.M., Huys, G., and Albert, M.J. (2007). Cystitis caused by *Aeromonas caviae*. J. Clin. Microbiol. *45*, 2348–2350.

Alperi, A., and Figueras, M.J. (2010). Human isolates of *Aeromonas* possess Shiga toxin genes (*stx1* and *stx2*) highly similar to the most virulent gene variants of *Escherichia coli*. Clin. Microbiol. Infect. *16*, 1563–1567.

Alperi, A., Figueras, M.J., Inza, I., and Martínez-Murcia, A.J. (2008). Analysis of 16S rRNA gene mutations in a subset of *Aeromonas* strains and their impact in species delineation. Int. Microbiol. *11*, 185–194.

Alperi, A., Martínez-Murcia, A.J., Ko, W., Monera, A., Saavedra, M.J., and Figueras, M.J. (2010a). *Aeromonas taiwanensis* sp. nov., and *Aeromonas sanarellii* sp. nov., two new clinical species from Taiwan. Int. J. Syst. Evol. Microbiol. *60*, 2048–2055.

Alperi, A., Martínez-Murcia, A.J., Monera, A., Saavedra, M.J., and Figueras, M.J. (2010b). *Aeromonas fluvialis* sp. nov., isolated from Spanish river. Int. J. Syst. Evol. Microbiol. *60*, 72–77.

Altwegg, M., Martinetti Lucchini, G., Lüthy-Hottenstein, J., and Rohrbach, M. (1991). *Aeromonas*-associated gastroenteritis after consumption of contaminated shrimp. Eur. J. Clin. Microbiol. Infect. Dis. *10*, 44–45.

Aravena-Román, M., Harnett, G.B., Riley, T.V., Inglis, T.J.J., and Chang, B.J. (2011a). *Aeromonas aquariorum* is widely distributed in clinical and environmental specimens and can be misidentified as *Aeromonas hydrophila*. J. Clin. Microbiol. *49*, 3006–3008.

Aravena-Román, M., Riley, T.V., Inglis, T.J.J., and Chang, B.J. (2011b). Phenotypic characteristics of human clinical and environmental *Aeromonas* in Western Australia. Pathology. *43*, 350–356.

Aravena-Román, M., Beaz-Hidalgo, R., Inglis, T.J., Riley, T.V., Martínez-Murcia, A.J., Chang, B.J., and Figueras, M.J. (2013). *Aeromonas australiensis* sp. nov., isolated from irrigation water in Western Australia. Int. J. Syst. Evol. Microbiol. *63*, 2270–2276.

Avolio, M., La Spisa, C., Moscariello, F., De Rosa, R., and Camporese, A. (2009). *Aeromonas* hydrophila ecthyma gangrenosum without bacteraemia in a diabetic man: the first case report in Italy. Infez. Med. *17*, 184–187.

Azzopardi, E.A., Azzopardi, S.M., Boyce, D.E., and Dickson, W.A. (2011). Emerging gram-negative infections in burn wounds. J. Burn Care Res. *32*, 570–576.

Beaz-Hidalgo, R., and Figueras, M.J. (2012). Molecular detection and characterization of furunculosis and other *Aeromonas* fish infections. In Health and Environment in Aquiculture, E.D. Carvalho, G.S. David, and R.J. da Silva, eds (InTech, Rijeka, Croatia), pp. 97–132. Available from: http://www.intechopen.com/books/health-and-environment-in-aquaculture/molecular-detection-and-characterization-of-furunculosis-and-other-aeromonas-fish-infections

Beaz-Hidalgo, R., and Figueras, M.J. (2013). *Aeromonas* spp. whole genomes and virulence factors implicated in fish disease. J. Fish Dis. *36*, 371–388.

Beaz-Hidalgo, R., Alperi, A., Figueras, M.J., and Romalde, J.L. (2009). *Aeromonas piscicola* sp. nov., isolated from diseased fish. Syst. Appl. Microbiol. *32*, 471–479.

Beaz-Hidalgo, R., Alperi, A., Buján, N., Romalde, J.L., and Figueras, M.J. (2010). Comparison of phenotypical and genetic identification of *Aeromonas* strains isolated from diseased fish. Syst. Appl. Microbiol. *33*, 149–153.

Beaz-Hidalgo, R., Shakèd, T., Laviad, S., Halpern, M., and Figueras, M.J. (2012). Chironomid egg masses harbour the clinical species *Aeromonas taiwanensis* and *Aeromonas sanarellii*. FEMS Microbiol. Lett. 337: 48–59.

Beaz-Hidalgo, R., Martínez-Murcia A.J., and Figueras, M.J. (2013). Reclassification of *Aeromonas hydrophila* subsp. *dhakensis* Huys *et al.* 2002 and *Aeromonas aquariorum* Martínez-Murcia *et al.* 2008 as *Aeromonas dhakensis* sp. nov. comb nov. and emendation of the species *Aeromonas hydrophila*. Syst. Appl. Microbiol. *36*, 171–176.

Behera, B., Bhoriwal, S., Mathur, P., Sagar, S., Singhal, M., and Misra, M.C. (2011). Post-traumatic skin and soft tissue infection due to *Aeromonas hydrophila*. Indian J. Crit. Care Med. *15*, 49–51.

Bomar, L., and Graf, J. (2012). Investigation into the physiologies of *Aeromonas veronii in vitro* and inside the digestive tract of the medicinal leech using RNA-seq. Biol. Bull. *223*, 155–166.

Borrell, N., Acinas, S.G., Figueras, M.J., and Martínez-Murcia, A.J. (1997). Identification of *Aeromonas* clinical isolates by restriction fragment length polymorphism of PCR amplified 16S rRNA genes. J. Clin. Microbiol. *35*, 1671–1674.

Bossi-Küpfer, M., Genini, A., Peduzzi, R., and Demarta, A. (2007). Tracheobronchitis caused by *Aeromonas veronii* biovar *sobria* after near-drowning. J. Med. Microbiol. *56*, 1563–1564.

Bravo, L., Fernández, A., Núñez, F.Á., Riverom, L.A., Ramírez, M., Aguila, A., Ledo, Y., Cruz, Y., and Hernández, J. (2012). [*Aeromonas* spp associated to acute diarrheic disease in Cuba: case–control study]. Rev. Chilena Infectol. *29*, 44–48.

Campo, C., Navarro, V., Pérez, C., Gutiérrez, I., and Alonso, R. (2001). *Aeromonas* spp bacteremia: study of 12 cases and review of the literature. Enferm. Infecc. Microbiol. Clin. *19*, 161–164.

Carmignani, L., Picozzi, S., Spinelli, M., Di Pierro, S., Mombelli, G., Negri, E., Tejada, M., Gaia, P., Costa, E., and Maggioni, A. (2012). Bacterial sepsis following prostatic biopsy. Int. Urol. Nephrol. *44*, 1055–1063.

Carnahan, A.M., Chakraborty, T., Fanning, G.R., Verma, D., Ali, A., Janda, J.M., and Joseph, S.W. (1991). *Aeromonas trota* sp. nov., an ampicillin-susceptible species isolated from clinical specimens. J. Clin. Microbiol. *29*, 1206–1210.

Castro-Escarpulli, G., Figueras, M.J., Aguilera-Arreola, G., Soler, L., Fernández-Rendón, E., Aparicio, G.O., Guarro, J., and Chacón, M.R. (2003). Characterisation of *Aeromonas* spp. isolated from frozen fish intended for human consumption in Mexico. Int. J. Food Microbiol. *84*, 41–49.

Chacón, M.R., Castro-Escarpulli, G., Soler, L., Guarro, J., and Figueras, M.J. (2002). A DNA probe specific for *Aeromonas* colonies. Diagn. Microbiol. Infect. Dis. *44*, 221–225.

Chacón, M.R., Soler, L., Groisman, E.A., Guarro, J., and Figueras, M.J. (2004). Type III secretion system genes in clinical *Aeromonas* isolates. J. Clin. Microbiol. *42*, 1285–1287.

Chan, F.K., Chin, J.K., Ling, T.K., Chung, S.C., and Sung, J.J. (2000). *Aeromonas* infections in acute suppurative cholangitis: review of 30 cases. J. Infect. *40*, 69–73.

Chang, C.F., Chen, T.L., Chen, T.W., Yang, W.C., and Lin, C.C. (2005). Recurrent dialysis-associated *Aeromonas hydrophila* peritonitis: reports of two cases and review of the literature. Perit. Dial. Int. *25*, 496–499.

Chao, C.M., Gau, S.J., and Lai, C.C. (2012a). *Aeromonas* genitourinary tract infection. J. Infect. *65*, 573–575.

Chao, C.M., Gau, S.J., and Lai, C.C. (2012b). Empyema caused by *Aeromonas* species in Taiwan. Am. J. Trop. Med. Hyg. *87*, 933–935.

Chao, C.M., Lai, C.C., Tang, H.J., Ko, W.C., and Hsueh, P.R. (2013a). Skin and soft-tissue infections caused by *Aeromonas* species. Eur. J. Clin. Microbiol. Infect. Dis. *32*, 543–547.

Chao, C.M., Lai, C.C., Tang, H.J., Ko, W.C., and Hsueh, P.R. (2013b). Biliary tract infections caused by *Aeromonas* species. Eur. J. Clin. Microbiol. Infect. Dis. *32*, 245–251.

Chao, C.M., Lai, C.C., Tsai, H.Y., Wu, C.J., Tang, H.J., Ko, W.C., and Hsueh, P.R. (2013c). Pneumonia caused by *Aeromonas* species in Taiwan, 2004–2011. Eur. J. Clin. Microbiol. Infect. Dis. *32*, 1069–1075.

Chen, P.L., Wu, C.J., Chen, C.S., Tsai, P.J., Tang, H.J., and Ko, W.C. (2013). A comparative study of clinical *Aeromonas dhakensis* and *Aeromonas hydrophila* isolates in southern Taiwan: *A. dhakensis* is more predominant and virulent. Clin. Microbiol. Infect. doi: 10.1111/1469-0691.12456.

Chelimilla, H., Widjaja, D., Carvajal, S., and Kumbum, K. (2013). Life-threatening manifestation of diarrhea. Case Rep. Infect. Dis. *2013*, 354697.

Choi, J.-P., Lee, S.O., Kwon, H.H., Kwak, Y.G., Choi, S.H., Lim, S.K., Kim, M.N., Jeong, J.Y., Choi, S.H., Woo, J.H., and Kim, Y.S. (2008). Clinical significance of spontaneous *Aeromonas* bacterial peritonitis in cirrhotic patients: a matched case–control study. Clin. Infect. Dis. *47*, 66–72.

Chu, Y.W., Wong, C.H., Tsang, G.K.L., Kwok, M S.W., Wong, R.K.O., Lo, J.Y.C., and Kam, K.M. (2006). Lack of association between presentation of diarrhoeal symptoms and fecal isolation of *Aeromonas* spp. amongst outpatients in Hong Kong. J. Med. Microbiol. *55*, 349–351.

Chuang, H.C., Ho, Y.H., Lay, C.J., Wang, L.S., Tsai, Y.S., and Tsai, C.C. (2011). Different clinical characteristics among *Aeromonas hydrophila*, *Aeromonas veronii* biovar *sobria* and *Aeromonas caviae* monomicrobial bacteraemia. J. Korean Med. Sci. *26*, 1415–1420.

Clark, N., and Chenoweth, C.E. (2003). *Aeromonas* infection of the hepatobiliary system: report of 15 cases and review of the literature. Clin. Infect. Dis. *37*, 506–513.

Coutinho, A.-S., de Morais, O.O., Gomes, C.M., and de Oliveira Carneiro da Motta, J. (2012). Cutaneous abscess leading to sepsis by *Aeromonas hydrophila*. Infection *41*, 595–596.

Crivelli, C., Demarta, A., and Peduzzi, R. (2001). Intestinal secretory immunoglobulin A (sIgA) response to *Aeromonas* exoproteins in patients with naturally acquired *Aeromonas* diarrhea. FEMS Immunol. Med. Microbiol. *30*, 31–35.

De Luca. F., Giraud-Morin, C., Rossolini, G.M., Docquier, J.D., and Fosse, T. (2010). Genetic and biochemical characterization of TRU-1, the endogenous class C beta-lactamase from *Aeromonas enteropelogenes*. Antimicrob. Agents Chemother. *54*, 1547–1554.

Demarta, A., Tonolla, M., Caminada, A., Beretta, M., and Peduzzi, R. (2000). Epidemiological relationships between *Aeromonas* strains isolated from symptomatic children and household environments as determined by ribotyping. Eur. J. Epidemiol. *16*, 447–453.

Demarta, A., Küpfer, M., Riegel, P., Harf-Monteil, C., Tonolla, M., Peduzzi, R., Monera, A., Saavedra, M.J., and Martínez-Murcia, A. (2008). *Aeromonas tecta* sp. nov., isolated from clinical and environmental sources. Syst. Appl. Microbiol. *31*, 278–286.

Danaher, P.J., and Mueller, W.P. (2011). *Aeromonas hydrophila* septic arthritis. Mil. Med. *176*, 1444–1446.

Elwitigala, J.P.-, Higgs, D.S., Namnyak, S., White, J.W., and Yaneza, A. (2005). Septic arthritis due to *Aeromonas hydrophila*: case report and review of the literature. Int. J. Clin. Pract. Suppl. *147*, 121–124.

Esser, B., Burnens, A.P., Lanfranchini, F.M., Somaruga, S.G.E., Von Vigier, R.O., Schaad, U.B., Aebi, C., and Bianchetti, M.G. (2000). Acute community-acquired diarrhea requiring hospital admission in Swiss children. Clin. Infect. Dis. *31*, 192–196.

Esteve, C., Alcaide, E., and Blasco, M.D. (2012). *Aeromonas hydrophila* subsp. *dhakensis* isolated from feces, water and fish in Mediterranean Spain. Microbes Environ. *27*, 367–373.

Fatima, A., Afridi, F.I., Qureshi, A., Farooqi, B.J., and Hussain, A. (2013). Bacteremia due to *Aeromonas hydrophila* in acute lymphoblastic leukemia. J Coll. Physicians Surg. Pak. *23*, 893–895.

Fica, A., Julio, V., Palma, E., Guinez, D., Porte, L., Braun, S., and Dabanch, J. (2013). Bacteremic colangitis in adults in a general hospital in Chile. Rev Chilena Infectol. *30*, 65–73.

Figueras, M.J. (2005). Clinical relevance of *Aeromonas*. Rev. Med. Microbiol. *16*, 145–153.

Figueras, M.J., and Beaz-Hidalgo, R. (2014). *Aeromonas:* introduction. In Encyclopedia of Food Microbiology, 2nd Ed., vol. 1, Batt, C.A., and Tortorello, M.L., eds, (Elsevier Ltd, Academic Press, Oxford, UK), pp. 24–30.

Figueras, M.J., and Borrego, J.J. (2010). New perspectives in monitoring drinking water microbial quality. Int. J. Environ. Res. Public Health. *7*, 4179–4202.

Figueras, M.J., Guarro, J., and Martínez-Murcia, A.J. (2000a). Clinically relevant *Aeromonas* species. Clin. Infect. Dis. *30*, 988–989.

Figueras, M.J., Soler, L., Chacón, M.R., Guarro, J., and Martínez-Murcia, A.J. (2000b). Extended method for discrimination of *Aeromonas* spp. by 16S rDNA RFLP analysis. Int. J. Syst. Evol. Microbiol. *50*, 2069–2073.

Figueras, M.J., Alperi, A., Guarro, J., and Martínez-Murcia, A.J., (2006). Genotyping of isolates included in the description of a novel species should be mandatory. Int. J. Syst. Evol. Microbiol. *56*, 1183–1184.

Figueras M.J., Aldea, M.J., Fernández, N., Aspíroz, C., Alperi, A., and Guarro, J. (2007a). *Aeromonas* haemolytic uremic syndrome. A case and a review of the literature. Diagn. Microbiol. Infect. Dis. *58*, 231–234.

Figueras, M.J., Alperi, A., and Guarro, J. (2007b). On the identification of clinical *Aeromonas* by a new restriction fragment length polymorphism of 16S rDNA method. Lett. Appl. Microbiol. *45*, 692–693.

Figueras, M.J., Horneman, A.J., Martinez-Murcia, A., and Guarro, J. (2007c). Controversial data on the association of *Aeromonas* with diarrhoea in a recent Hong Kong study. J. Med. Microbiol. *56*, 996–998.

Figueras, M.J., Alperi, A., Saavedra, M.J., Ko, W.C., Gonzalo, N., Navarro, M., and Martínez-Murcia, A. (2009). Clinical relevance of the recently described *Aeromonas aquariorum*. J. Clin. Microbiol. *47*, 3742–3746.

Figueras, M.J., Alperi, A., Beaz-Hidalgo, R., Stackebrandt, E., Brambilla, E.M., Monera, A., and Martínez-Murcia, A.J. (2011a). *Aeromonas rivuli* sp. nov., isolated from the upstream region of a karst water rivulet in Germany. Int. J. Syst. Evol. Microbiol. *61*, 242–248.

Figueras, M.J., Beaz-Hidalgo, R., Collado, L., and Martínez-Murcia, A.J. (2011b). Recommendations for a new bacterial species description based on analyses of the unrelated genera *Aeromonas* and Arcobacter. Bull. BISMiS *2*, 1–16.

Figueras, M.J., Beaz-Hidalgo, R., and Paredes, K. (2011c). Furunculosis y otras infecciones producidas por *Aeromonas*. In Enfermedades infecciosas del cultivo de Salmónidos en Chile y el Mundo, Avendaño-Herrera, R. ed. (Niva Chile S.A., Puerto Varas, Chile), pp. 285–374.

Figueras, M.J., Beaz-Hidalgo, R., Senderovich, Y., Laviad, S., and Halpern, M. (2011d). Re-identification of *Aeromonas* isolates from chironomid egg masses as the potential pathogenic bacteria *Aeromonas aquariorum*. Environ. Microbiol. Rep. *3*, 239–244.

Figueira, V., Vaz-Moreira, I., Silva, M., and Manaia, C.M. (2011). Diversity and antibiotic resistance of *Aeromonas* spp. in drinking and waste water treatment plants. Water Res. *45*, 5599–5611.

Finkelstein, R., and Oren, I. (2011). Soft tissue infections caused by marine bacterial pathogens: epidemiology, diagnosis, and management. Curr. Infect. Dis.Rep. *13*, 470–477.

Fontes, M.C., Saavedra, M.J., Monera, A., Martins, C., and Martínez-Murcia, A. (2010). Phylogenetic identification of *Aeromonas simiae* from a pig, first isolate since species description. Vet. Microbiol. *142*, 313–316.

Fontes, M.C., Saavedra, M.J., Martins, C., and Martínez-Murcia, A. (2011). Phylogenetic identification of *Aeromonas* from pigs slaughtered for consumption in slaughterhouses at the North of Portugal. Int. J. Food Microbiol. *146*, 118–122.

Fraisse, T., Lechiche, C., Sotto, A., and Lavigne, J.P. (2008). [*Aeromonas* spp. infections: retrospective study in Nîmes University Hospital, 1997–2004]. [Article in French] Pathol. Biol. (Paris) *56*, 70–76.

García-Lozano, T., Aznar Oroval, E., Cruz Mojarrieta, J., Messeguer Badía, F. (2012). [Vesiculobullous lesions in an immunocompromised cancer patient]. [Article in Spanish] Rev. Argent. Microbiol. *44*, 61.

Ghenghesh, K.S., Ahmed, S.F., El-Khalek, R.A., Al-Gendy, A., and Klena, J. (2008). *Aeromonas*-associated infections in developing countries. J. Infect. Dev. Ctries. 2, 81–98.

Giltner, C.L., Bobenchik, A.M., Uslan, D.Z., Deville, J.G., and Humphries, R.M. (2013). Ciprofloxacin-resistant *Aeromonas hydrophila* cellulitis following leech therapy. J.Clin. Microbiol. *51*, 1324–1326.

von Graevenitz, A. (2007). The role of *Aeromonas* in diarrhea: a review. Infection 35, 59–64.

Graf, J. (2000). Symbiosis of *Aeromonas* and *Hirudo medicinalis*, the medicinal leech. ASM News *66*, 147–153.

Grim, C.J., Kozlova, E.V., Sha, J., Fitts, E.C., van Lier, C.J., Kirtley, M.L., Joseph, S.J., Read, T.D., Burd, E.M., Tall, B.D. *et al.* (2013). Characterization of *Aeromonas hydrophila* wound pathotypes by comparative genomic and functional analyses of virulence genes. MBio. *4*, e00064–13.

Grimes, J. (2006). Koch's Postulates-Then and Now. Microbe *1*, 223–228.

Gröbner, S., Bissinger, A.L., Raible, A., Heeg, P., Autenrieth, I.B., and Schmidt, S.M. (2007). Severe diarrhoea caused by *Aeromonas veronii* biovar *sobria* in a patient with metastasised GIST. Pol. J. Microbiol. *56*, 277–279.

Harf-Monteil, C., Le Flèche, A., Riegel, P., Prèvost, G., Bermond, D., Grimont, P.A.D., and Monteil, H. (2004). *Aeromonas simiae* sp. nov., isolated from monkey faeces. Int. J. Syst. Evol. Microbiol. *54*, 481–485.

Hasan N.A., Choi, S.Y., Eppinger, M., Clark, P.W., Chen, A., Alam, M., Haley, B.J., Taviani, E., Hine, E., Su, Q., *et al.* (2012). Genomic diversity of 2010 Haitian cholera outbreak strains. Proc. Natl. Acad. Sci. U.S.A. *109*, E2010–7.

Hiransuthikul, N., Tantisiriwat, W., Lertutsahakul, K., Vibhagool, A., and Boonma, P. (2005). Skin and soft-tissue infections among tsunami survivors in southern Thailand. Clin. Infect. Dis *41*, e93–6.

Hochedez, P., Hope-Rapp, E., Olive, C., Nicolas, M., Beaucaire, G., and Cabié, A. (2010). Bacteremia caused by *Aeromonas hydrophila* complex species [corrected] complex in the Caribbean Islands of Martinique and Guadeloupe. Am. J. Trop. Med. Hyg. *83*, 1123–1127. Erratum in: Am. J. Trop. Med. Hyg. 2011. *84*, 842.

Hochedez, P., Hope-Rapp, E., Olive, C., Nicolas, M., Beaucaire, G., and Cabié, A. (2011). Bacteremia caused by *Aeromonas* species [corrected] in the Caribbean Islands of Martinique and Guadeloupe. Am. J. Trop. Med. Hyg. 2011. *84*, 842.

Horneman, A.J., and Ali, A. (2011). *Aeromonas.* In Manual of Clinical Microbiology, 10th edn, Versalovic, J. ed. (ASM Press, Washington, DC), pp. 658–665.

Hua, H.T., Bollet, C., Tercian, S., Drancourt, M., and Raoult, D. (2004). *Aeromonas popoffii* urinary tract infection. J. Clin. Microbiol. *42*, 5427–5428.

Huys, G., Kämpfer, P., Albert, M.J., Kühn, I., Denys, R., and Swings, J. (2002). *Aeromonas hydrophila* subsp. *dhakensis* subsp. nov., isolated from children with diarrhoea in Bangladesh, and extended description of *Aeromonas hydrophila* subsp. *hydrophila* (Chester 1901) Stanier 1943 (approved lists 1980). Int. J. Syst. Evol. Microbiol. 52, 705–712.

Huys, G., Coopman,V., Van Varenbergh, D., and Cordonnier, J. (2012). Selective culturing and genus-specific PCR detection for identification of *Aeromonas* in tissue samples to assist the medico-legal diagnosis of death by drowning. Forensic Sci. Int. *221*, 11–15.

Igbinosa, I.H., Igumbor, E.U., Aghdasi, F., Tom, M., and Okoh, A.I. (2012). Emerging *Aeromonas* species infections and their significance in public health. Scientific World J. 2012: 625023.

Issa, N., and Napolitano, L.M. (2011). *Aeromonas* pneumonia in a trauma patient requiring extracorporeal membrane oxygenation for severe acute respiratory distress syndrome: case report and literature review. Surg. Infect. (Larchmt) *12*, 241–245.

Janda, J.M., and Abbott, S.L. (1996). Human pathogens. In The Genus *Aeromonas*, B. Austin, M. Altwegg, P.J. Gosling, and S. Josep, eds (John Wiley & Sons, Inc., New York, NY), pp. 151–174.

Janda, J.M., and Abbott, S.L. (1998). Evolving concepts regarding the genus *Aeromonas*: an expanding panorama of species, disease presentation, and unanswered questions. Clin. Infect. Dis. *27*, 332–344.

Janda, J.M., and Abbott, S.L. (2007). 16S rRNA gene sequencing for bacterial identification in the diagnostic laboratory: pluses, perils, and pitfalls. J. Clin. Microbiol. *45*, 2761–2764.

Janda, J.M., and Abbott, S.L. (2010). The genus *Aeromonas*, taxonomy, pathogenicity, and infection. Clin. Microbiol. Rev. *23*, 35–73.

Juan, H.J., Tang, R.B., Wu, T.C., and Yu, K.W. (2000). Isolation of *Aeromonas hydrophila* in children with diarrhea. J. Microbiol. Immunol. Infect. *33*, 115–117.

Khalil, M.A., Rehman, A., Kashif, W.U., Rangasami, M., and Tan, J. (2013). A rare case of *Aeromonas hydrophila* catheter related sepsis in a patient with chronic kidney disease receiving steroids and dialysis:

a case report and review of *Aeromonas* infections in chronic kidney disease patients. Case Rep Nephrol. 2013, 735194.

Kao, H.T., Huang, Y.C., and Lin, T.Y. (2003). Fatal bacteremic pneumonia caused by *Aeromonas hydropila* in a previously healthy child. J. Microbiol. Immunol. Infect. *36*, 209–211.

Kao, T.L., and Kao, M.L. (2012). A fatal case of necrotizing *Aeromonas schubertii* fasciitis after penetrating injury. Am. J. Emerg. Med. *30*, 258.e3–5.

Khajanchi, B.K., Fadl, A.A., Borchardt, M.A., Berg, R.L., Horneman, A.J., Stemper, M.E., Joseph, S.W., Moyer, N.P., Sha, J., and Chopra, A.K. (2010). Distribution of virulence factors and molecular fingerprinting of *Aeromonas* species isolates from water and clinical samples: suggestive evidence of water-to-human transmission. Appl. Environ. Microbiol. *76*, 2313–2325.

Kimura, M., Araoka, H., and Yoneyama, A. (2013). *Aeromonas caviae* is the most frequent pathogen amongst cases of *Aeromonas* bacteraemia in Japan. Scand. J. Infect. Dis. *45*, 304–309.

Ko, W.C., Lee, H.C., Chuang, Y.C., Liu, C.C., and Wu, J.J. (2000). Clinical features and therapeutic implications of 104 episodes of monomicrobial *Aeromonas* bacteraemia. J. Infect. *40*, 267–273.

Koth, K., Boniface, J., Chance, E.A., and Hanes, M.C. (2012). *Enterobacter asburiae* and *Aeromonas hydrophila*: soft tissue infection requiring debridement. Orthopedics 35, e996–9.

Kotloff, K.L., Nataro, J.P., Blackwelder, W.C., Nasrin, D., Farag, T.H., Panchalingam, S., Wu, Y., Sow. S.O., Sur, D., Breiman, R.F. *et al.* (2013). Burden and aetiology of diarrhoeal disease in infants and young children in developing countries (the Global Enteric Multicenter Study, GEMS): a prospective, case–control study. Lancet *382*, 209–222.

Kumar, S., Mukhopadhyay, P., Chatterjee, M., Bandyopadhyay, M.K., Bandyopadhyay, M., Ghosh, T., and Samaddar, D. (2012). Necrotizing fasciitis caused by *Aeromonas caviae*. Avicenna J Med. *2*, 94–96.

Lamy, B., Kodjo, A., and Laurent, F. (2009). Prospective nationwide study of *Aeromonas* infections in France. J. Clin. Microbiol. *47*, 1234–1237.

Lamy, B., Laurent, F., and Kodjo, A. (2010a). Validation of a partial *rpoB* gene sequence as a tool for phylogenetic identification of aeromonads isolated from environmental sources. Can. J. Microbiol. *56*, 217–228.

Lamy, B., Laurent, F., Verdier, I., Decousser, J.W., Lecaillon, E., Marchandin, H., Roger, F., Tigaud, S., de Montclos H; colBVH Study Group, Kodjo, A. (2010b). Accuracy of 6 commercial systems for identifying clinical *Aeromonas* isolates. Diagn. Microbiol. Infect. Dis. *67*, 9–14.

Lau, S.M., Peng, M.Y., and Chang, F.Y. (2000). Outcomes of *Aeromonas* bacteremia in patients with different types of underlying disease. J. Microbiol. Immunol Infec. *33*, 241–247.

Lay, C.J., Zhuang, H.J., Ho, Y.H., Tsai, Y.S., Wang, L.S., and Tsai, C.C. (2010). Different clinical characteristics between polymicrobial and monomicrobial *Aeromonas* bacteremia a study of 216 cases. Intern Med. *49*, 2415–2421.

Lete, A.L., Suñé, N.M., Pellicer, J.A., and Blanco, J.J. (2013). *Aeromonas* diarrhea and Behçet's disease: more than a mere association. Rheumatol. Int. *33*, 543–544.

Levine, S.M., Frangos, S.G., Hanna, B., Colen, K., and Levine, J.P. (2010). *Aeromonas* septicemia after medicinal leech use following replantation of severed digits. Am. J. Crit. Care. *19*, 469–471.

Liakopoulos, V., Arampatzis, S., Kourti, P., Tsolkas, T., Zarogiannis, S., Eleftheriadis, T., Giannopoulou, M., and Stefanidis, I. (2011). *Aeromonas hydrophila* as a causative organism in peritoneal dialysis-related peritonitis: case report and review of the literature. Clin. Nephrol. *75*, 65–68.

Lin, W.T., Su, S.Y., Lai, C.C., Tsai, T.C., Gau, S.J., and Chao, C.M. (2013). Peritonitis caused by *Aeromonas* species at a hospital in southern Taiwan. Intern Med. *52*, 2517–2521.

Liu, J., Janiki, L., and Houpt, E. (2012). Precaution for using nucleic acid-based methods to detect *Aeromonas*. J. Clin. Microbiol. *50*, 1829.

Llopis, F., Grau, I., Tubau, F., Cisnal, M., and Pallares, R. (2004). Epidemiological and clinical characteristics of bacteraemia caused by *Aeromonas* spp. as compared with *Escherichia coli* and *Pseudomonas aeruginosa*. Scand. J. Infect. Dis. *36*, 335–341.

Maetz, B., Abbou, R., Andreoletti, J.B., and Bruant-Rodier, C. (2012). Infections following the application of leeches: two case reports and review of the literature. J. Med. Case Rep. *6*, 364.

Mahabeer, Y., Khumalo, A., Kiratu, E., and Mlisana, K. (2014). Posttraumatic brain abscess caused by *Aeromonas hydrophila*. J. Clin. Microbiol. *52*, 1796–1797.

Mandal, J., Dhodapkar, R., Acharya, N.S., Sastry, A., and Parija, S.C. (2010). Urinary tract infection due to *Aeromonas* spp., a lesser known causative bacterium. J. Infect. Dev. Ctries *4*, 679–681.

Mansour, A.M., Abd Elkhalek, R., Shaheen, H.I., El Mohammady, H., Refaey, S., Hassan, K., Riddle, M., Sanders, J.W., Sebeny, P.J., Young, S.Y., and Frenck, R. (2012). Burden of *Aeromonas hydrophila*-associated diarrhea among children younger than 2 years in rural Egyptian community. J. Infect. Dev. Ctries. *6*, 842–846.

Martin-Carnahan, A., and Joseph, S.W. (2005). Family I. *Aeromonadaceae* Colwell, MacDonell and DeLey 1986. In Bergey's Manual of Systematic Bacteriology, 2nd edn, vol. 2, D.J. Brennan, N.R. Krieg, J.T. Staley, and G.N. Garrity, eds (Springer-Verlag, New York, NY), pp. 556–578.

Martín Delgado, M., Matute Cruz, P., Nuñez Gallo, D., Fernández González, C., López González-Coviella, N., Valladares Hernández, B., Pita Toledo, M.L., and Martínez Navarro, F. (2001). Diarrea del viajero asociada a *Aeromonas hydrophyla*. Rev. Salut Ambient. *1*, 30–35.

Martínez-Murcia, A.J., Benlloch, S., and Collins, M.D. (1992). Phylogenetic interrelationships of members of the genera *Aeromonas* and *Plesiomonas* as determined by 16S ribosomal DNA sequencing: lack of congruence with results of DNA–DNA hybridizations. Int. J. Syst. Bacteriol. *42*, 412–421.

Martínez-Murcia, A.J., Borrell, N., and Figueras, M.J. (2000). Typing of clinical and environmental *Aeromonas veronii* strains based on the 16S-23S rDNA spacers. FEMS Immunol. Med. Microbiol. *28*, 225–232.

Martínez-Murcia, A.J., Soler, L., Saavedra, M.J., Chacón, M.R., Guarro, J., Stackebrandt, E., and Figueras, M.J. (2005). Phenotypic, genotypic, and phylogenetic discrepancies to differentiate *Aeromonas salmonicida* from *Aeromonas bestiarum*. Int. Microbiol. *8*, 259–269.

Martínez-Murcia, A.J., Saavedra, M.J., Mota, V.R., Maier, T., Stackebrandt, E., and Cousin, S. (2008). *Aeromonas aquariorum* sp. nov., isolated from aquaria of ornamental fish. Int. J. Syst. Evol. Microbiol. *58*, 1169–1175.

Martínez-Murcia, A., Monera, A., Alperi, A., Figueras, M.J., and Saavedra, M.J. (2009). Phylogenetic evidence suggests that strains of *Aeromonas hydrophila* subsp. *dhakensis* belong to the species *Aeromonas aquariorum* sp. nov. Curr. Microbiol. *58*, 76–80.

Martínez-Murcia, A., Morena, A., Saavedra, M.J., Ocina, R., López-Álvarez, M., Lara, E., and Figueras, M.J. (2011). Multilocus phylogenetic analysis of the genus *Aeromonas*. Syst. Appl. Microbiol. *34*, 189–199.

Martínez-Murcia, A., Beaz-Hidalgo, R., Saavedra, M.J., Figueras, M.J., and Sedlacek, I. (2013). *Aeromonas cavernicola* sp. nov., isolated from fresh water of a brook in a cavern. Curr. Microbiol. *66*, 197–204.

Martino, M.E., Fasolato, L., Montemurro, F., Rosteghin, M., Manfrin, A., Patarnello, T., Novelli, E., and Cardazzo, B. (2011). Determination of microbial diversity of *Aeromonas* strains on the basis of multilocus sequence typing, phenotype, and presence of putative virulence genes. Appl. Environ. Microbiol. *77*, 4986–5000.

Mendes-Marques, C.L., Nascimento, L.M., Theophilo, G.N., Hofer, E., Melo Neto, O.P., and Leal, N.C. (2012). Molecular characterization of *Aeromonas* spp. and *Vibrio cholerae* O1 isolated during a diarrhea outbreak. Rev. Inst. Med. Trop. Sao. Paulo. *54*, 299–304.

Miñana-Galbis D., Farfan, M., Loren, J.G., and Fusté, M.C. (2010). Proposal to assign *Aeromonas diversa* sp. *nov*., as a novel species designation for *Aeromonas* group 501. Syst. Appl. Microbiol. 33, 15–19.

Monaghan, S.F., Anjaria, D., Mohr, A., and Livingston, D.H. (2008). Necrotizing fasciitis and sepsis caused by *Aeromonas hydrophila* after crush injury of the lower extremity. Surg. Infect. (Larchmt) *9*, 459–467.

Morinaga, Y., Yanagihara, K., Araki, N., Harada, Y., Yamada, K., Akamatsu, N., Matsuda, J., Nishino, T., Hasegawa, H., Izumikawa, K. *et al.* (2011). Clinical characteristics of seven patients with *Aeromonas* septicemia in a Japanese hospital. Tohoku J. Exp. Med. *225*, 81–84.

Morinaga, Y., Yanagihara, K., Latif Eugenin, F.L., Beaz-Hidalgo, R., and Figueras, M.J. (2013). Identification errors of *Aeromonas* species: the considerable virulence of the new species *Aeromonas aquariorum*. Diag. Microbiol. Infect. Dis. *76*, 106–109.

Mukhopadhyay, C., Bhargava, A., and Ayyagari, A. (2003). *Aeromonas hydrophila* and aspiration pneumonia: a diverse presentation. Yonsei Med. J. *44*, 1087–1090.

Mukhopadhyay, C., Chawla, K., Sharma, Y., and Bairy, I. (2008). Emerging extra-intestinal infections with *Aeromonas hydrophila* in coastal region of southern Karnataka. J. Postgrad. Med. *54*, 199–202.

Mumcuoglu, K.Y. (2014). Recommendations for the use of leeches in reconstructive plastic surgery. Evid. Based Complement Alternat. Med. 2014, 205929.

Mumcuoglu, K.Y., Huberman, L., Cohen, R., Temper, V., Adler, A., Galun, R., and Block, C. (2009). Elimination of symbiotic *Aeromonas* spp. from the intestinal tract of the medicinal leech, *Hirudo medicinalis*, using ciprofloxacin feeding. Clin. Microbiol. Infect. *16*, 563–567.

Murata, H., Yoshimoto, H., Masuo, M., Hitoshi, T., Kitamura, S., Otsuka, Y., and Miura, Y. (2001). Fulminant pneumonia due to *Aeromonas hydrophila* in a man with chronic renal failure and liver cirrhosis. Intern Med. *40*, 118–123.

Nagar, V., Shashidhar, R., and Bandekar, J.R. (2013). Characterization of *Aeromonas* strains isolated from Indian foods using *rpoD* gene sequencing and whole cell protein analysis.World J. Microbiol. Biotechnol. *29*, 745–752.

Nagata, K., Takeshima, Y., Tomii, K., and Imai, Y. (2011). Fulminant fatal bacteremic pneumonia due to *Aeromonas hydrophila* in a non-immunocompromised woman. Intern. Med. *50*, 63–65.

Nakasome, I., Yamane, N., Miyazaki, T., Onaga, S., and Higa, M. (2001). *Aeromonas* species infection with severe clinical manifestation in Okinawa, Japan-association with gas gangrene. Rinsho Biseibutshu Jinsoku Shindan Kenkyukai Shi *12*, 15–21.

Nawaz, M., Sung, K., Khan, S.A., Khan, A.A., and Steele, R. (2006). Biochemical and molecular characterization of tetracycline-resistant *Aeromonas veronii* isolates from catfish. Appl. Environ. Microbiol. 27, 6461–6466.

Nzeako, B., and Okafor, N. (2002). Bacterial enteropathogens and factors associated with seasonal episodes of gastroenteritis in Nsukka, Nigeria. Br. J. Biomed. Sci. *59*, 76–79.

Ojeda-Vargas, M., González-Fernández, M.A., Alfonso-Rodríguez, O., and Monzón-Moreno, C. (2005). [Urinary tract infection caused by bacteria of the genus *Aeromonas*]. [Article in Spanish]. Enferm. Infecc. Microbiol. Clin. *23*, 181–182.

Okumura, K., Shoji, F., Yoshida, M., Mizuta, A., Makino, I., and Higashi, H. (2011). Severe sepsis caused by *Aeromonas hydrophila* in a patient using tocilizumab: a case report. J. Med. Case Rep. *5*, 499.

Ørmen, Ø., Granum, P.E., Lassen, J., and Figueras, M.J. (2005). Lack of agreement between biochemical and genetic identification of *Aeromonas* spp. APMIS. *113*, 203–207.

Ottaviani, D., Leoni, F., Rocchegiani, E., Santarelli, S., Masini, L., D'Annibale, M.L., Pianetti, A., and Carraturo, A. (2013). A severe case of *Aeromonas veronii* biovar *sobria* travellers' diarrhoea characterized by *Vibrio parahaemolyticus* co-isolation. J. Med. Microbiol. *62*, 161–164.

Ottaviani, D., Parlani, C., Citterio, B., Masini, L., Leoni, F., Canonico, C., Sabatini, L., Bruscolinim, F., and Pianetti, A. (2011). Putative virulence properties of *Aeromonas* strains isolated from food, environmental and clinical sources in Italy: a comparative study. Int. J. Food Microbiol. *144*, 538–545.

Ouderkirk, J.P., Bekhor, D., Turett, G.S., and Murali, R. (2004). *Aeromonas* meningitis complicating medicinal leech therapy. Clin. Infect. Dis. 38, e36–7.

Pablos, M., Remacha, M.A., Rodríguez-Calleja, J.M., Santos, J.A., Otero, A., and García-López, M.L. (2010). Identity, virulence genes, and clonal relatedness of *Aeromonas* isolates from patients with diarrhoea and drinking water. Eur. J. Clin. Microbiol. Infect. Dis. *29*, 1163–1172.

Pablos, M., Huys, G., Cnockaert, M., Rodríguez-Calleja, J.M., Otero, A., Santos, J.A., and García-López. M.L. (2011). Identification and epidemiological relationships of *Aeromonas* isolates from patients with diarrhea, drinking water and foods. Int. J. Food Microbiol. *147*, 203–210.

Pampín, F., Bou, G., Galeiras, R., Freire, D., Bouza, M.T., and Zúñiga, M. del C. (2012). [*Aeromonas* and meningitis: An unusual presentation]. [Article in Spanish] Neurocirugia (Astur). *23*, 200–202.

Papadakis, V., Poniros, N., Katsibardi, K., Charissiadou, A.E., Anastasopoulos, J., and Polychronopoulou, S. (2012). Fulminant *Aeromonas hydrophila* infection during acute lymphoblastic leukemia treatment. J. Microbiol. Immunol. Infect. 45, 154–157.

Park, T.S., Oh, S.H., Lee, E.Y., Lee, T.K., Park, K.H., Figueras, M.J., and Chang, C.L. (2003). Misidentification of *Aeromonas veronii* biovar sobria as *Vibrio alginolyticus* by the Vitek system. Lett. Appl. Microbiol. *37*, 349–353.

Parker, J.L., and Shaw, J.G. (2011). *Aeromonas* spp. clinical microbiology and disease. J. Infect. *62*, 109–118.

Patel, K.M., Svestka, M., Sinkin, J., and Ruff, P. 4th. (2013). Ciprofloxacin-resistant *Aeromonas hydrophila* infection following leech therapy: A case report and review of the literature. J. Plast. Reconstr. Aesthet. Surg. 66, e20–2.

Pérez-Valdespino, A., Celestino-Mancera, M., Villegas-Rodríguez, V.L., and Curiel-Quesada, E. (2014). Characterization of mercury-resistant clinical *Aeromonas* species. Braz. J. Microbiol. *44*, 1279–1283.

Presley, S.M., Rainwater, T.R., Austin, G.P., Platt, S.G., Zak, J.C., Cobb, G.P., Marsland, E.J., Tian, K., Zhang, B., Anderson, T.A., *et al.* (2006). Assessment of pathogens and toxicants in New Orleans, LA following Hurricane Katrina. Environ. Sci. Technol. *40*, 468–474.

Puah, S.M., Puthucheary, S.D., Liew, F.Y., and Chua, K.H. (2013). *Aeromonas aquariorum* clinical isolates: antimicrobial profiles, plasmids and genetic determinants. Int. J. Antimicrob. Agents *41*, 281–284.

Puthucheary, S.D., Puah, S.M., and Chua, K.H. (2012). Molecular characterization of clinical isolates of *Aeromonas* species from Malaysia. Plos One *7*, e30205.

Ramalivhana, J.N., Obi, C.L., Samie, A., Labuschagne, C., and Weldhagen, G.F. (2010). Random amplified polymorphic DNA typing of clinical and environmental *Aeromonas hydrophila* strains from Limpopo province, South Africa. J. Health Popul. Nutr. *28*, 1–6.

Roberts, M.T., Enoch, D.A., Harris, K.A., and Karas, J.A. (2006). *Aeromonas veronii* biovar *sobria* bacteraemia with septic arthritis confirmed by 16S rDNA PCR in an immunocompetent adult. J. Med. Microbiol. *55*, 241–243.

Roger, F., Lamy, B., Jumas-Bilak, E., Kodjo, A., colBVH study group and Marchandin, H. (2012a). Ribosomal multi-operon diversity: an original perspective on the genus *Aeromonas*. PLoS One. *7*, e46268.

Roger, F., Marchandin, H., Jumas-Bilak, E., Kodjo, A., colBVH study group, and Lamy, B. (2012b). Multilocus genetics to reconstruct aeromonad evolution. BMC Microbiol. *12*, 62.

Saavedra, M.J., Perea, V., Fontes, M.C., Martins, C., and Martínez-Murcia, A. (2007). Phylogenetic identification of *Aeromonas* strains isolated from carcasses of pig as new members of the species *Aeromonas* allosaccharophila. Antonie Van Leeuwenhoek *91*, 159–167.

Sahin, I., and Barut, H.S. (2010). Quinolone-resistant *Aeromonas hydrophila* peritonitis in a CAPD patient. Clin. Nephrol. *73*, 241–243.

Sánchez-Céspedes, J., Figueras, M.J., Aspiroz, C., Aldea, M.J., Toledo, M., Alperi, A., Marco, F., and Vila, J. (2009). Development of imipenem resistance in an *Aeromonas veronii* biovar *sobria* clinical isolate recovered from a patient with cholangitis. J. Med. Microbiol. *58*, 451–455.

Sartor C., Bornet, C., Guinard, D., and Fournier, P.E. (2013). Transmission of *Aeromonas hydrophila* by leeches. Lancet *381*, 1686.

Sartor, C., Limouzin-Perotti, F., Legre, R., Casanova, D., Bongrand, M.-C., Sambuc, R., and Drancourt, M. (2002). Nosocomial infections with *Aeromonas hydrophila* from leeches. Clin. Infect. Dis. *35*, e1–e5.

Schnabl, S.M., Kuntz, C., Unglaub, F., Polykandriotis, E., Horch, R.E., and Dragu, A. (2010). Acute postoperative infection with *Aeromonas hydrophila* after using medical leeches for treatment of venous congestion. Arch. Orthop. Trauma Surg. *130*, 1323–1328.

Sedláček, I., Krejčí, E., Andělová, A., Sedláčková, M., Porazilová, I., and Holochová, P. (2012). *Aeromonas hydrophila* subsp. *dhakensis* a causative agent of gastroenteritis imported into the Czech Republic. Ann. Agric. Environ. Med. *19*, 409–413.

Senderovich, Y., Gershtein, Y., Halewa, E., and Halpern, M. (2008). *Vibrio cholerae* and *Aeromonas*: do they share a mutual host? ISME J. *2*, 276–283.

Senderovich, Y., Shifra, K.D., Vainblat, I., Blau, D., Izhaki, I., and Halpern, M. (2012). A molecular study on the prevalence and virulence potential of *Aeromonas* spp. recovered from patients suffering from diarrhoea in Israel. PLoS ONE 7, e30070.

Shak, J.R., Whitaker, J.A., Ribner, B.S., and Burd, E.M. (2011). Aminoglycoside-resistant *Aeromonas hydrophila* as part of a polymicrobial infection following a traumatic fall into freshwater. J. Clin. Microbiol. *49*, 1169–1170.

Sherry, N.L., Padiglione, A.A., Spelman, D.W., and Cleland, H. (2013). Microbiology of wildfire victims differs significantly from routine burns patients: Data from an Australian wildfire disaster. Burns *39*, 331–334.

Shin, G.W., You, M.J., Cho, H.S., Yi, S.W., and Lee, C.S. (2013). Severe sepsis due to *Aeromonas aquariorum* in a patient with liver cirrhosis.Jpn. J. Infect. Dis. *66*, 519–522.

Shizuma, T., Tanaka, C., Mori, H., and Fukuyama, N. (2011). Investigation of Bacteremia due to *Aeromonas* species and comparison with that due to enterobacteria in patients with liver cirrhosis. Gastroenterol. Res. Pract. *2011*, 930826.

Soler, L., Figueras, M.J., Chacon, M.R., Vila, J., Marco, F., Martínez-Murcia, A.J., and Guarro, J., (2002). Potential virulence and antimicrobial susceptibility of *Aeromonas popoffii* recovered from freshwater and seawater. FEMS Immunol. Med. Microbiol. *32*, 243–247

Soler, L., Figueras, M.J., Chacon, M.R., Guarro, J., and Martinez-Murcia, A.J. (2003a). Comparison of three molecular methods for typing *Aeromonas popoffii* isolates. Antonie Van Leeuwenhoek. *83*, 341–349.

Soler, L., Marco, F., Vila, J., Chacon, M.R., Guarro, J., and Figueras, M.J. (2003b). Evaluation of two miniaturized systems, MicroScan W/A and BBL Crystal E/NF, for identification of clinical isolates of *Aeromonas* spp. J. Clin. Microbiol. *41*, 5732–5734.

Soler, L., Yañez, M.A., Chacón, M.R., Aguilera-Arreola, M.G., Catalán, V., Figueras, M.J., and Martínez-Murcia, A.J. (2004). Phylogenetic analysis of the genus *Aeromonas* based on two housekeeping genes. Int. J. Syst. Evol. Microbiol. *54*, 1511–1519.

Soltan Dallal, M.M., and Moezardalan, K. (2004). *Aeromonas* spp. associated with children's diarrhoea in Tehran: a case–control study. Ann. Trop. Paediatr. *24*, 45–51.

Stano, F., Brindicci, G., Monno, R., Rizzo, C., Ghezzani, F., Carbonara, S., Guaglianone, E., Donelli, G., and Monno, L. (2009). *Aeromonas sobria* sepsis complicated by rhabdomyolysis in an HIV-positive patient: case report and evaluation of traits associated with bacterial virulence. Int. J. Infect. Dis. *13*, e113–8.

Steer, A., Daley, A.J., and Curtis, N. (2005). Suppurative sequelae of symbiosis. Lancet *365*, 188.

Su, S.Y., Chao, C.M., and Lai, C.C. (2012). *Aeromonas* bacteremia in southern Taiwan. Scand. J. Infect. Dis. 45:728.

Su, S.Y., Lin, W.T., and Chao, C.M. (2013). Is it really *Aeromonas hydrophila*? Infection *41*, 279.

Surek, M., Vizzotto, B.S., Souza, E.M., Pedrosa, Fde. O., Dallagassa, C.B., Farah, S.M., and Fadel-Picheth, C.M. (2010). Identification and antimicrobial susceptibility of *Aeromonas* spp. isolated from stool samples of Brazilian subjects with diarrhoea and healthy controls. J. Med. Microbiol. *59*, 373–374.

Svenungsson, B., Lagergren, A., Ekwall, E., Evengård, B., Hedlund, K.O., Kärnell, A., Löfdahl, S., Svensson, L., and Weintraub, A. (2000). Enteropathogens in adult patients with diarrhea and healthy control subjects: a 1-year prospective study in a Swedish clinic for infectious diseases. Clin. Infect. Dis. *30*, 770–778.

Taneja, N., Khurana, S., Trhan, A., Marwaha, R.K., and Sharma, M. (2004). An outbreak of hospital acquired diarrhea due to *Aeromona sobria*. Indian Pediatr. *41*, 912–916.

Tang, H.J., Lai, C.C., Lin, H.L., and Chao, C.M. (2014). Clinical manifestations of bacteremia caused by *Aeromonas* species in southern Taiwan. PLoS One. *9*, e91642.

Tena, D., González-Praetorius, A., Gimeno, C., Pérez-Pomata, M.T., and Bisquert, J. (2007). Extraintestinal infection due to *Aeromonas* spp.: review of 38 cases. Enferm. Infecc. Microbiol. Clin. *25*, 235–241.

Tena, D., Aspíroz, C., Figueras, M.J., González-Praetorius, A., Aldea, M.J., Alperí, A., and Bisquert, J. (2009). Surgical site infection due to *Aeromonas* species: report of nine cases and literature review. Scand. J. Infect. Dis. *41*, 164–170.

Tsai, Y.H., Hsu, R.W., Huang, T.J., Hsu, W.H., Huang, K.C., Li, Y.Y., and Peng, K.T. (2007). Necrotizing soft-tissue infections and sepsis caused by *Vibrio vulnificus* compared with those caused by *Aeromonas* species. J. Bone Joint Surg. Am. *89*, 631–636.

Tsai, Y.H., Huang, K.C., Huang, T.J., and Hsu, W.H. (2009). Case reports: fatal necrotizing fasciitis caused by *Aeromonas sobria* in two diabetic patients. Clin. Orthop. Relat. Res. *467*, 846–849.

Tsai, M.S., Kuo, C.Y., Wang, M.C., Wu, H.C., Chien, C.C., and Liu, J.W. (2006). Clinical features and risk factors for mortality in *Aeromonas* bacteremic adults with hematologic malignancies. J. Microbiol. Immunol. Infect. *39*, 150–154.

Turner, P., Willemse, C., Phakaudom, K., Zin, T.W., Nosten, F., and McGready, R. (2012). *Aeromonas* spp. bacteraemia in pregnant women, Thailand-Myanmar border. Emerg. Infect. Dis. *18*, 1522–1523.

Tomás, J.M. (2012). The main *Aeromonas* pathogenic factors. ISRN Microbiol. *4*, 2012:256261.

Vally, H., Whittle, A., Cameron, S., Dowse, G.K., and Watson, T. (2004). Outbreak of *Aeromonas hydrophila* wound infections associated with mud football. Clin. Infect. Dis. *38*, 1084–1089.

Vila, J., Ruiz, J., Gallardo, F., Vargas, M., Soler, L., Figueras, M.J., and Gascon, J. (2003). *Aeromonas* spp. and traveler's diarrhea: clinical features and antimicrobial resistance. Emerg. Infect. Dis. *9*, 552–555.

Wang, J.H., Wang, C.Y., Chi, C.Y., Ho, M.W., Ho, C.M., and Lin, P.C. (2009). Clinical presentations, prognostic factors, and mortality in patients with *Aeromonas sobria* complex bacteremia in a teaching hospital: a 5-year experience. J. Microbiol. Immunol. Infect. *42*, 510–515.

Whitaker, I.S., Josty, I.C., Hawkins, S., Azzopardi, E., Naderi, N., Graf, J., Damaris, L., Lineaweaver, W.C., and Kon, M. (2011). Medicinal leeches and the microsurgeon: a four-year study, clinical series and risk benefit review. Microsurgery 31, 281–287.

Wu, C.J., Wu, J.J., Yan, J.J., Lee, H.C., Lee, N.Y., Chang, C.M., Shih, H.I., Wu, H.M., Wang, L.R., and Ko, W.C. (2007). Clinical significance and distribution of putative virulence markers of 116 consecutive clinical *Aeromonas* isolates in southern Taiwan. J. Infect. *54*, 151–158.

Wu, C.J., Lee, H.C., Chang, T.T., Chen, C.Y., Lee, N.Y., Chang, C.M., Sheu, B.S., Cheng, P.N., Shih, H.I., and Ko, W.C. (2009). *Aeromonas* spontaneous bacterial peritonitis: a highly fatal infectious disease in patients with advanced liver cirrhosis. J. Formos Med. Assoc. *108*, 293–300.

Wu, C.J., Chen, P.L., Tang, H.J., Chen, H.M., Tseng, F.C., Shih, H.I., Hung, Y.P., Chung, C.H., and Ko, W.C. (2012a). Incidence of *Aeromonas* bacteraemia in southern Taiwan: Vibrio and Salmonella bacteraemia as comparators. J. Microbiol. Immunol. Infect. *47*, 145–148.

Wu, C.J., Chen, P.L., Wu, J.J., Yan, J.J., Lee, C.C., Lee, H.C., Lee, N.Y., Chang, C.M., Lin, Y.T., Chiu, Y.C., and Ko, W.C. (2012b). Distribution and phenotypic and genotypic detection of a metallo-β-lactamase, CphA, among bacteraemic *Aeromonas* isolates. J. Med. Microbiol. *61*, 712–719.

Wu, C.J., Tsai, P.J., Chen, P.L., Wu, I.C., Lin, Y.T., Chen, Y.H., Wang, L.R., and Ko, W.C. (2012c). *Aeromonas* aquariorum septicemia and enterocolitis in a cirrhotic patient. Diagn. Microbiol. Infect. Dis. *74*, 406–408.

Wu, C.J., Wang, H.C., Chen, C.S., Shu, H.Y., Kao, A.W., Chen, P.L., and Ko, W.C. (2012d). Genome sequence of a novel human pathogen, *Aeromonas* aquariorum. J. Bacteriol. *194*, 4114.

Wu, C.J., Wang, H.C., Chen, P.L., Chang, M.C., Sunny Sun, H., Chou, P.H., and Ko, W.C. (2013). AQU-1, a chromosomal class C β-lactamase, among clinical *Aeromonas* dhakensis isolates: distribution and clinical significance. Int. J. Antimicrob. Agents. *42*, 456–461.

Yang, X., Yang, Q.Q., Guo,Q.Y., Yi, C.Y., Mao, H.P., Lin, J.X., Jiang, Z.P., and Yu, X.Q. (2008). *Aeromonas salmonicida* peritonitis after eating fish in a patient undergoing CAPD. Perit. Dial. Int. *28*, 316–317.

Yánez, M.A., Catalán, V., Apráiz, D., Figueras, M.J., and Martínez-Murcia, A.J. (2003). Phylogenetic analysis of members of the genus *Aeromonas* based on *gyrB* gene sequences. Int. J. Syst. Evol. Microbiol. *53*, 875–883.

Yasti, A.C., Otan, E., Doganay, M., and Kama, N.A. (2009). *Aeromonas hydrophila* in a burn patient. J. Burn Care Res. *30*, 898–900.

Yi, S.W., You, M.J., Cho, H.S., Lee, C.S., Kwon, J.K., and Shin, G.W. (2013). Molecular characterization of *Aeromonas* species isolated from farmed eels (*Anguilla japonica*).Vet. Microbiol. *164*, 195–200.

Zhang, Q., Shi, G.Q., Tang, G.P., Zou, Z.T., Yao, G.H., and Zeng, G. (2012). A foodborne outbreak of *Aeromonas hydrophila* in a college, Xingyi City, Guizhou, China, 2012. Western Pac. Surveill Response J. 3, 39–43.

Aeromonas, a Multifaceted Microbe: Beneficial Associations with Animals

5

Joerg Graf

Abstract

One fascinating aspect of *Aeromonas* spp. is that they are not only pathogens but also form commensal and beneficial associations with different animals (Graf, 1999; Janda and Abbott 2010; Cheesman *et al.*, 2011; Roeselers *et al.*, 2011). Indeed even one species can be found in association with a wide range of animals that are commensal, beneficial or pathogenic, which depends on the host is colonized (Silver *et al.*, 2007a). This diversity of associations provides an opportunity for investigating the underlying mechanisms of non-pathogenic microbe–host interactions and for gaining insights into the evolution of virulence (Silver *et al.*, 2011). Among the Aeromonads, the best-studied symbiotic interaction is that of *Aeromonas* spp. with medicinal leeches, e.g. *Hirudo verbana* (Graf, 2000; Graf *et al.*, 2006; Nelson and Graf, 2012; Nyholm and Graf, 2012). Recent work has revealed that *Aeromonas* spp. induce the proper development of the gut in Zebrafish (Bates *et al.*, 2007; Cheesman *et al.*, 2011), which is in stark contrast to the ulcerative syndrome and other diseases in fishes (Cai *et al.*, 2012) and suggests the importance of the location in determining the outcome of the associations. There is also circumstantial evidence that *Aeromonas* spp. form non-pathogenic associations with different species of mosquitoes and also vampire bats (Muller *et al.*, 1980; Pidiyar *et al.*, 2002, 2004).

Medicinal leeches

Medicinal leeches are exclusive blood feeders and can consume in a single meal over five times their body weight (Sawyer, 1986). This feat is accomplished by the secretion of powerful vasodilators and anticoagulants, which stimulate the blood flow, and the active removal of blood by the animal through rhythmic contractions. This ability has led to the use of medicinal leeches in modern medicine to treat venous congestions after reconstructive, microvascular surgery (Foucher *et al.*, 1981; Whitlock *et al.*, 1983; Whitaker *et al.*, 2011). A complication of this therapy are wound infections by *Aeromonas hydrophila* (Lineaweaver *et al.*, 1992; Whitaker *et al.*, 2011). These infections were prevented for years by the prophylactic administration of ciprofloxacin, but in recent years, infections have been caused by ciprofloxacin resistant bacteria (Levine *et al.*, 2010; Mumcuoglu *et al.*, 2010; Giltner *et al.*, 2013; Patel *et al.*, 2013).

The medicinal leech feeds exclusively on vertebrate blood that is deficient in vitamins (Sawyer, 1986). In other blood-feeding animals, it has been demonstrated that certain vitamins such as vitamin B12 are provided by symbiotic bacteria (Nogge, 1976). Similarly, it has been proposed that *Aeromonas* spp. provide leeches with vitamin B12 (Graf, 2002). Other hypotheses are that *Aeromonas* spp. are required for the digestion of the ingested blood meal (Büsing, 1951; Wilde, 1975) and that they could detoxification of haem. A separate role has been proposed in contributing to the unusual specificity (Büsing *et al.*, 1953). None of these functions have been convincingly demonstrated experimentally (Graf, 2002, Graf *et al.*, 2006). The lack of data is due in part to the longevity of leeches, which feed only a few times per year and can live for many years in captivity, which makes such functional studies very time-consuming.

Early, culture-based research suggested that the microbiota residing in the leech gut was unusually simple, consisting only of *Aeromonas hydrophila* leading to the proposal that this bacterium releases antimicrobial compounds preventing the growth of other bacteria (Lehmensick, 1941; Büsing, 1951; Wilde, 1975). More recent data utilizing culture-independent approaches such as sequencing 16S rRNA genes and fluorescent *in situ* hybridization has shown that the microbiota is more complex (Graf, 1999; Worthen *et al.*, 2006; Kikuchi and Graf, 2007; Laufer *et al.*, 2008). In *Hirudo verbana* it is dominated by two microorganisms, an *Aeromonas* species and a *Rikenella*-like bacterium, which is a member of the Bacteroidetes (Worthen *et al.*, 2006). In addition, other microorganisms are present at lower levels and determining the core microbiome requires additional studies utilizing more animals and next generation sequencing approaches.

The simple question of the identity of the leech symbiont has been challenging. Early reports consistently identified the symbionts as *A. hydrophila* but this was only based on biochemical tests (Jennings and van der Lande, 1967; Mackay *et al.*, 1999). The first study that used a combination of biochemical and molecular tests to identify the symbionts found *A. veronii* in animals that were reported to be *Hirudo medicinalis* (Graf, 1999). Current data suggest, that this confusion was caused not just by changes in the taxonomy of *Aeromonas* and but also of the medicinal leeches themselves. These contradictions were clarified by the discovery that *H. medicinalis* was comprised out of at least three species, *H. medicinalis*, *H. orientalis* and *H. verbana* (Siddall *et al.*, 2007). Siddall's study suggests *H. medicinalis* is rare and not typically sold by medical suppliers and that *H. medicinalis* is colonized by *A. hydrophila* (Siddall *et al.*, 2007, 2011). The eastern medicinal leech, *H. orientalis*, is colonized by *A. jandaei* and *A. veronii* (Laufer *et al.*, 2008). While the most commonly used medicinal leech is *H. verbana*, which is colonized by *A. veronii*. However the continuous reports of patients becoming infected with *A. hydrophila* after leech therapy suggests the final answer to this interesting question has not yet been written and requires an approach that utilizes greater in depth sampling of the microbial community or identifying the leeches using molecular approaches (Snower *et al.*, 1989; Lineaweaver *et al.*, 1992; Patel *et al.*, 2013).

Culture-independent approaches revealed the presence of a second dominant symbiont, the *Rikenella*-like bacterium in the digestive tract of leeches and suggest the presence of a simple microbial community as compared to the very complex digestive tract communities of mammals (Worthen *et al.*, 2006; Kikuchi and Graf, 2007; Bomar *et al.*, 2011). The data also suggest that a stable, long-term association is formed and that other bacteria, which must enter the digestive tract, fail to establish themselves. This raises the question of what factors contribute to the specificity. One factor appears to be the vertical or semi-vertical transmission of the symbionts from parent to offspring. Leeches lay cocoons that harbour multiple embryos with each one contained within a membrane. Using diagnostic PCR, that the developing embryos harboured *Aeromonas* already inside the cocoon (Rio *et al.*, 2009).

How the cocoon is inoculated is unclear as it is secreted from pores on the dorsal surface and shaped with the anterior sucker. But a vertical transmission could provide the community of digestive-tract symbionts an early competitive advantage. But other factors must play a role as the microbiome of animals is not static but changes over time.

The ability to culture and genetically manipulate *Aeromonas* strains provides important insight into the genetic requirements for colonizing the leech (Nelson and Graf, 2012; Nyholm and Graf, 2012). A study by Silver *et al.* (2011) compared the ability of *A. veronii* strains obtained from leeches, humans or veterinary samples to colonize the leech and found that leech isolates outcompeted the strains that were isolated from other sources. This indicated that strains cultured from leeches were adapted specifically to the leech environment. A multilocus sequencing study of housekeeping genes and virulence factors did not detect any phylogenetic signal that correlated with the source of isolation. Instead a surprising degree of horizontal gene transfer not just of genes but of gene fragments was observed. Whole genome comparisons might provide insight into factors that contribute to the specialization but the initial approach was to identify genes that are required for colonization.

One approach to identify mechanisms that contribute to the specificity is to identify mutants that have a reduced ability to colonize the digestive tract of the leech, thus providing clues of the not just the genetic requirements but also molecular mechanisms involved. The leech symbiont *A. veronii* Hm21 can be manipulated genetically and colonization assays was established that tested the ability of the strain to proliferate in the leech digestive tract either by itself or in competition with other strains.

A signature-tagged transposon mutant screen revealed a number of mutants that had a decreased ability to colonize including one mutant that had an inactivated type three-secretion system, T3SS (Silver *et al.* 2007a,b). The T3SS is a well-described virulence factor in *A. salmonicida* and *A. hydrophila*. The role in the extracellular colonization of a digestive tract was surprising until fluorescent *in situ* hybridization micrographs revealed the presence of phagocytic immune cells or leech haemocytes. The haemocytes phagocytosed the T3SS mutants but not the parent strain. Interestingly the T3SS is required by the same strain for virulence in a mouse septicaemia model. An analysis of the Hm21 genome revealed the presence of a second, unrelated T3SS in the leech symbiont. Its role has yet to be determined. T3SS elicit their effect on host cells by injecting effector molecules. The strains isolated from the leech contained the well characterized AexT homologue that was first described in *A. salmonicida* and also AexU that was found in *A. hydrophila* strain SSU (Sha *et al.*, 2007). A PCR survey found that all of the *A. veronii* strains examined possessed the T3SS and these effectors (Silver and Graf, 2009). Interestingly, the genome sequence analysis revealed that an *A. veronii* strain isolated in the sediments of a carp pond did not possess the T3SS which the leech isolated HM21 possessed two T3SS, indicating that this is not a universal characteristic (Li *et al.*, 2011; Bomar *et al.*, 2013).

One common characteristic of *A. veronii* strains isolated from the leech is that they are beta haemolytic. This capability seemed surprising as the erythrocytes inside the leech gut appear to by physically intact even after months after feeding. In a different transposon mutagenesis, mutants were screened for the loss of beta-haemolysis. In this screen of 18,000 mutants only one mutant was identified and it had an insertion in *exeM*, a component of the type two secretion system, T2SS (Maltz and Graf, 2011). The T2SS is used to secrete a large number of exoenzymes and haemolysins in *Aeromonas* spp. and has been well studied in other *Aeromonas* strains. The *exeM* mutant had a reduced ability to colonized during the initial growth inside the medicinal leech after the animal fed on a blood meal. When this mutant was grown together with a wild type strain, it reached similar levels when inoculated at 1:1 ratio, but even at a 1:3 ratio, the mutant reached lower colonization levels than the wild type strain. These results are consistent with a secreted protein being responsible for the colonization defect. This defect could be levitated by lysing the erythrocytes of the blood prior to feeding or complementing the mutant with *exeMN*. These data indicated that the ability to lyse or partially lyse erythrocytes is critical for *A. veronii* to colonize *H. verbana*. The lysis of the erythrocytes would not make the membranes more accessible as a nutrient but instead release haemoglobin. It is possible that *A. veronii* utilizes the iron from haem for growth during the first few hours of colonization. Thus iron acquisition using siderophores or haem receptors could serve as an additional specificity determinant.

Several lines of evidence suggest that *Aeromonas* encounters membrane stress inside the leech and that this stress arises from different sources (Indergand and Graf, 2000; Braschler *et al.*, 2003; Silver *et al.*, 2007b; Bomar and Graf, 2012). A study with LPS mutants revealed that an intact LPS is required for successfully colonizing the leech. Inactivating the complement system of the ingested blood meal allowed the mutant to colonize to similar levels as wild type strain, suggesting that the complement system of the ingested vertebrate blood remained active inside the leech and killed sensitive bacteria (Braschler *et al.*, 2003). In a subsequent STM study, a large number of mutants were found to have to have become sensitive to serum, reinforcing the importance of complement resistance in the colonization process (Silver *et al.*, 2007b). Another mutant had a defective Brown's lipoprotein and this mutant had a colonization defect when fed to the leech in heat-inactivated blood. This mutant was more sensitive to SDS and as the Brown's lipoprotein tethers the outer membrane to the peptidoglycan layer this suggests the presence of another membrane stress causing compound that is independent of the complement system.

In addition to mutant studies, insights into the physiology of the bacteria gained through transcriptomics analysis can provide important insight into the environment the bacteria encounter in their habitat. One study compared the transcripts from *Aeromonas* inside the leech gut to those from *A. veronii* grown in a nutrient-rich broth in the laboratory (Bomar and Graf, 2012). Genes encoding the alternative sigma factor, RpoE, and phage shock proteins were dramatically up-regulated inside the leech crop. RpoE is induced when bacteria encounter membrane stress due to factors such as osmolarity, antimicrobial compounds, or reactive oxygen species. Similarly, the phage shock protein encoding genes, *pspA*, *pspB* and *pspC*, were more than 50-fold inside the leech crop as compared to *in vitro*. The *psp* genes have been shown to be induced by osmotic stress, loss of proton motive force, exposure to organic solvents, stationary phase growth, and filamentous phage infection.

The study of *Aeromonas* as digestive-tract symbiont of the medicinal leech has revealed a number of adaptations that required for *A. veronii* to be a successful symbiont. While many different *A. veronii* strains can colonize the leech crop, strains isolated from the leech crop outcompete these other strains suggesting that the leech isolates possess properties unique to them. An initial phylogenetic analysis of *A. veronii* strains did not show the leech isolates to be monophyletic, indicating that multiple lineage may possess these properties (Silver *et al.* 2011). Comparative genomic approaches may provide insight into the underlying molecular mechanisms of this competitive advantage.

Zebrafish

While it is well established that several *Aeromonas* strains are pathogens of fishes, it was recently discovered that *Aeromonas hydrophila* and *A. veronii* are parts of the digestive tract microbiome of zebra fish (Rawls *et al.*, 2004; Roeselers *et al.*, 2011). This association extends far beyond simply growing in the digestive tract of zebra fish; the microbiome and *Aeromonas* affect the development of the fish gut. In germ-free zebrafish, the alkaline phosphatase activity of the brush border is absent and there are fewer goblet and enteroendocrine cells present (Bates *et al.*, 2006). Exposing the fish to the normal microbiota or even a pure culture of *A. veronii* or *A. hydrophila* reverses all of these phenotypes. This suggests that digestive tract inhabitants of zebrafish, including *Aeromonas* species influence the proper development of these vertebrates.

Other associations with blood-feeding animals

In addition to the above leeches and fish, *Aeromonas* spp. are also thought to be commensal or beneficial symbionts of other blood feeding animals. A single study examined the faeces of vampire bats and was able to culture *A. hydrophila* (Muller *et al.*, 1980). When the genome of the mosquito *Anopheles gambiae* was published intriguingly it also contained sequences that apparently were of bacterial origin, including some that were most similar to *A. veronii*, suggesting that this bacterium inhabited the digestive tract of the mosquitoes that were sequenced (Holt *et al.*, 2002). Several studies have since examined the microbiome of mosquito digestive tracts from both the aquatic larvae and adults (Pidiyar *et al.*, 2002, 2004).

References

Bates, J.M., Akerlund, J., Mittge, E., and Guillemin, K. (2007). Intestinal alkaline phosphatase detoxifies lipopolysaccharide and prevents inflammation in zebrafish in response to the gut microbiota. Cell Host Microbe 2, 371–382.

Bates, J.M., Mittge, E., Kuhlman, J., Baden, K.N., Cheesman, S.E., and Guillemin, K. (2006). Distinct signals from the microbiota promote different aspects of zebrafish gut differentiation. Dev. Biol. *297*, 374–386.

Bomar, L., and Graf, J. (2012). Investigation into the physiologies of *Aeromonas veronii in vitro* and inside the digestive tract of the medicinal leech using RNA-seq. Biol. Bull. *223*, 155–166.

Bomar, L., Maltz, M., Colston, S., and Graf, J. (2011). Directed culturing of microorganisms using metatranscriptomics. MBio *2*, e00012–e00011.

Bomar, L., Stephens, Z., Nelson, M.C., Velle, K., Guillemin, K., and Graf, J. (2013). Draft genome of *Aeromonas veronii* Hm21, a leech symbiont. Genome Annunc. *1*, e00800-13.

Braschler, T.R., Merino, S., Tomas, J.M., and Graf, J. (2003). Complement resistance is essential for colonization of the digestive tract of Hirudo medicinalis by *Aeromonas* strains. Appl. Environ. Microbiol. *69*, 4268–4271.

Büsing, K.-H. (1951). *Pseudomonas hirudinis*, ein bakterieller Darmsymbiont des Blutegels (*Hirudo officinalis*). Zentralbl. Bakteriol. *157*, 478–485.

Büsing, K.-H., Döll, W., and Freytag, K. (1953). Die Bakterienflora der medizinischen Blutegel. Arch. Mikrobiol. *19*, 52–86.

Cai, S.H., Wu, Z.H., Jian, J.C., Lu, Y.S., and Tang, J.F. (2012). Characterization of pathogenic *Aeromonas veronii* bv. veronii associated with ulcerative syndrome from chinese longsnout catfish (Leiocassis longirostris Gunther). Braz. J. Microbiol. *43*, 382–388.

Cheesman, S.E., Neal, J.T., Mittge, E., Seredick, B.M., and Guillemin, K. (2011). Epithelial cell proliferation in the developing zebrafish intestine is regulated by the Wnt pathway and microbial signaling via Myd88. Proc. Natl. Acad. Sci. U.S.A. *108* (Suppl. 1), 4570–4577.

Foucher, G., Henderson, H.R., Maneau, M., Merle, M., and Braun, F.M. (1981). Distal digital replantation: one of the best indications for microsurgery. Int. J. Microsurg. *3*, 263–270.

Giltner, C.L., Bobenchik, A.M., Uslan, D.Z., Deville, J.G., and Humphries, R.M. (2013). Ciprofloxacin-resistant *Aeromonas hydrophila* cellulitis following leech therapy. J. Clin. Microbiol. *51*, 1324–1326.

Graf, J. (1999). Symbiosis of *Aeromonas veronii* biovar sobria and *Hirudo medicinalis*, the medicinal leech: a novel model for digestive tract associations. Infect. Immun. *67*, 1–7.

Graf, J. (2000). The symbiosis of *Aeromonas* and *Hirudo medicinalis*, the medicinal leech. ASM News *66*, 147–153.

Graf, J. (2002). The effect of the symbionts on the physiology of *Hirudo medicinalis*, the medicinal leech. Invert. Reprod. Biol. *41*, 269–275.

Graf, J., Kikuchi, Y., and Rio, R.V. (2006). Leeches and their microbiota: naturally simple symbiosis models. Trends Microbiol. *14*, 365–371.

Holt, R.A., Subramanian, G.M., Halpern, A., Sutton, G.G., Charlab, R., Nusskern, D.R., Wincker, P., Clark, A.G., Ribeiro, J.M., Wides, R., *et al.* (2002). The genome sequence of the malaria mosquito *Anopheles gambiae*. Science *298*, 129–149.

Indergand, S., and Graf, J. (2000). Ingested blood contributes to the specificity of the symbiosis of *Aeromonas veronii* biovar sobria and *Hirudo medicinalis*, the medicinal leech. Appl. Environ. Microbiol. *66*, 4735–4741.

Janda, J.M., and Abbott, S.L. (2010). The genus *Aeromonas*: taxonomy, pathogenicity, and infection. Clin. Microbiol. Rev. *23*, 35–73.

Jennings, J.B., and van der Lande, V.M. (1967). Histochemical and bacteriological studies on digestion in nine species of leeches (Annelidia: Hirudinea). Biol. Bull. *33*, 166–183.

Kikuchi, Y., and Graf, J. (2007). Spatial and temporal population dynamics of a naturally occurring two-species microbial community inside the digestive tract of the medicinal leech. Appl. Environ. Microbiol. *73*, 1984–1991.

Laufer, A.S., Siddall, M.E., and Graf, J. (2008). Characterization of the digestive-tract microbiota of *Hirudo orientalis*, a European medicinal leech. Appl. Environ. Microbiol. *74*, 6151–6154.

Lehmensick, R. (1941). Ueber einen neuen bakteriellen Symbionten im Darm von *Hirudo officinalis* L. Zentralbl. Bakteriol. *147*, 317–321.

Levine, S.M., Frangos, S.G., Hanna, B., Colen, K., and Levine, J.P. (2010). *Aeromonas* septicemia after medicinal leech use following replantation of severed digits. Am. J. Crit. Care *19*, 469–471.

Li, Y., Liu, Y., Zhou, Z., Huang, H., Ren, Y., Zhang, Y., Li, G., Zhou, Z., and Wang, L. (2011). Complete genome sequence of *Aeromonas veronii* strain B565. J. Bacteriol. *193*, 3389–3390.

Lineaweaver, W.C., Hill, M.K., Buncke, G.M., Follansbee, S., Buncke, H.J., Wong, R.K., Manders, E.K., Grotting, J.C., Anthony, J., and Mathes, S.J. (1992). *Aeromonas hydrophila* infections following use of medicinal leeches in replantation and flap surgery. Ann. Plast. Surg *29*, 238–244.

Mackay, D.R., Manders, E.K., Saggers, G.C., Banducci, D.R., Prinsloo, J., and Klugman, K. (1999). *Aeromonas* species isolated from medicinal leeches. Ann. Plast. Surg. *42*, 275–279.

Maltz, M., and Graf, J. (2011). The type II secretion system is essential for erythrocyte lysis and gut colonization by the leech digestive tract symbiont *Aeromonas veronii*. Appl. Environ. Microbiol. *77*, 597–603.

Muller, H.E., Pinus, M., and Schmidt, U. (1980). [*Aeromonas hydrophila* as a normal intestinal bacterium of the vampire bat (*Desmodus rotundus*)]. Zentralbl. Veterinarmed B *27*, 419–424.

Mumcuoglu, K.Y., Huberman, L., Cohen, R., Temper, V., Adler, A., Galun, R., and Block, C. (2010). Elimination of symbiotic *Aeromonas* spp. from the intestinal tract of the medicinal leech, *Hirudo medicinalis*, using ciprofloxacin feeding. Clin. Microbiol. Infect. *16*, 563–567.

Nelson, M.C., and Graf, J. (2012). Bacterial symbioses of the medicinal leech Hirudo verbana. Gut Microbes 3, 322–331.

Nogge, G. (1976). Sterility in tsetse flies (*Glossina morsitans* Westwood) caused by loss of symbionts. Experientia *32*, 995–996.

Nyholm, S.V., and Graf, J. (2012). Knowing your friends: invertebrate innate immunity fosters beneficial bacterial symbioses. Nat. Rev. Microbiol. *10*, 815–827.

Patel, K.M., Svestka, M., Sinkin, J., and Ruff, P. 4th (2013). Ciprofloxacin-resistant *Aeromonas hydrophila* infection following leech therapy: a case report and review of the literature. J. Plast. Reconstr. Aesthet. Surg. *66*, e20–22.

Pidiyar, V., Kaznowski, A., Narayan, N.B., Patole, M., and Shouche, Y.S. (2002). *Aeromonas culicicola* sp. nov., from the midgut of Culex quinquefasciatus. Int. J. Syst. Evol. Microbiol. 52, 1723–1728.

Pidiyar, V.J., Jangid, K., Patole, M.S., and Shouche, Y.S. (2004). Studies on cultured and uncultured microbiota of wild *Culex quinquefasciatus* mosquito midgut based on 16S ribosomal RNA gene analysis. Am. J. Trop. Med. Hyg. *70*, 597–603.

Rawls, J.F., Samuel, B.S., and Gordon, J.I. (2004). Gnotobiotic zebrafish reveal evolutionarily conserved responses to the gut microbiota. Proc. Natl. Acad. Sci. U.S.A. *101*, 4596–4601.

Rio, R.V., Maltz, M., McCormick, B., Reiss, A., and Graf, J. (2009). Symbiont succession during embryonic development of the European medicinal leech, *Hirudo verbana*. Appl. Environ. Microbiol. 75, 6890–6895.

Roeselers, G., Mittge, E.K., Stephens, W.Z., Parichy, D.M., Cavanaugh, C.M., Guillemin, K., and Rawls, J.F. (2011). Evidence for a core gut microbiota in the zebrafish. ISME J. *5*, 1595–1608.

Sawyer, R.T. (1986). Leech Biology and Behavior (Clarendon Press, Oxford, UK).

Sha, J., Wang, S.F., Suarez, G., Sierra, J.C., Fadl, A.A., Erova, T.E., Foltz, S.M., Khajanchi, B.K., Silver, A.C., Graf, J., *et al.* (2007). Further characterization of a type III secretion system (T3SS) and of a new effector protein 3 from a clinical isolate of *Aeromonas hydrophila*-Part I. Patho. Microbiol. *43*, 127–146.

Siddall, M.E., Min, G.S., Fontanella, F.M., Phillips, A.J., and Watson, S.C. (2011). Bacterial symbiont and salivary peptide evolution in the context of leech phylogeny. Parasitology *138*, 1815–1827.

Siddall, M.E., Trontelj, P., Utevsky, S.Y., Nkamany, M., and Macdonald, K.S. (2007). Diverse molecular data demonstrate that commercially available medicinal leeches are not *Hirudo medicinalis*. Proc. Biol. Sci. *274*, 1481–1487.

Silver, A.C., and Graf, J. (2009). Prevalence of genes encoding the type three secretion system and the effectors AexT and AexU in the *Aeromonas veronii* group. DNA Cell Biol. *28*, 383–388.

Silver, A.C., Kikuchi, Y., Fadl, A.A., Sha, J., Chopra, A.K., and Graf, J. (2007a). Interaction between innate immune cells and a bacterial type III secretion system in mutualistic and pathogenic associations. Proc. Natl. Acad. Sci. U.S.A. *104*, 9481–9486.

Silver, A.C., Rabinowitz, N., Küffer, S., and Graf, J. (2007b). Identification of *Aeromonas veronii* genes required for colonization of the medicinal leech, *Hirudo verbana*. J. Bacteriol. *189*, 6763–6722.

Silver, A.C., Williams, D., Faucher, J., Horneman, A.J., Gogarten, J.P., and Graf, J. (2011). Complex evolutionary history of the *Aeromonas veronii* group revealed by host interaction and DNA sequence data. PLoS One 6, e16751.

Snower, D.P., Ruef, C., Kuritza, A.P., and Edberg, S.C. (1989). *Aeromonas hydrophila* infection associated with the use of medicinal leeches. J. Clin. Microbiol. 27, 1421–1422.

Whitaker, I.S., Josty, I.C., Hawkins, S., Azzopardi, E., Naderi, N., Graf, J., Damaris, L., Lineaweaver, W.C., and Kon, M. (2011). Medicinal leeches and the microsurgeon: a four-year study, clinical series and risk benefit review. Microsurgery *31*, 281–287.

Whitlock, M.R., O'Hare, P.M., Sanders, R., and Morrow, N.C. (1983). The medicinal leech and its use in plastic surgery: a possible cause for infection. Br. J. Plast. Surg *36*, 240–244.

Wilde, V. (1975). Untersuchungen zum Symbioseverhältnis zwischen *Hirudo officinalis* und Bakterien. Zool. Anz. *195*, 289–306.

Worthen, P.L., Gode, C.J., and Graf, J. (2006). Culture-independent characterization of the digestive-tract microbiota of the medicinal leech reveals a tripartite symbiosis. Appl. Environ. Microbiol. 72, 4775–4781.

Aeromonad Antigenic Structures

6

Jennifer Parker and Jonathan Shaw

Abstract

Within their cell envelope bacteria of the genus *Aeromonas* have numerous components that help them colonize and survive within diverse environments. Both polar and lateral flagella help in the initial colonization process, with polar flagella being required in liquid surroundings and lateral flagella for the migration across and interaction with surfaces. Glycosylation of the polar flagellar filament is essential for its function, however the full biological role for this is not yet understood. Following the initial interaction with a surface through flagella-mediated motility, filamentous and non-filamentous adherence factors are employed, with the MSHA type IV pilus system playing a major role. However, in non-filamentous strains a number of outer membrane proteins have been proposed to be important in the adherence process. Following colonization, more invasive isolates of the aeromonads that are frequently isolated from cases of bacteraemia or wound infections are able to employ a series of other cell envelope factors that protect them against the host's immune response. These factors include lipopolysaccharide, capsule and the S-layer.

Introduction

Aeromonas spp. are able to colonize many different environments, they can be found as free living planktonic bacteria or in complex biotic or abiotic biofilms. Like many Gram-negative bacteria the aeromonads have many antigenic structures that are present on the external leaflet of the cell envelope that help them inhabit these differing locations. The majority of these structures have been implicated in the colonization process of either the host or abiotic surfaces leading to the formation of biofilms. For aeromonad–host interactions, they occur either through direct binding of host cells or matrix components, or protecting the bacterium from the host's defence mechanisms. Therefore for the gastroenteritis producing aeromonad strains, flagella, pili and other adherence structures are probably more important, whereas strains that cause bacteraemia and wound infections, protection from the host by LPS, capsule or S-layers are likely to be more important. Since the publication of the previous book on the genus over fifteen years ago a number of aeromonad genome sequences have become available and have helped expand our understanding of the cell envelope of this bacterium. Below we will outline the major features and recent findings in this ever-expanding area.

Flagella

Bacterial flagella are complex nano-machines in which a 10–15 μm helical filament that extends from the cell surface and is anchored to a rotating basal body that spans the bacterial cell envelope (Terashima *et al.*, 2008). The helical filament is composed of repeating subunits known as flagellins. The main role of the flagella is motility in liquid environments and on solid surfaces, to move towards more favourable environments and nutrients and away from less favourable environments, however it is also an important antigenic structure involved in successful adhesion and colonization. Flagellin is a direct target of Toll-like receptor-5 (TLR-5) leading to increased production of interleukin-6 in mammals (Hayashi *et al.*, 2001). Approximately 60% of mesophilic *Aeromonas* spp. possess dual flagella systems, in that they produce a constitutively expressed polar flagellum for motility in liquid environments, as well as an inducible distinct lateral flagellar system for movement across solid surfaces or viscous environments (Gavin *et al.*, 2002; Kirov *et al.*, 2002; Rabaan *et al.*, 2001).

The mesophilic aeromonads such as *A. caviae* and *A. hydrophila* are motile in liquid environments (in broth culture) by the action of a single unsheathed polar flagellum that is constitutively expressed and is encoded by over 50 genes located within 6 major gene clusters on the chromosome. The helical filament of the polar flagellum of *Aeromonas* is usually made up of two repeating flagellin subunits known as FlaA and FlaB (Canals *et al.*, 2006b; Rabaan *et al.*, 2001). In the laboratory, polar flagella-mediated motility of aeromonads can be assayed by the assessment of their ability to swim through low percentage agar (0.25–0.3%).

The well-documented discrepancies regarding *Aeromonas* taxonomy issues has been largely avoided with *A. salmonicida* due to its distinct role as a fish pathogen, and its apparent lack of motility making it simpler to identify than other species. It was previously defined as both aflagellated and non-motile. However, McIntosh and Austin (1991) demonstrated that *A. salmonicida* was capable of polar flagellated motility when grown at supra-optimal temperature (30–37°C). The first genetic evidence that *A. salmonicida* were in fact capable of producing flagella was provided by Umelo and Trust in 1997. In this study the flagellin genes (*flaA* and *flaB*) of *A. salmonicida* A449 were identified, cloned and expressed in *E. coli.* The study was extended through electron microscopy analysis of *A. salmonicida* A449 which confirmed that it expresses an unsheathed polar flagella at extremely low frequencies when grown in liquid culture under normal laboratory growth conditions (Umelo and Trust, 1997), providing an explanation as to why it had been previously undiscovered. Genome sequencing of *A. salmonicida* A449 demonstrated the presence of all the genes required for the biosynthesis of a polar flagellum. However, several of the genes contained disruptions through nucleotide deletions (*flrA* and *maf1*) or nucleotide duplications (*flgL*) thus resulting in mutations that abrogated the synthesis of a functional flagellum (Reith *et al.*, 2008).

Most work on the *Aeromonas* flagellar systems has concentrated on the 'motile' mesophilic species. The aeromonad polar flagellum has two flagellin subunits (FlaA and FlaB) that are exported as monomers via a dedicated type III protein secretion system. Following secretion, the flagellin monomers polymerize under the capping protein FlaH and assemble in to filaments forming the polar flagellum that elongates as more monomers of flagellin are secreted. Flagellin proteins are composed of three main domains that are required for function: the N-terminal region (approximately amino acids 1–50) of the protein, which contains a putative signal sequence required for export of the protein via the dedicated type III protein secretion system; the variable central D2/D3 antigenic domain (approximately amino acids 146–231) which is modified with the addition of sugar residues in a large number of bacteria including aeromonads; and the C-terminal region that is (approximately amino acids 262-end), required for recognition by the dedicated flagellin specific chaperone (annotated in aeromonads as FlaJ). The conserved regions at the C- and N-termini are the portions of flagellin that are recognized by the innate immune system receptor Toll-like receptor 5 (TLR5). The flagellin monomers, fold in such a way that the variable central D2/D3 domain is exposed to the outer surface of the filament, while concealing the parts of the flagellin that interact with TLR-5 inside the structure of the polymerized flagellar filament (Smith *et al.*, 2003).

Glycosylation of flagellins, the major subunits of the flagellar filament, has been described for a number of bacteria (Schirm *et al.*, 2004; Verma *et al.*, 2006; Thibault *et al.*, 2001; Josenhans *et al.*, 2002; Schirm *et al.*, 2003; Leclerc *et al.*, 1998) and was first described in *A. caviae* in 2001 (Rabaan *et al.*, 2001), where it was observed that FlaA and FlaB showed aberrant migration when analysed by SDS-PAGE. In a later study, mass spectrometry analysis of purified flagellin of *A. caviae* indicated that FlaA and FlaB of *Aeromonas* are homogeneously glycosylated with between six and eight pseudaminic acid (Pse5Ac7Ac) moieties of 317 Da that are linked on to serine and threonine residues in the central immunogenic D2/D3 domains of the flagellin via O-linked glycosylation (Tabei *et al.*, 2009). Pseudaminic acid is a nine carbon non-ulonosonic acid sugar related to sialic acid. Tabei and co-workers demonstrated that five of the genes *flmA*, *flmB*, *neuB*, *flmD* and *neuA* were involved in the synthesis of Pse5Ac7Ac (Tabei *et al.*, 2009) and that these genes were essential for motility through the production of flagella. In order for Pse5Ac7Ac to be transferred onto the flagellin it has to be activated with the addition of CMP by a NeuA paralogue PseF to give Pse5Ac7Ac-CMP at the time of writing the glycosyltransferase that carries out this process has yet to be determined. However, it has been predicted that this role is likely to be carried out by the Maf family of proteins (Parker *et al.*, 2012). Genes encoding putative Maf proteins have been shown to be required for the formation of a fully functional flagellum in *A. caviae* (Parker *et al.*, 2012) and *A. hydrophila* (Canals *et al.*, 2007). In contrast to the glycosylated polar flagellin of *A. caviae,* the polar flagellin of *A. hydrophila* AH-3 is modified with a heterogeneous glycan that consists of a heptasaccharide linked to a pseudaminic acid derivative of 376 Da (Wilhelms *et al.*, 2012). In addition the flagellin glycosylation island of *A. hydrophila* AH-3 is more complex with the presence of several extra genes that could possibly account for the heterogeneous glycan or the modified pseudaminic acid derivative (Canals *et al.*, 2007). Currently the exact cellular function of the flagellin glycosylation is unknown but it is thought to be necessary for flagella filament assembly via its dedicated Type III secretion apparatus with unglycosylated flagellin accumulating in the cytoplasm (Josenhans *et al.*, 2002). Because unglycosylated flagellins are not exported and do not produce a functional filament, it is difficult to determine whether or not the modifications themselves have a precise role in pathogenicity or infection. Additional hypotheses for the role of flagellin glycosylation include the idea that glycosylation acts as a signal for secretion, increases the solubility of the flagellin or may be required to evade the immune system by preventing (or decreasing) recognition by TLR-5 (Logan, 2006). Evidence from *A. caviae* Sch3 suggests the glycosylated flagellin is recognized better by anti-sera than the unglycosylated form (Parker *et al.*, 2012). As well as motility in liquid environments, polar flagella have also been identified as important virulence factors with roles in adherence and colonization of hosts. The polar flagella of *A. caviae* have been shown to play a role in adherence to

HEp-2, Henle407 and Caco-2 cells, and in the formation of biofilms on borosilicate glass, with both flagellins (FlaA and FlaB) being required (Rabaan *et al.*, 2001; Kirov *et al.*, 2004). The *A. hydrophila* polar flagellum is also seen as a good target for vaccine development for the protection of channel catfish. Yeh and Klesius have overexpressed and purified 19 *A. hydrophila* flagellar proteins and reported strongly reacting sera to FlgK in experimentally infected catfish (Yeh and Klesius, 2011).

Other components of the aeromonad polar flagellum that are of interest is the flagellar motor that is composed of the rotor and stator. The rotary component (rotor) comprises the proteins FliG, FliM and FliN that make up the C-ring and in *E. coli* and *Salmonella* the stationary component (stator) is made up of the proteins MotA and MotB. In *E. coli* and *Salmonella* MotA and MotB couple the flow of protons down an ion gradient to flagella rotation. However, the stator complex for the polar flagellum in *Aeromonas*, like those of *Shewanella oneidensis* and *Vibrio parahaemolyticus* appear to be more complex. In contrast to the systems of *E. coli* and *Salmonella*, here the stator complex is driven by sodium ions rather than protons and require four proteins PomA, PomB, MotX and MotY, where PomA and PomB are the paralogues of the proton-driven MotA and MotB (Wilhelms *et al.*, 2009; Molero *et al.*, 2011). MotX and MotY are not present in the *E. coli* flagellar paradigm, but in the *Vibrio* polar flagellum they form the T-ring component of the basal body. Furthermore, aeromonads have two sets of the Pom–motor proteins designated PomA, PomB and PomA2, PomB2 (Wilhelms *et al.*, 2009). Both sets are sodium-coupled stator complexes and neither set is essential on its own, but they do have variations in their sensitivities to sodium ions (Wilhelms *et al.*, 2009).

Shimada and co-workers using electron microscopy originally reported that mesophilic aeromonads when grown in liquid medium had a single polar flagellum, but growth on surfaces induced 20–30% of the strains tested to produce peritrichous flagella (Shimada *et al.*, 1985). Subsequent work has demonstrated that approximately 50–60% of mesophilic aeromonads use an entirely distinct peritrichous lateral flagella system for movement across a solid surface (known as swarming motility) (Kirov *et al.*, 2002). The complete gene set of 38 genes required for lateral flagella biosynthesis are found in a single chromosomal region (Canals *et al.*, 2006a). This gene cluster is also present in *A. salmonicida*, although *lafA* is disrupted by the insertion sequence *IS3* and *lfgD* has a nucleotide deletion resulting in the lateral flagella negative phenotype of this bacterium (Merino *et al.*, 2003; Reith *et al.*, 2008). Lateral flagella filaments on aeromonads are composed of repeating lateral flagellin known as LafA, which form unsheathed flagella filaments. In 2002 a comprehensive study showed that 27 of 55 (49%) of clinical isolates tested contained *laf* genes and the strains containing these genes could swarm successfully over the surface of solid media (Kirov *et al.*, 2002). Electron microscopy analysis of *Aeromonas* grown under laboratory conditions showed that > 60% of an *A. caviae* Sch3 population expressed lateral flagella when grown on solid medium at 37°C for 6 hours (Kirov *et al.*, 2002). The lateral flagella of a number of *Aeromonas* species were investigated further under varying growth conditions, and found that no lateral flagella could be identified in *A. caviae*, *A. veronii* biovar sobria, and *A. hydrophila* when they were grown in liquid medium (Kirov *et al.*, 2002). Unlike the polar flagellar stator the lateral flagellar stator complex is composed of the proteins LafT and LafU and is thought to be proton driven (Gavin *et al.*, 2002; Wilhelms *et al.*, 2009).

Most of the sequenced mesophilic aeromonads encode a single lateral flagellin gene (known as *lafA*). The genome of *A. caviae* Sch3 and a minority of other aeromonad strains encode two lateral flagellin genes (*lafA1* and *lafA2*) with studies showing that the lateral flagella filament of *A. caviae* Sch3 are comprised of two distinct lateral flagellins known as LafA1 and LafA2 (Gavin *et al.*, 2002). The first indication that the lateral flagellins of *A. caviae* and *A. hydrophila* were post-translationally modified via glycosylation was found by Gavin *et al* (2002) where aberrant migration of the lateral flagellins was observed when analysing by SDS-PAGE. More recently, findings have shown that the lateral flagellin of *A. hydrophila* is modified via O-linked glycosylation on to serine and threonine residues. *A. hydrophila* lateral flagellins are glycosylated exclusively with a 376 Da derivative of pseudaminic acid that is also used to modify the polar flagellin (Wilhelms *et al.*, 2012). To date it has yet to be determined whether or not glycosylation of the lateral flagellin of *A. hydrophila* is essential for swarming motility over surfaces, or indeed what the role of this modification is at all. Like polar flagella, the lateral flagella system of *Aeromonas* has also been identified as an important virulence factor, involved in adherence to host cells. Gavin *et al.* (2002) confirmed that the distinct lateral flagella systems of *A. caviae* and *A. hydrophila* are essential for the formation of biofilms to both polystyrene and glass and for adherence to HEp-2 epithelial cells (Gavin *et al.*, 2002; Gavin *et al.*, 2003; Kirov *et al.*, 2004). The possession of both flagellar systems but especially lateral flagella has been greatly associated with the ability to form biofilms in *A. caviae* strains (Santos *et al.*, 2011).

Flagella systems are metabolically expensive structures to produce and because of this the regulation of expression of the genes involved in polar and lateral flagella biosynthesis occurs as a hierarchical cascade that is tightly regulated. The best studied *Aeromonas* system is that of *A. hydrophila* AH-3 where the genetic regulation of the polar (Altarriba *et al.*, 2003; Canals *et al.*, 2006b; Wilhelms *et al.*, 2011) and the lateral (Canals *et al.*, 2006a) systems have been studied in detail. The constitutively expressed fully functional polar flagellum involves the expression of over 50 genes that are organized in to clusters and located in six different chromosomal regions, with timings of expression coupled to protein export and flagella formation. As well as the structural components of the flagellum, these expressed genes encode a number of regulators as well as proteins involved in chemotaxis (the movement towards or away from a certain chemical). The transcriptional hierarchy of the genes required for and involved in polar flagellum formation in *A. hydrophila* AH-3 has been elucidated. In 2011, a comprehensive study by, Wilhelms *et al.* (2011) confirmed that transcription of the polar-flagellum class I genes is σ^{70} dependent and is organized in four transcriptional levels (classes I to IV), where each level contains the activator for the subsequent transcriptional level. The class I genes encodes the σ^{70} dependent response regulator-like protein FlrA. Cells where the gene for *flrA* is disrupted are aflagellate, but lateral flagella formation is unaffected. Expression of *flrA* is σ^{28} and σ^{54} independent suggesting that FlrA is the master regulator of the polar-flagellum regulon. Transcription of the class II and III genes are mainly σ^{54} and FlrA dependent and encode many of the structural components of the basal body, the hook, the export/assembly apparatus, and chemotaxis. The regulatory protein encoding genes *flrBC*, and the gene encoding σ^{28} (*fliA*) are also in this class of the genetic hierarchy. Class IV make up the late flagellar genes which encode components such as the flagellins (FlaA and FlaB), motor components, filament cap (FlaH), flagellin chaperone (FlaJ) and the anti-sigma factor FlgM. The class IV genes are σ^{28} (FliA) dependent (Wilhelms *et al.*, 2011).

In contrast to constitutively expressed polar flagellin genes, the genes required for lateral flagella biosynthesis are expressed only under certain conditions, and are environmentally regulated (Merino *et al.*, 2006). Expression of the lateral flagella system is induced when grown on surfaces or in a more viscous environment. All of the genes required for lateral flagella synthesis in *A. hydrophila* are located together in one region of the chromosome (Canals *et al.*, 2006a). The regulatory cascade of the polar and lateral flagella system separates genes in to early, middle and late. For lateral flagella expression LafK is the master regulator. The expression of the middle genes of the induced lateral flagella genes of *A. hydrophila* is controlled by σ^{54} whereas findings suggest that in most cases the late gene expression is controlled by the σ^{28} flagella-specific factor, and its cognate anti-σ factor FlgM (Canals *et al.*, 2006a). The signal that activates LafK and how *Aeromonas* senses a surface is unknown. However, it appears to be different from the mechanism described for *Vibrio parahaemolyticus,* this bacterium uses its polar flagellum as a sensor dynamometer and mutations in the polar flagellum cause constitutive expression of lateral flagella (Merino *et al.*, 2006). This is not the case for *Aeromonas,* as mutations in the polar flagellum do not affect lateral flagella expressions therefore the polar flagellum is not the sensor (Merino *et al.*, 2006; Altarriba *et al.*, 2003).

Pili/fimbriae

For gastrointestinal pathogens adherence to the host is an essential step in their colonization process as it stops the bacterium being washed away and it is required for both efficient delivery of toxins and invasion of deeper tissues. The adherence mechanisms of aeromonads are complex, differences in adherence patterns have been observed for single strains that change upon growth phase, temperature and media (Thornley *et al.*, 1996). This suggests that the aeromonad adherence mechanism is complicated and multifactorial and it is a process that is still relatively poorly understood. However, the advent of whole genome sequencing has helped improve our understanding of the possible mechanisms, giving a system wide view of what factors are potentially expressed by different *Aeromonas* strains.

Early work on adherence in *Aeromonas hydrophila* concentrated on the haemagglutination (HA) of erythrocytes (Atkinson and Trust, 1980) and described over 40 different HA patterns. However, haemagglutination has been associated with both the possession of filamentous and non-filamentous adhesins. This was followed by work showing the association of filamentous structures with adherence (Kirov and Sanderson, 1995) and that the mechanical removal of these structures could result in a great reduction the adhesive ability of aeromonad strains (Kirov and Sanderson, 1996). Furthermore, fimbrial differences have been reported between environmental and clinical strains, with the clinical strains displaying a small number (<10 per cell) of long-wavy (L/W) pili and environmental strains possessing a larger number of short-rigid (S/R) pili (Carello *et al.*, 1988). Suggesting the L/W pili are more relevant to clinical infections and pathogenesis. Following ultra structural studies on a strain of *A. hydrophila* the major pilin of a straight rigid (S/R) pilus was isolated, it had a molecular weight of 17 kDa and its amino acid sequence homology was similar to the *Escherichia coli* type 1 pilus and Pap pili (Ho *et al.*, 1990). More, recently genome sequencing has revealed the presence of a type I pilus locus in the genomes of both *Aeromonas salmonicida* and several mesophilic aeromonads (Reith *et al.*, 2008; Seshadri *et al.*, 2006; J. Shaw unpublished). The *A. salmonicida* type I pilus locus has been mutated and a Fim-deficient strain was not attenuated in live challenges of Atlantic salmon, but it was demonstrated that Fim acts as an adhesin and it improves the efficiency of *A. salmonicida* infection of its host (Dacanay *et al.*, 2010).

Most investigations on fimbrial-associated adherence have concentrated on the long-wavy (L/W) pili of the mesophilic *Aeromonas* species these strains have been shown to be able to form filamentous networks or bundles (Kirov *et al.*, 1995). Furthermore removal of these structures, or competition from the purified filaments can greatly reduce adherence of the parental bacterial strain in competition assays (Iwanaga and Hokama, 1992). The first L/W pilus to be genetically characterized was the so, called 'mini-pilin' of *A. hydrophila*, this pilus was identified by ultra-structural studies on *A. hydrophila* strain AH26 that produced both S/R and L/W pili (Ho *et al.*, 1990). The 'mini-pilin' was found to be environmentally regulated, being produced at 22°C and not at 37°C and when the availability of free iron was reduced by an iron-chelator. The gene for this flexible pilus (*fxp*) was found on a 7.6 kb plasmid, this encoded a 4 kDa pilin composed of 46 amino acids (Ho *et al.*, 1990). The pilin subunit did not share any similarity with any known pilin sequences, nor was the *fxp* gene present in any other of the 66 *Aeromonas* strains tested. Moreover, the 'mini-pilin' Fxp shares homology to Cep the virion capsid protein of the *V. cholerae* phage CTX (Waldor and Mekalanos, 1996), leading to the suggestion that the 'mini-pilin' does not encode a flexible pilus but a filamentous phage particle.

Most true L/W pili have diameters of 4–7 nm and have been isolated from a number of strains, the molecular masses of their pilins tend to be between 19 and 23 kDa and they are immunologically and electrophoretically distinct although all have N-terminal sequence homology to type IV pilins (Kirov and Sanderson, 1996). The major pilin of the bundle-forming pilus from *A. veronii* biovar sobria strain BC88 has been purified and its N-terminal sequence showed most sequence homology to the type IV pilin the mannose-sensitive haemagglutinin (MSHA) of *Vibrio cholerae* (Kirov and Sanderson 1996). However, this group of pilins belong to the classical non-bundle-forming type IVa family, whereas true bundle-forming pili are found in the type IVb family. Expression of the bundle-forming pilus appeared to be environmentally regulated with more pili seen on the surface of the bacterium when grown at 22°C in liquid as opposed to 37°C on solid media (Kirov and Sanderson, 1996). More detailed investigation of the bundle-forming pilus strain *A. veronii* biovar sobria BC88 led to the isolation of the genes encoding another type-IV pilus system (Tap). This followed the cloning of the biogenesis cluster (*tapABCD*), these were most similar to the *piliABCD* genes of *Pseudomonas aeruginosa* and not the MSHA pilus (Pepe *et al.*, 1996). The role of the Tap-pilus is unknown, as mutation of *tapA* the gene encoding the major pilin had no effect on adherence to both biotic and abiotic surfaces (Kirov *et al.*, 2000). The equivalent pilus system has been shown to be present in all the aeromonad strains that have had their genomes sequenced. The aeromonad Tap pilus system is encoded by 22 genes (*tapABCD* plus eighteen other), these are scattered around the chromosome in several loci varying from one gene to seven (Boyd *et al.*, 2008; Seshadri *et al.*, 2006; J.G. Shaw unpublished). The investigations of Kirov and co-workers demonstrated that *A. veronii* biovar sobria BC88 encoded at least two distinct type IV pilus families Tap and Bfp (MSHA) whose major pilins differ in their N-terminal amino acid sequences and molecular weights (Barnett *et al.*, 1997).

Genome sequencing has indeed confirmed that most aeromonads encode three type IV pilus systems in their genomes in addition to the Fim-like type I pilus system (Reith *et al.*, 2008; Seshadri *et al.*, 2006). Genome sequencing of *A. salmonicida* A449 revealed the genes encoding three type-IV pilus systems, *flp*, *msh* and *tap*. The Tap system was similar to the one previously described in both mesophilic *Aeromonas* spp. and *A. salmonicida*. The Flp system was similar to the Flp pilus of *Actinobacillus actinomycetemcomitans* and *Caulobacter crecentus* (Boyd *et al.*, 2008). The Flp pilus is encoded by 13 genes located in a single gene cluster (Reith *et al.*, 2008; Boyd *et al.*, 2008). The locus encoding the MSHA pilus (similar to the *A. veronii* biovar sobria bundle-forming pilus) was thought to be non-functional as it contained a large internal deletion of several genes. Mutation of the Flp system suggested that it played no part in the virulence of *A. salmonicida* when tested in the Atlantic salmon (Boyd *et al.*, 2008). Whereas, mutation of the *A. salmonicida* Tap system provided evidence that it performed a role in the infection of Atlantic salmon and thus virulence, unlike the Tap system of mesophilic aeromonads (Boyd *et al.*, 2008). In *A. salmonicida* the genes for the Tap pilus appear to be constitutively expressed whereas the Flp pilus genes are up-regulated when the bacterium is grown *in vivo* and under low iron conditions (Boyd *et al.*, 2008).

As stated above the predominant type IV pilus isolated from the mesophilic aeromonads are related to the bundle forming pilus described by Kirov and co-workers. This group of pili are thought to be the major aeromonad adhesin as their removal by mechanical shearing, competition with purified filaments or inhibition by antibody had a great effect at reducing the adherence of *Aeromonas* to host cells (Kirov and Sanderson, 1996; Kirov *et al.*, 1999). The N-terminal sequence of the *A. veronii* biovar sobria BC88 bundle-forming pilus showed closest homology to the MSHA of *Vibrio cholerae* El Tor (Kirov and Sanderson, 1996; Marsh and Taylor, 1999). The bundle-forming pilus/MSHA encoding locus *msh* from *A. veronii* biovar sobria BC88 has recently been characterized (Hadi *et al.*, 2012). The bundle-forming pilus was encoded on a single 22 kb locus that contained seventeen genes, these were organized into two large operons *mshI-F* and *mshB-Q*. The locus had four genes that encoded pre-pilin proteins that are thought to form the pilus filament, one encoded protein MshA exactly matched the N-terminal amino acid sequence of the Bfp-major pilin protein. All four of the pre-pilin proteins appear to be essential as their mutation results in over a 90% reduction of adherence to both biotic and abiotic surfaces (Hadi *et al.*, 2012). This provided the first genetic evidence using isogenic mutants that the bundle forming pilus is the major pilus system required for adherence of *A. veronii* biovar sobria BC88 and possibly other mesophilic aeromonads.

This study also demonstrated the importance of one of the Tap pilus proteins. One of the genes found in the *tapABCD* gene cluster encodes TapD that is the only aeromonad pre-pilin peptidase found in the aeromonad genome. This enzyme is required for processing of both the pre-pilin proteins of type IV pili, but also the pseudopilins of the type-II secretion system (T2SS). Mutation of the *tapD* gene had a pleiotropic effect that resulted not only in the loss of the MSHA bundle-forming pilus and therefore adherence, but also the secretion of a number of extracellular enzymes such as protease, DNase and haemolysin (aerolysin) that are secreted by the T2SS (Hadi *et al.*, 2012). Located upstream of the *msh* gene cluster, is a conserved gene *mshH* that encodes a potential regulatory protein. The gene *mshH* is conserved in this position in all aeromonad strains that have had their genomes sequenced, as well as *Vibrio*, *Shewanella* and *Photobacterium*. This gene encodes the protein MshH that contains diguanylate cyclase (GGDEF) and phosphodiesterase (EAL) domains that are involved in the synthesis and degradation of cyclic-di-GMP respectively. Cyclic-di-GMP is involved in bacterial lifestyle choices, such as the change between sessile and motile existence especially biofilm formation. This gene was therefore thought to encode a candidate protein that regulated the production of the bundle-forming pilus. However, Hadi and co-workers demonstrated that the *mshH* was not transcriptionally linked to the rest of the *msh* gene cluster and that its mutation had little effect on expression of the bundle-forming pilus and thus adherence (Hadi *et al.*, 2012).

Lipopolysaccharide (LPS)

Like other Gram-negative bacteria the aeromonads Lipopolysaccharide (LPS) is organized into three separate domains; the conserved and toxic lipid A; the highly variable O-antigen polysaccharide and the core oligosaccharide that links the two. Sakazaki and Shimada (1984) identified 44 O-antigen serogroups among 307 aeromonad strains. An additional 52 new serogroups were isolated by Thomas and colleagues when examining a further 1255 strains. Even with this extended scheme a significant number of strains did not belong to any serogroup suggesting the presence of yet more unidentified serogroups in the genus (Thomas *et al.*, 1990). This work demonstrated that as a genus *Aeromonas* has a highly variable population of O-antigen serogroups. However, the serogroups O:3, O:11, O14, O:16 and O:34 predominate among the human isolates (Sakazaki and Shimada, 1984; Thomas *et al.*, 1990).

Aeromonad LPS has been designated as a possible colonization factor, Franki and Chang (1994) demonstrated a 61% reduction in adhesion of an *A. veronii* biovar sobria strain in stationary phase when no LPS was detectable. LPS has been shown to play in important role in the colonization of the germ free chicken by *A. hydrophila* of serogroup O:34 (Merino *et al.*, 1996b) and the medicinal leech by *A. veronii* biovar sobria (Braschler *et al.*, 2003). In addition smooth (S) LPS has been associated with adherence to HEp-2 cells of *A. hydrophila*, and the presence of smooth LPS when *A. hydrophila* O:34 strains were gown at 20°C resulted in them being more virulent in fish and mice than their rough (R) counterparts grown at 37°C (Merino *et al.*, 1992). LPS plays an important role in aeromonad resistance to non-immune serum and thus to complement mediated killing (Merino *et al.*, 1991). Transposon mutants devoid of O-Antigen and LPS core structure were found to be sensitive to complement mediated lysis in contrast to the smooth parent strain (Merino *et al.*, 1996a). It has been suggested that enterotoxigenic gastroenteritis associated aeromonad strains would not require an extensive O-Antigen to initiate colonization of mucosal surfaces, whereas invasive strains would benefit from a smooth LPS due to their ability to resist complement killing once inside the host, causing bacteraemic infections (Tso and Dooley, 1995).

Three main electrophoretic patterns have been identified for the O-Antigen of *Aeromonas* LPS: a predominant smooth (S) ladder-like pattern commonly seen for the enterobacteria; a homogeneous pattern showing some O-Antigen chains sometimes in doublets; and a rough (R) or semi-rough pattern exhibiting one main band representing the core with no detectable O-Antigen bands (Kokka *et al.*, 1990; Tso and Dooley, 1995; Khashe *et al.*, 1996). As one would expect from the large variety of serogroups the chemical composition of the O-Antigen is very diverse, however, the structure and chemical compositions of the O-Antigen from a number of *Aeromonas* species have been elucidated. One of the first aeromonad O-Antigen structures to be determined was that of an *A. caviae* gastrointestinal isolate, this was a branched pentasaccharide O-Antigen unit composed of L-Rha, D-Gal, D-ManNAc and D-GalNAc (Linnerborg *et al.*, 1996). Biosynthesis of LPS is complex and requires over 40 genes that encode proteins involved in the synthesis of the sugars and lipids in addition to transport and assembly proteins. One of the best studied aeromonad LPS systems is O:34 of *A. hydrophila* AH-3, where the genetics and the regulation of the O-antigen encoding *wb** gene cluster has been characterized (Jimenez *et al.*, 2008b). Previous work by the group of Tomás has determined the structure of the O:34 antigen and have shown it to contain D-Man, D-GalNAc and 6-deoxy-L-talose (Knirel *et al.*, 2002). The $wb^{*}_{O:34}$ gene cluster is 18.1 kb in size, encodes 17 open reading frames and it has a lower G+C content (44.3%) than the *Aeromonas* average consistent with lateral gene transfer (Jimenez *et al.*, 2008b). The genes within the cluster encode proteins consistent with the LPS O-antigen structure, required for the synthesis of D-Man, L-Rha and 6-deoxy-L-talose. The presence of genes encoding Wzy and Wzx suggest the LPS O-antigen is exported in a Wzy-dependent manner. The *wzz* gene of the $wb^{*}_{O:34}$ gene cluster was demonstrated to be regulated by temperature, showing greater transcription at 20°C when compared to 37°C (Jimenez *et al.*, 2008b), this also agreed with their previous observations that as well as osmolarity, temperature also regulated LPS production in this strain (Aguilar *et al.*, 1997). The LPS core structure has also been determined for *A. hydrophila* O:34 (Knirel *et al.*, 2004) and the inner core was shown to be highly conserved with that of *A. salmonicida* (Wang *et al.*, 2006). For the *A. hydrophila* strain AH-3 of serogroup O:34, three gene clusters have been isolated involved in LPS core biosynthesis (Jimenez, *et al.*, 2008a). Region 1 contained seven genes, region 2 four genes and region 3 two genes, along with the core structural information this allowed the putative designation of all LPS core biosynthetic functions (Jimenez *et al.*, 2008a). Recently a comprehensive review has been published detailing the known structures of numerous aeromonad carbohydrate antigens including both O-Antigen and core-oligosaccharides for the reader interested in more structural information I would direct them there (Nazarenko *et al.*, 2011).

In *A. caviae* Sch3 the laboratory of Shaw and coworkers isolated a series of transposon mutants in a locus that affected adherence, motility by the loss of flagella and the absence of the LPS O-Antigen band (Gryllos *et al.*, 2001). These mutations were mapped to the *flm* locus, a cluster of seven genes that appeared to have a dual role in both flagella and LPS biosynthesis (Gryllos *et al.*, 2001; Tabei *et al.*, 2009). Subsequent work demonstrated that five of the genes *flmA, flmB, neuB, flmD* and *neuA* encoded proteins involved in the synthesis of the nine-carbon sialic acid related sugar pseudaminic acid (Pse5Ac7Ac) (Tabei *et al.*, 2009). As described previously in the flagella section of this chapter, this sugar also decorates the polar flagellin proteins of this strain. These genes were essential for both motility through the production of flagella and the LPS O-Antigen, thus having a dual role in both of these systems. The other two genes in the gene cluster, *lst* and *lsg* encoded a glycosyltransferase and a LPS O-Antigen flippase respectively, mutation of either of these genes did not affect flagellin glycosylation or motility, but caused the loss of the LPS O-Antigen, suggesting these genes were specific for LPS production only. Mass spectrometry showed that the polar flagellins of this strain were glycosylated with pseudaminic acid and that at least two pseudaminic acid residues were present in the LPS O-Antigen of *A. caviae* Sch3 along with two other un-identified sugars (Tabei *et al.*, 2009).

Capsule

Capsules are hydrated repetitions of a molecule usually a monosaccharide, that covers the external surface of the bacterium. The genus *Aeromonas* has been described as non-capsulated, however, there have been several reports of the presence of capsular polysaccharides in both motile and non-motile species. The first demonstration of the production of capsules by aeromonads was in *A. salmonicida* 80204-1 grown *in vitro* in glucose rich media (Garrote *et al.*, 1992) or during *in vivo* growth in fish implants (Garduno and Kay, 1995). Subsequently, capsules were found to be produced by *A. hydrophila* serogroups O:11 and O:34 (Martinez *et al.*, 1995). Interestingly, there are differences in the chemical composition of the aeromonad capsules with glucose dominating in most strains. However, the *A. hydrophila* O:34 and O:11 capsules were composed of D-glucose, D-mannose, L-rhamnose, D-mannuronic acid and acetic acid (Martinez *et al.*, 1995). Whereas the *A. salmonicida* capsule lacked acetic acid but did contain *N*-acetylmannosamine (Garrote *et al.*, 1992).

It would be most likely that capsules would confer an advantage on the more invasive *Aeromonas* strains that are involved in bacteraemia by conferring on them serum resistance by keeping the membrane attack complex of complement away from the bacterial cell membranes. Furthermore, aeromonad capsules have been shown to decrease opsonization and thus phagocytosis (Merino *et al.*, 1996a). In addition, capsules have been found to increase adherence and invasiveness of fish cell lines by *A. hydrophila* O:34 (Merino *et al.*, 1997). Aguilar and co-workers isolated two genes from *A. hydrophila* serogroup O:34 strain that conferred serum resistance on *E. coli* DH5α and were thought necessary for the production of the colanic acid capsule in the *E. coli* strain (Aguilar *et al.* 1999). The *A. hydrophila* genes were named *orf1* and *wcaJ*, and unlike the wild-type, mutation of the latter gene *wcaJ* resulted in a *A. hydrophila* strain that was unable to produce a capsule in media rich in glucose or in autolysates of fish (Aguilar *et al.*, 1999). The first complete capsule locus was isolated from *A. hydrophila* PPD134/91, this comprised of a 17.5 kb locus of 13 ORFs organized into three distinct regions representative of group II capsules (Zhang *et al* 2002). The genes for regions I and III encoded the transport genes while region II was required for capsule biosynthesis. The purified capsular polysaccharide from the virulent *A. hydrophila* strain PPD134/91 was able to confer serum resistance on the avirulent strain PPD35/85 without interfering with the adherence of this strain (Zhang *et al.*, 2002). Ten out of 33 *A. hydrophila* strains were shown to be able to possess group II capsules when examined by electron microscopy and genetic means (Zhang *et al.*, 2003). Using a PCR-based system these ten were able to be sub-divided in to group IIA and group IIB capsules. Group IIA capsules were mainly found in *A. hydrophila* O:18 and O;34 sergroups, whereas group IIB capsules were associated with O:21 and O:27 serogroups (Zhang *et al.* 2003). The possession of a group II capsule greatly increased the survivability of *A. hydrophila* when challenged with either serum or phagocytes (Zhang *et al.*, 2003). However, information is still limited and the serogroups and strains that possess capsules has not been fully studied and needs to be established.

S-layer

Strains within the genus *Aeromonas* possess a paracrystalline tetragonal array of a surface protein that forms a layer around the bacterium (S-layer), the layer is composed of identical surface array protein (SAP) subunits and forms the predominant surface antigen. Possession of an S-layer (originally called A-layer) was first demonstrated in *A. salmonicida* (Kay *et al.*, 1981) and it was recognized as one of the major aeromonad virulence factors as it was associated with the ability of *A. salmonicida* to cause septicaemia in fish. The S-layer is less common in the mesophilic strains of *Aeromonas*, but it has been found in *A. hydrophila* and *A. veronii* biovar sobria serogroup O:11, O:14 and O:81 (Kokka *et al.*, 1990; Esteve *et al.*, 2004), but the S-layer is absent from *A. caviae*. The surface array proteins are thought to be tethered onto the LPS O-Antigen of homogeneous length, as LPS mutants devoid of a smooth LPS O-Antigen do not assemble an S-layer and secrete SAP into the medium, whereas partial loss of O-Antigen results in a disrupted S-layer (Kokka *et al.*, 1990; Griffiths and Lynch 1990). For the interaction with LPS, the C-terminal domain of the S-layer protein was found to be essential (Thomas *et al.*, 1992). The gene *vapA* encoding the S-layer protein in *A. salmonicida* can be deleted when the bacterium is grown at over 25°C, resulting in loss of the S-layer (Belland and Trust, 1987)

The S-layer proteins of aeromonads contain a signal sequence of between 19–21 amino acids and are thought to be secreted via the classical signal sequence route of type two secretion (T2SS). In fact in the genome of *A. salmonicida* A449 the S-layer protein gene *vapA* lies downstream from an operon encoding components of a VapA specific T2SS (Reith *et al.*, 2008). This T2SS was shown to be specific and required for S-layer secretion as mutation of two of the genes *spsD* or *sps*E resulted in accumulation of the S-layer protein in the periplasm but did not affect the secretion of other extracellular proteins (Noonan and Trust, 1995b). In *A. salmonicida* A449 the *vapA* gene and secretion operon are in fact found in a larger gene cluster that encodes numerous carbohydrate biosynthesis and modification proteins. These are thought to be involved in LPS biosynthesis, and LPS anchors the S-layer protein to the cell surface (Reith *et al.*, 1995). The G + C content of this chromosomal region is unusually low for *Aeromonas* and this cluster is thought to have been laterally acquired (Reith *et al.*, 2008). The structural SAP subunits are usually between 50–52 kDa (Kokka *et al.*, 1990; Kay *et al.*, 1981), although the predicted molecular weights from the genes encoding the SAPs would suggest that AshA the *A. hydrophila* SAP is post-translationally modified whereas the *A. salmonicida* SAP VapA is not (Noonan and Trust, 1997). Post-translational modifications for *A. hydrophila* S-layer proteins have been reported; phosphorylation of tyrosine residues have been identified that decreased the PI of the S-layer protein from 6.7 to 4.6 (Thomas and Trust 1995). For the S-layer protein to cover the surface of the bacterium, it must represent 10–15% of the total cellular protein, therefore the gene that encodes the S-layer protein must be efficiently transcribed and translated. Two promoters have been identified upstream of the *vapA* gene in *A. salmonicida* (Chu *et al.*, 1993), one of which is most active during the early stages of growth (Noonan and Trust 1995a)

S-layers have been associated with the ability of *A. salmonicida* (Kay *et al.*, 1981) and *A. hydrophila* (Murray *et al.*, 1988) to cause disease in fish and have been suggested to act by protecting the cells from complement mediated killing (Merino *et al.*, 1996a). Both the SAPs from *A. salmonicida* and *A. hydrophila* are resistant to proteolytic degradation and it is suggested that the S-layer could render the bacteria resistant to proteolytic degradation once inside the phagolysosome of the macrophage (Noonan and Trust, 1997). In addition the S-layers of *A. hydrophila* fish and human isolates bind to the extracellular matrix proteins collagen, fibronectin, laminin and to murine macrophages (Garduno *et al.*, 1992). Although the role of the S-layers of *A. hydrophila* and *A. veronii* biovar *sobria* have not been fully determined, their importance is not thought to be as great as for *A. salmonicida* in fish disease (Merino *et al.*, 1995). Differences in the SAPs, such as post-translational modifications exist between the *A. salmonicida* and mesophilic aeromonad proteins. Moreover, *A. hydrophila* S-layer mutants did not exhibit increased susceptibility to rabbit serum (Noonan and Trust, 1997), or loss of virulence in mice (Kokka *et al.*, 1991). Therefore the full role of S-layers in the virulence of the mesophilic *Aeromonas* strains still remains to be determined.

Outer membrane proteins (OMPs)

Studies on outer membrane protein (OMPs) have mainly concentrated on the mesophilic species of *A. hydrophila*, *A. veronii* biovar sobria and *A. caviae*. The outer membrane proteins (OMPs) described show variation in size with most exhibiting OMPs of 30–54 kDa (Kokka *et al.*, 1990). The 39 kDa porin of *A. hydrophila* was shown to bind C1q and thus activate the classical complement pathway (Merino *et al.*, 1998). However, in strains that have smooth LPS, complement binding is protected by the extended LPS O-Antigen.

Temperature dependent OMP expression between 10–40°C has been observed in motile *Aeromonas* species (Tso and Dooley, 1995). Similar results have been observed for *A. caviae* grown between 4°C and 43°C, where the changes in OMP expression at 43°C coincided with the loss of haemagglutination (J.G. Shaw unpublished). The loss of the 43 kDa OMP in a *A. hydrophila* clinical isolate was seen when it was grown at 43°C instead of 37°C, this also coincided with the loss of haemagglutination (Atkinson *et al.*, 1987). Since this was the OMP that was lost upon temperature increase this protein was proposed to be a haemagglutinin. Two haemagglutinating carbohydrate-reactive OMPs (CROMPs) were subsequently isolated (Quinn *et al.*, 1994). The CROMPs were found to be outer membrane porins similar to those of *Escherichia coli*. Unfortunately, the role if any of haemagglutinin in *in vitro* adherence was not determined.

Outer membrane proteins have been proposed to mediate adherence of non-piliated human isolates of *Aeromonas* to cell lines (Nandapalan and Chang, 1989; Bartkova and Ciznar, 1994). A number of studies have looked at the differential expression of *A. hydrophila* OMPs during growth *in vitro* and *in vivo* (Rahman and Kawai, 2000; Poobalane *et al.*, 2008). In *A. hydrophila* a unique OMP band of 55 kDa was seen in four virulent and two avirulent strains when grown *in vivo* when compared to *in vitro* suggesting a potential vaccine target (Poobalane *et al.*, 2008). Interest has also been shown in OmpW of *A. hydrophila*, as the equivalent protein in *Vibrio cholerae* is conserved and thought to be a good vaccine candidate (Maiti *et al.*, 2009). The protein for OmpW has been overexpressed, purified and antisera raised against the recombinant protein. Western blotting of a number of *Aeromonas* strains using the anti-OmpW sera showed that its expression was dependent on a high salt concentration (Maiti *et al.*, 2009). The OmpW protein was found to highly immunogenic in fish and was suggested to be useful as a vaccine. Another protein, OmpTS that shares a high degree of homology to the aeromonad adhesin Aha offers protection to fish (Khushiramani *et al.*, 2006). Recently, another recombinant OMP has been expressed (rOmp-G); again this protein is very similar to the Aha adhesin (Guan *et al.*, 2011), and thus OmpTS, suggesting that they are all in fact the same protein. This recombinant protein offers protection to eels against challenge by either *A. hydrophila* or *A. sobria* (Guan *et al.*, 2011). However, the roles of these proteins in aeromonad virulence has not been established, and it needs to be determined if OmpTS, Aha and OmpG are in fact the same protein.

In a non-piliated strain of *A. caviae* Ae398 that demonstrated aggregative adherence a 43 kDa OMP was thought to be the major adhesion. This *A. caviae* strain showed maximum adherence at 22°C and the 43 kDa OMP isolated from this strain bound specifically to HEp-2 cells (Rocha-de-Souza *et al.*, 2001). Furthermore, antibody raised against this protein was able to inhibit the adhesion of the parental bacterium to host cells (Rocha-de-Souza *et al.*, 2001). Two OMPs of 37 kDa and 48 kDa respectively were identified in *A. veronii* (Vazquez-Juarez *et al.*, 2004). Determination of the N-terminal amino acid sequence of the 48 kDa protein lead to its assignment in the Omp48 family similar to LamB. The purified *A. veronii*

Omp48 was shown to bind to a series of extracellular matrix proteins such as collagen and fibronectin as well as mucin and lactoferrin (Vazquez-Juarez *et al.*, 2004). Treatment of host cells with sugars could inhibit the interaction with Omp48, suggesting it has a lectin-like binding domain; furthermore, treatment of *A. veronii* with anti-Omp48 antibody could inhibit *A. veronii*, *A. hydrophila* and *A. caviae* adhesion to HeLa cells (Vazquez-Juarez *et al.*, 2004). Omp48 appears to be present in the vast majority of *Aeromonas* strains and again has been suggested to be a good vaccine candidate (Khushiramani *et al.*, 2008).

The outer membrane protein profile of *A. salmonicida* has been analysed, using two-dimensional electrophoresis followed by liquid-chromatography and mass spectrometry (LC/MS/MS). A total of 76 unique proteins were identified of which the surface array protein VapA dominated accounting for 60% of the OMPs (Ebanks *et al.*, 2005). In addition 10 porins and various nutrient acquisition proteins were identified although the role of these in pathogenesis is unknown (Ebanks *et al.*, 2005). Furthermore, *A. salmonicida* iron limitation has been shown to alter OMP expression in this bacterium and four iron-restricted OMPs (IROMPS) have been shown to be expressed *in vivo* and to be immunogenic in the Atlantic salmon (Hirst and Ellis, 1994; Ebanks *et al.*, 2004).

Certain strains of aeromonads have been reported to cause a dysentery-like illness, suggesting that these strains have the ability to invade the host cells of the gut epithelium. Strains of *Aeromonas* spp. have been reported to be invasive in several different cells lines (Gray *et al.*, 1990; Shaw *et al.*, 1995; Shaw *et al.*, 1997). However, nothing is known about the possible mechanisms of how this would occur. The published genome sequence of *A. hydrophila* ATCC7966^{T} (Seshadri *et al.*, 2006) and the draft genome of *A. caviae* Sch3 (unpublished) a strain reported to be invasive, have genes encoding or annotated as potential adhesins or invasins. In *A. hydrophila* ATCC 7966^{T} these are found in the locus AHA1064-AHA1066, and in *A. caviae* it is the genes *aigA* and *aigB*. However, similar genes have not been reported in other *Aeromonas* strains yet. The proteins encoded by *aigA*, *aigB* and *AHA1066* all have several bacterial immunoglobulin-like (Big) domains. Examples of proteins that also have these domains would be the invasin protein of *Yersinia* spp. and intimin of enteropathogenic *E. coli* (Hamburger *et al.*, 1999; Luo *et al.*, 2000), these proteins are involved in invasion and intimate adherence to host cells respectively. These proteins tend to have a β-barrel domain at their N-terminus that anchors the protein in the outer membrane with other domains of the protein presented externally. The N-terminus is followed by a number of Big-domains (3–7) attached to a receptor domain at the C-terminus (Hamburger *et al.*, 1999; Luo *et al.*, 2000). The *Aeromonas* Big-like proteins (AigA, AigB and AHA1066) are different as they do not have the β-barrel domain and they appear to be lipo-proteins as they contain a lipidation sequence at their N-terminus. Moreover, instead of being presented externally, lipoproteins usually face into the periplasm from the inner leaflet of the outer-membrane, and would not be ideal as an adhesin or invasin, as they would not be in the correct orientation to be on the outer surface of the bacterium. In terms of sequence homology the aeromonad proteins are most similar to the Lig proteins of *Leptospira interrogans*, these are also Big-domain containing lipoproteins; however, they are thought to be on the outer leaflet of the outer membrane of this bacterium as they play an essential role in this organisms adherence and pathogenesis (Palaniappan *et al.*, 2002; Choy *et al.*, 2007). The Lig proteins of *L. interrogans* have been shown to bind a whole series of extracellular matrix proteins including, fibronectin, laminin, collagen and fibrinogen (Choy *et al.*, 2007). However the role of the Big-domain proteins in aeromonad virulence is unknown, it would be intriguing to known their role in *Aeromonas* biology. However, homologues of AigA and AigB have recently appeared in the unpublished genome sequence databases of *A. media* WS and *A. veronii* AMC34, respectively.

Conclusion

As one would expect from a bacterium that interacts with both diverse hosts and environments, various strains of *Aeromonas* have a plethora of surface structures that facilitate this process. We have summarized the external antigenic structures found on aeromonad strains that are thought to have roles in its pathogenesis in Fig. 6.1. In this chapter we have concentrated on the recent findings of these major structures. Since the previous book on *Aeromonas* over fifteen years ago there have been a number of aeromonad genomes sequenced that have been a great asset in helping laboratories to unravel the biology of this enigmatic pathogen. However, as with most things in life more answers give rise to yet more questions. Major in-roads have been taken in understanding the motility or lack of motility in some aeromonad strains. The genetics of both the polar and lateral flagella systems has been elucidated. *Aeromonas* have been shown to glycosylate their flagellar filament with the sugar pseudaminic acid, a process shared by other pathogens such as *Campylobacter* and *Helicobacter*. Although, the role of this glycosylation process is as yet unknown. We have also gained information on the aeromonads' ability to produce capsules and different types of LPS. This work has been complemented with recent findings about their multiple type IV and other pilus systems with regard to their adherence mechanisms. Isogenic mutants have shown the importance of the Bundle-forming MSHA type IV pilus system in adherence to both biotic and abiotic surfaces. But we do not yet understand the role of the other type-IV pilus system present in some strains of mesophilic aeromonads, such as the Flp-pilus system. Furthermore, more attention needs to be directed to the role of the type-1 Fim-like pilus system that has not been investigated for around twenty years. Other areas investigating the aeromonad cell envelope appear to have also stalled, little work has been

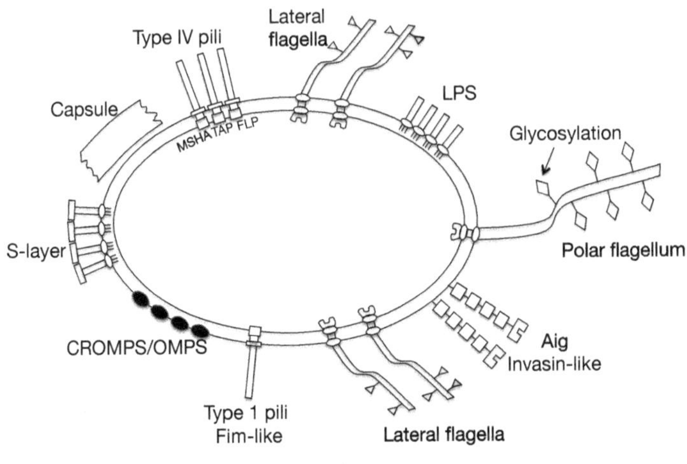

Figure 6.1 Schematic diagram of a stylized *Aeromonas* strain depicting the various cell envelope antigenic structures that have been implicated or suggested to be involved in its pathogenesis.

carried out on the S-layer in the last decade apart from investigating its potential as a vaccine candidate. With regards to the S-layer there are still many questions regarding the potential of post-translational modifications of the Surface protein and the biology and biochemistry associated with this process. The roles of the Omps and CROMPS need to be investigated genetically to be able to fully understand their roles in adherence and colonization. Furthermore, the new outer membrane protein candidates of the Lig-like proteins AigA and AigB that contain Big-domains have been recognized through genome analysis these have given us new targets to investigate for adherence and invasion as nothing is known about the roles of these proteins in *Aeromonas* biology. As more aeromonad genome sequences become available, the genes encoding these proteins are being recognized and found more frequently. We are sure by the time of the next book on the genus *Aeromonas* many of these areas of investigation will provide answers and many new questions.

References

Aguilar, A., Merino, S., Rubires, X., and Tomàs, J.M. (1997). The influence of osmolarity on lipopolysaccharide and virulence of *Aeromonas hydrophila* serotype O:34 strains grown at 37°C. Infect. Immun. *65*, 1245–1250.

Aguilar, A., Merino, S., Mercé, M., Regué, M., and Tomàs, J.M. (1999). Two genes from the capsule of *Aeromonas hydrophila* (serotype O:34) confer serum resistance to *Escherichia coli* K12 strains. Res. Microbiol. *150*, 395–402.

Altarriba, M., Merino, S., Gavin, R., Canals, R., Rabaan, A., Shaw, J.G., and Tomàs, J.M. (2003). A polar flagella operon (*flg*) of *Aeromonas hydrophila* contains genes required for lateral flagella expression. Microb. Pathog. *34*, 249–259.

Atkinson, H.M., and Trust, T.J. (1980). Haemagglutination properties and adherence ability of *Aeromonas hydrophila*. Infect. Immun. *27*, 938–946.

Atkinson, H.M., Adams, D., Savvas, R.S., and Trust, T.J. (1987). *Aeromonas* adhesin antigens. Experientia *43*, 372–374.

Barnett T.C., Kirov, S.M., Strom, M.S., and Sanderson, K. (1997). *Aeromonas* spp. possess at least two distinct type IV pilus families. Microb. Pathog. *23*, 241–247.

Bartkova, G., and Ciznar, I. (1994). Adherence pattern of non-piliated *Aeromonas hydrophila* strains to tissue cultures. Microbios. *77*, 47–55.

Belland, R.J., and Trust, T.J. (1987). Cloning of the gene for the surface array protein of *Aeromonas salmonicida* and evidence linking loss of expression to genetics deletion. J. Bacteriol. *169*, 4086–4091

Boyd J.M., Dacanay, A., Knickle, L.C., Touhami, A., Brown, L.L., Jericho, M.H., Johnson, S.C., and Reith, M. (2008). Contribution of Type IV pili to the virulence of *Aeromonas salmonicida* subsp. *salmonicida* in Atlantic salmon (*Salmo salar* L.). Infect. Immun. *76*, 1445–1455.

Braschler, T.R., Merino, S., Tomàs, J.M., and Graf, J. (2003). Complement resistance is essential for colonization of the digestive tract of *Hirudo medicinalis* by *Aeromonas* strains. Appl. Env. Microbiol. *69*, 4268–4271.

Canals, R., Altarriba, M., Vilches, S., Horsburgh, G., Shaw, J.G., Tomàs, J.M., and Merino, S. (2006a). Analysis of the lateral flagellar gene system of *Aeromonas hydrophila* AH-3. J. Bacteriol. *188*, 852–862.

Canals, R., Ramirez, S., Vilches, S., Horsburgh, G., Shaw, J.G., Tomàs, J.M., and Merino, S. (2006b). Polar flagellum biogenesis in *Aeromonas hydrophila*. J. Bacteriol. *188*, 542–555.

Canals, R., Vilches, S., Wilhelms, M., Shaw, J.G., Merino, S., and Tomàs, J.M. (2007). Non-structural flagella genes affecting both polar and lateral flagella-mediated motility in *Aeromonas hydrophila*. Microbiology *153*, 1165–1175.

Carello, A., Silburn, K.A., Budden, J.R., and Chang, B.J. (1988). Adhesion of clinical and environmental *Aeromonas* isolates to HEp-2 cells. J. Med. Microbiol. *26*, 19–27.

Choy, H.A., Kelley, M.M., Chen, T.L., Møller, A.K., Matsunaga, J., and Haake, D.A. (2007). Physiological osmatic induction of *Leptospira interrogans* adhesion: LigA and LigB bind extracellular matrix proteins and fibrinogen. Infect. Immum. *75*, 2441–2450.

Chu, S., Gustafson, C.E., Feutrier, J., Cavaignac, S., and Trust, T.J. (1993). Transcriptional analysis of the *Aeromonas salmonicida* S-layer protein gene *vapA*. J. Bacteriol. *175*, 7968–7975.

Dacanay, A., Boyd, J.M., Fast, M.D., Knickle, L.C., and Reith, M.E. (2010). The *Aeromonas salmonicida* type I pilus system contributes to host colonization but not invasion. Dis. Aquat. Organ. *88*, 199–206.

Ebanks, R.O., Dacanay, A., Goguen, M., Pinto, D.M., and Ross, N.W. (2004). Differential proteomic analysis of *Aeromonas salmonicida* outer membrane proteins in response to low iron and *in vivo* growth conditions. Proteomics *4*, 1074–1085.

Ebanks, R.O., Goguen, M., McKinnon, S., Pinto, D.M., and Ross, N.W. (2005). Identification of the major outer-membrane proteins of *Aeromonas salmonicida*. Dis. Aquat. Org. *68*, 29–38.

Esteve, C., Alcaide, E., Canals, R., Merino, S., Blasco, D., Figueras, M.J., and Tomás, J.M. (2004). Pathogenic *Aeromonas hydrophila* serogroup O: 14 and O: 81 strains with an S layer. Appli. Env. Microbiol. *70*, 5898–5904.

Franki, K.T., and Chang, B.J. (1994). Variable expression of O-antigen and the role of lipopolysaccharide as an adhesion in *Aeromonas sobria*. FEMS Microbiol. Lett. *122*, 97–102.

Garduno, R.A., and Kay, W.W. (1995). Capsulated cells of *Aeromonas salmonicida* grown *in vitro* have different functional properties than capsulated cells grown *in vivo*. Can. J. Microbiol. *41*, 941–945.

Garduno, R.A., Lee, E.J.Y., and Kay, W.W. (1992). S-layer-mediated association of *Aeromonas salmonicida* with murine macrophages. Infect. Immun. *60*, 4373–4382.

Garrote, A., Bonet, R., Merino, S., Simon-Pujol, M.D., and Gongregado, F. (1992). Occurrence of capsule in *Aeromonas salmonicida*. FEMS Microbiol. Lett. *95*, 127–132.

Gavin, R., Rabaan, A.A., Merino, S., Tomás, J.M., Gryllos, I., and Shaw, J.G. (2002). Lateral flagella of *Aeromonas* species are essential for epithelial cell adherence and biofilm formation. Mol. Microbiol. *43*, 383–397.

Gavin R., Merino, S.,Altarriba, M., Canals, R., Shaw, J.G., and Tomás, J.M. (2003). Lateral flagella are required for increased cell adherence, invasion and biofilm formation by *Aeromonas* spp. FEMS Microbiol. Lett. *224*, 77–83.

Gray, S.J., Stickler, D.J., and Bryant, T.N. (1990). The incidence of virulence factors in mesophilic *Aeromonas* species isolated from farm animals and their environment. Epidemiol. Infect. *105*, 277–294.

Griffiths, S.G., and Lynch, W.H. (1990). Characterization of *Aeromonas salmonicida* variants with altered cell surfaces and their use in studying surface protein assembly. Arch. Microbiol. *154*, 308–312.

Gryllos, I., Shaw, J.G., Gavin, R., Merino, S., and Tomás, J.M. (2001). Role of *flm* locus in mesophilic *Aeromonas* adherence. Infect. Immun. *69*, 65–74.

Guan, R., Xiong, J., Huang, W., and Guo, S. (2011). Enhancement of protective immunity in European eel (*Angulla anguilla*) against *Aeromonas hydrophila* and *Aeromonas sobria* by recombinant *Aeromonas* outer membrane protein. Acta Biochim. Biophys. Sin. *43*, 79–88.

Hadi, N., Yang, Q., Barnett, T.C., Tabei, S.M.B., Kirov, S.M., and Shaw, J.G. (2012). Bundle-forming pilus locus of *Aeromonas veronii* bv sobria. Infect. Immun. *80*, 1351–1360.

Hamburger, Z.A., Brown, M.S., Isberg, R.R., and Bjorkman, P.J. (1999). Crystal structure of Invasin: bacterial integrin-binding protein. Science *286*, 291–295

Hayashi, F., Smith, K.D., Ozinsky, A., Hawn, T.R., Yi, E.C., Goodlett, D.R., Eng, J.K., Akira, S., Underhill, D.M., and Aderem, A. (2001). The innate immune response to bacterial flagellin is mediated by Toll-like receptor 5. Nature *410*, 1099–1103.

Hirst, I.D., and Ellis, A.E. (1994). Iron-regulated outer membrane proteins of *Aeromonas salmonicida* are important protective antigens in Atlantic salmon against furunculosis. Fish Shellfish Immunol. *4*, 29–45.

Ho, A.S.Y., Miietzner, T.A., Smith, A.J., and Schoolnik, G.K. (1990). The pili of *Aeromonas hydrophila*: identification of an environmentally regulated 'mini pili'. J. Exp. Med. *172*, 795–806.

Iwanaga, M., and Hokama, A. (1992). Characterisation of *Aeromonas sobria* TAP13 pili: a possible new colonization factor. J. Gen. Microbiol. *138*, 1913–1919.

Jimenez, N., Canals, R., Lacasta, A., Kondakova, A.N., Lindner, B., Knirel, Y.A., Merino, S., Regué, M., and Tomás, J.M. (2008a). Molecular analysis of three *Aeromonas hydrophila* AH-3 (serotype O:34) lipopolysaccharide core biosynthesis gene cluster. J. Bacteriol. *190*, 3176–3184.

Jimenez, N., Canals, R., Saló, M.T., Vilches, S., Merino, S., and Tomás, J.M. (2008b). The *Aeromonas hydrophila wb**O:34 gene cluster: genetics and temperature regulation. J. Bacteriol. *190*, 4198–4209.

Josenhans, C., Vossebein, L., Friedrich, S., and Suerbaum, S. (2002). The *neuA/flmD* gene cluster of *Helicobacter pylori* is involved in flagellar biosynthesis and flagellin glycosylation. FEMS Microbiol. Lett. *210*, 165–172.

Kay, W.W., Buckley, J.T., Ishiguro, E.E., Phipps, B.M., Monette, J.P.L., and Trust, T.J. (1981). Purification and disposition of a surface protein associated with virulence of *Aeromonas salmonicida*. J. Bacteriol. *147*, 1077–1084.

Khashe, S., Hill, W., and Janda, M.J. (1996). Characterisation of *Aeromonas hydrophila* strains of clinical, animal and environmental origin expressing the O:34 antigen. Curr. Microbiol. 33, 104–108.

Khushiramani, R., Girisha, S.K., Karunasagar, I., and Karunasagar, I. (2006). Protective efficacy of recombinant OmpTS protein of *Aeromonas hydrophila* in Indian major carp.Vaccine 25, 1157–1158.

Khushiramani, R., Girisha, S.K., Bhowmick, P.P., Karunasagar, I., and Karunasagar, I. (2008). Prevalence of different outer membrane proteins in isolates of *Aeromonas* species. World. J. Microbiol. Biotechnol. *24*, 2263–2268.

Kirov, S.M., and Sanderson, K. (1995). *Aeromonas* cell line adhesion, surface structures and *in vivo* models of intestinal colonization. Med. Microbiol. Lett. *4*, 305–315.

Kirov S.M., and Sanderson, K. (1996). Characterisation of a type IV bundle-forming pilus (SFP) from a gastroenteritis-associated strain of *Aeromonas veronii* biovar sobria. Microb. Pathog. *21*, 23–34.

Kirov, S.M., Jacobs, I., Hayward, L.J., and Hapin, R.H. (1995). Electron-microscopic examination of factors influencing the expression of filamentous surface structures on clinical and environmental isolates of *Aeromonas veronii* biotype sobria. Microbiol. Immunol. 39, 329–338.

Kirov S.M., O'Donovan, L.A., and Sanderson, K. (1999). Functional characterisation of type IV pili expressed on diarrhea-associated isolates of *Aeromonas* species. Infect. Immun. *67*, 5447–5454.

Kirov, S.M., Barnett, T.C., Pepe, C.M., Strom, M.S., and Albert, M.J. (2000). Investigation of the role of type IV *Aeromonas* pilus (Tap) in the pathogenesis of *Aeromonas* gastrointestinal infection. Infect. Immun. *68*, 4040–4048.

Kirov, S.M., Tassell, B.C., Semmler, A.B., O'Donovan, L.A., Rabaan, A.A., and Shaw, J.G. (2002). Lateral flagella and swarming in *Aeromonas* species. J. Bacteriol. *184*, 547–555.

Kirov S.M., Castrisios, M., and Shaw, J.G. (2004). *Aeromonas* flagella (polar and lateral) are enterocyte adhesins that contribute to biofilm formation on surfaces. Infect. Immun. *72*, 1939–1945.

Knirel, Y.A., Shaskov, A.S., Senchenkova, S.N., Merino, S., and Tomás, J.M. (2002). Structure of the O-polysaccharide of *Aeromonas hydrophila* O:34: a case of random O-acetylation of 6-deoxy-L-talose. Carbohydr. Res. 337, 1381–1386

Knirel, Y.A., Vinogradov, E., Jimenez, N., Merino, S., and Tomás, J.M. (2004). Structural studies on the R-type lipopolysaccharide of *Aeromonas hydrohila*. Carbohydr. Res. *339*, 787–793.

Kokka, R.P., Vedros, N.A., and Janda, M.J. (1990). Electrophoretic analysis of the surface components of autoagglutinating surface array protein-positive and surface array protein-negative *Aeromonas hydrophila* and *Aeromonas sobria*. J. Clin Microbiol. *28*, 2240–2247.

Kokka, R.P., Vedros, N.A., and Janda, M.J. (1991). Characterisation of classic and atypical serogroup O:11 *Aeromonas*: evidence that the surface array protein is not directly involved in mouse pathogenicity. Microb. Pathog. *10*, 71–79

LeClerc, G., Wang, S., and Ely, B. (1998). A new class of *Caulobacter crescentus* flagellar genes. J. Bacteriol. *180*, 5010–5019.

Linnerborg, M., Widmalm, G., Rahman, M.M., Jansson, P., Holme, T., Qadri, F., and Albert, M.J. (1996). Structural studies of the O-antigenic polysaccharide from an *Aeromonas* caviae strain. Carbohydr. Res. *291*, 165–174.

Logan, S.M., (2006). Flagellar glycosylation – a new component of the motility repertoire? Microbiology *152*, 1249–1262.

Luo, Y., Frey, E.A., Pfuetzner, R.A., Greagh, A.L., Knoechel, D.G., Haynes, C.A., Finlay, B.B., and Strynadka, N.C.J. (2000). Crystal structure of enteropathogenic *Escherichia coli* intimin–receptor complex. Nature *405*, 1073–1077.

McIntosh, D., and Austin, B. (1991). Atypical characteristics of the salmonid pathogen *Aeromonas salmonicida*. J. Gen. Microbiol. *137*, 1341–1343.

Maiti, P., Raghunath, P., Karunasagar, I., and Karunasagar, I. (2009). Cloning and expression of an outer membrane protein OmpW of *Aeromonas hydrophila* and study of its distribution in *Aeromonas* spp. J. Appl. Microbiol. *107*, 1157–1167.

Marsh, J.W., and Taylor, R.K. (1999). Genetic and transcriptional analyses of the *Vibrio cholerae* mannose-sensitive hemagglutinin type 4 pilus locus. J. Bacteriol. *181*, 1110–1117.

Martinez, M.J., Pujol, D.S., Congregado, F., Merino, S., Rubires, X., and Tomás, J.M. (1995). The presence of capsular polysaccharide in mesophilic *Aeromonas hydrophila* O:11 and O:34. FEMS. Microbiol. Lett. *128*, 69–74.

Merino, S., Camprubi, S., and Tomás, J.M. (1991). The role of lipopolysaccharide in complement-killing of *Aeromonas hydrophila* strains of serotype O:34. J. Gen. Microbiol. *137*, 1583–1590.

Merino, S., Camprubi, S., and Tomás, J.M. (1992). Effect of growth temperature on outer membrane components and virulence of *Aeromonas hydrophila* strains of serotype O:34. Infect. Immun. *60*, 4343–4349.

Merino, S., Rubires, X., Konkel, S., and Tomás, J.M. (1995). Emerging pathogens: *Aeromonas* spp. Inter. J. Food. Microbiol. *28*, 157–168.

Merino, S., Rubires, X., Aguilar, A., Alberti, S., Hernandez-Alles, S., Benedi, V.J., and Tomás, J.M. (1996a). Mesophilic *Aeromonas* sp. serogroup O:11 resistance to complement-mediated killing. Infect. Immun. *64*, 5302–5309.

Merino, S., Rubires, X., Aguilar, A., Guillot, J.F., and Tomás, J.M. (1996b). The role of the O-antigen lipopolysaccharide on the colonization *in vivo* of the germ free chicken gut by *Aeromonas hydrophila* serogroup O:34. Microb. Pathog. *20*, 325–333.

Merino, S., Aguilar, A., Rubires, X., Abitiu, N., Regue, M., and Tomás, J.M. (1997). The role of the capsular polysaccharide of *Aeromonas hydrophila* serogroup O:34 in the adherence to and invasion of fish cell lines. Res. Microbiol. *148*, 625–631.

Merino, S., Nogueras, M.M., Aguilar, A., Rubires, X., Alberti, S., Beneti, V.J., and Tomás, J.M. (1998). Activation of the complement classical pathway (C1q binding) by mesophilic *Aeromonas hydrophila* outer membrane protein. Infect. Immun. *66*, 3825–3831.

Merino S., Gavin, R., Vilches, S., Shaw, J.G., and Tomás, J.M. (2003). A colonization factor (production of lateral flagella) of mesophilic *Aeromonas* spp. is inactive in *Aeromonas salmonicida* strains. Appl. Env. Microbiol. *69*, 663–667

Merino, S., Shaw, J.G., and Tomás, J.M. (2006). Bacterial lateral flagella: an inducible flagella system. FEMS Microbiol. Lett. *263*, 127–135.

Molero, R., Wilhelms, M., Infanzón., Tomás, J.M., and Merino, S. (2011). *Aeromonas hydrophila motY* is essential for polar flagellum function and requires coordinate expression of *motX* and Pom proteins. Microbiol. *157*, 2772–2784.

Murray, R.G.E., Dooley, J.S.G., Whippey, P.W., and Trust, T.J. (1988). Structure of an S layer on a pathogenic strain of *Aeromonas hydrophila*. J. Bacteriol. *170*, 2625–2630.

Nandapalan, N., and Chang, B.J. (1989). Production and characterization of monoclonal antibodies to *Aeromonas sobria* surface antigens. FEMS Microbiol. Immunol. *47*, 515–524.

Nazarenko, E.L., Crawford, R.J., and Ivanova, E.P. (2011). The structural diversity of carbohydrate antigens of selected Gram-negative marine bacteria. Mar. Drugs *9*, 1914–1954.

Noonan, B., and Trust, T.J. (1995a). The leucine zipper of *Aeromonas salmonicida* AbcA is required for the transcriptional activation of the P2 promoter of the surface-layer structural gene, *vapA*, in *Escherichia coli*. Mol. Microbiol. *17*, 379–386.

Noonan, B., and Trust, T.J. (1995b). Molecular analysis of an A-protein secretion mutant of *Aeromonas salmonicida* reveals a surface layer-specific protein secretion pathway. J. Mol. Biol. *248*, 316–327.

Noonan, B., and Trust, T.J. (1997). The synthesis, secretion and role in virulence of the paracrystalline surface layers of *Aeromonas salmonicida* and *A. hydrophila*. FEMS Microbiol. Lett. *154*, 1–7.

Palaniappan, R.U.M., Chang, Y.-F., Jusuf, S.S.D., Artiushin, S., Timoney, J.F., McDonough, S.P., Barr, S.C., Divers, T.J., Simpson, K.W., McDonough, P.L., and Mohammed, H.O. (2002). Cloning and molecular characterization of an immunogenic LigA protein of *Leptospira interrogans*. Infect. Immun. *70*, 5924–5930.

Parker, J.L., Day-Williams, M.J., Tomás, J.M., Stafford, G.P., and Shaw, J.G. (2012). Identification of a putative glycosyltransferase responsible for the transfer of pseudaminic acid onto the polar flagellin of *Aeromonas caviae* Sch3N. Microbiologyopen *1*, 149–160.

Pepe, C.M., Eklund, M.W., and Strom, M.S. (1996). Cloning of an *Aeromonas hydrophila* type IV pilus biogenesis gene cluster: complementation of pilus assembly functions and characterisation of a type IV leader peptidase/N-methyltransferase required for extracellular secretion. Mol. Microbiol. *19*, 857–869.

Poobalane, S., Thompson, K.D., Diab, A., László, A., Jeney, G., and Adams, A. (2008). Protein expression by *Aeromonas hydrophila* during growth *in vitro* and *in vivo*. Microb. Pathog. *45*, 60–69.

Quinn, D.F., Atkinson, H.M., Bretag, A.H., Tester, M., Trust, T.J., Wong, C.Y.F., and Flower, R.L.P. (1994). Carbohydrate-reactive pore-forming outer membrane proteins of *Aeromonas hydrophila*. Infect. Immun. *62*, 4054–4058.

Rabaan, A.A., Gryllos, I., Tomás, J.M., and Shaw, J.G. (2001). Motility and the polar flagellum are required for *Aeromonas caviae* adherence to HEp-2 cells. Infect. Immun. *69*, 4257–4267.

Rahman, M.H., and Kawai, K. (2000). Outer membrane proteins of *Aeromonas hydrophila* induce protective immunity in goldfish. Fish Shellfish Immunol. *10*, 379–982

Reith M.E., Singh R.K., Curtis, B., Boyd, J.M., Bouevitch, A., Kimball, J., Munholland, J., Murphy, C., Sarty, D., Williams, J., *et al.* 2008). The genome of *Aeromonas salmonicida* subsp. *salmonicida* A449: insights into the evolution of a fish pathogen. BMC Genomics 9, 427–441.

Rocha-de-Souza, C.M., Colombo, A.V., Hirata, R., Mattos-Guaraldi, A.L., Monteiro-Leal, L.H., Previato, J.O., Freitas, A.C., and Andrade, A.F.B. (2001). Identification of a 43-kDa outer-membrane protein as an adhesin in *Aeromonas caviae*. J. Med. Microbiol. *50*, 313–319.

Sakazaki, R., and Shimada, T. (1984). O-serogrouping scheme for mesophilic *Aeromonas* strains. Japan. J. Med. Sci. Biol. *37*, 247–255.

Santos, P.G., Santos, P.A., Bello, A.R., and Freitas-Almeida A.C. (2011). Association of *Aeromonas* caviae polar and lateral flagella with biofilm formation. Lett. Appl. Microbiol. *52*, 49–55.

Schirm, M., Soo, E.C., Aubry, A.J., Austin, J., Thibault, P., and Logan, S.M. (2003). Structural, genetic and functional characterization of the flagellin glycosylation process in *Helicobacter pylori*. Mol. Microbiol. *48*, 1579–1592.

Schirm, M., Arora, S.K., Verma, A., Vinogradov, E., Thibault, P., Ramphal, R., and Logan, S.M. (2004). Structural and genetic characterization of glycosylation of type a flagellin in *Pseudomonas aeruginosa*. J. Bacteriol. *186*, 2523–2531.

Seshadri, R., Joseph, S., Chopra, A., Sha, J., Shaw, J., Graf. J., Haft, G., Wu, M., Ren, Q., Rosovitz, M.J., *et al.* (2006). Genome sequence of *Aeromonas hydrophila* ATCC 7966^{T}: Jack of all trades. J. Bacteriol. *188*, 8272–8282.

Shaw, J.G., Thornley, J.P., Palmer, I., and Geary, I. (1995). Invasion of tissue culture cells by *Aeromonas caviae*. Med. Microbiol. Lett. *4*, 324–331

Shaw, J., Thornley, J., and Eley, A. (1997). Adherence and invasion of *Aeromonas caviae* to monolayer cells. Adv. Exp. Med. Biol. *412*, 217–219.

Shimada, T., Sakazaki, R., and Suzuki, K. (1985). Peritrichous flagella in mesophilic strains of *Aeromonas*. Jpn. J. Med. Sci. Biol. *38*, 141–145.

Smith, K.D., Andersen-Nissen, E., Hayashi, F., Strobe, K., Bergman, M.A., Barrett, S.L., Cookson, B.T., and Aderem, A. (2003). Toll-like receptor 5 recognizes a conserved site on flagellin required for protofilament formation and bacterial motility. Nat. Immunol. *4*, 1247–1253.

Tabei, S.M.B., Hitchen, P.G., Day-Williams, M.J., Merino, S., Vart, R., Pang, P.-C., Horsburgh, G.J., Viches, S., Wilhelms, M., Tomás, J.M., *et al.* (2009). An *Aeromonas caviae* genomic island is required for both O-antigen lipopolysaccharide biosynthesis and flagellin glycosylation. J. Bacteriol. *191*, 2851–2863.

Terashima, H., Kojima, S., and Homma, M. (2008). Flagellar motility in bacteria structure and function of flagellar motor. Int. Rev. Cell Mol. Biol. *270*, 39–85.

Thibault, P., Logan, S.M., Kelly, J.F., Brisson, J.-R., Ewing, C.P., Trust, T.J., and Guerry, P. (2001). Identification of the carbohydrate moieties and glycosylation motifs in *Campylobacter jejuni* flagellin. J. Biol. Chem. *276*, 34862–34870.

Thomas, S.R., and Trust, T.J. (1995). A specific PulD homolog is required for the secretion of paracrystalline surface array subunits in *Aeromonas hydrophila*. J. Bacteriol. *177*, 3932–3939.

Thomas, L.V., Gross, R.J., Cheasty, T., and Rowe, B. (1990). Extended serogrouping scheme for motile mesophilic *Aeromonas* species. J. Clin. Microbiol. *28*, 980–984.

Thomas, S., Austin, J.W., McCubbin, W.D., Kay, C.D., and Trust, T.J. (1992). Roles of structural domains in the morphology and surface anchoring of the tetragonal paracrystalline array of *Aeromonas hydrophila*. J. Mol. Biol. *228*, 652–661.

Thornley, J.P., Shaw, J.G., Gryllos, I.A., and Eley, A. (1996). Adherence of *Aeromonas caviae* to human cell lines HEp-2 and Caco-2. J. Med. Microbiol. *45*, 445–451.

Tso, M.D., and Dooley, J.S.G. (1995). Temperature-dependent protein and lipopolysaccharide expression in clinical *Aeromonas* isolates. J. Med. Microbiol. *42*, 32–38.

Umelo, E., and Trust, T.J. (1997). Identification and molecular characterization of two tandemly located flagellin genes from *Aeromonas salmonicida* A449. J. Bacteriol. *179*, 5292–5299.

Vázquez-Juárez, R.C., Romero, M.J., and Ascencio, F. (2004). Adhesive properties of a LamB-like outer-membrane proteins and its contribution to *Aeromonas veronii* adhesion. J. Appl. Microbiol. *96*, 700–708.

Verma, A., Schirm, M., Arora, S.K., Thibault, P., Logan, S.M., and Ramphal, R. (2006). Glycosylation of b-type flagellin of *Pseudomonas aeruginosa*: structural and genetic basis. J. Bacteriol. *188*, 4395–4403.

Waldor, M.K., and Mekalanos, J.J. (1996). Lysogenic conversion by a filamentous phage encoding cholera toxin. Science 272, 1910–1914.

Wang, Z., Li, J., Vingradov, E., and Altman, E. (2006). Structural studies of the core region of *Aeromonas salmonicida* subsp. salmonicida lipolysaccharide. Carbohydr. Res. *341*, 109–117.

Wilhelms, M., Vilches, S., Molero, R., Shaw, J.G., Tomás, J.M., and Merino, S. (2009). Two redundant sodium-driven stator motor proteins are involved in *Aeromonas hydrophila* polar flagellum rotation. J. Bacteriol. *191*, 2206–2217.

Wilhelms, M., Molero, R., Shaw, J.G., Tomás, J.M., and Merino, S. (2011). Transcriptional hierarchy of *Aeromonas hydrophila* polar-flagellum genes. J. Bacteriol. *193*, 5179–5190.

Wilhelms, M., Fulton, K.M., Twine, S.M., Tomás, J.M., and Merino, S. (2012). Differential glycosylation of polar and lateral flagellins in *Aeromonas hydrophila* AH-3. J. Biol. Chem. *287*, 27851–27862.

Yeh, H.Y., and Klesius, P.H. (2011). Over-expression, purification and immune responses to *Aeromonas hydrophila* AL09–73 flagellar proteins. Fish and Shellfish Immunol. *31*, 1278–1283

Zhang, Y.L., Arakawa, E., and Leung, K.Y. (2002). Novel *Aeromonas hydrophila* PPD134/91 genes involved in O-antigen and capsule biosynthesis. Infect. Immun. *70*, 2326–2335.

Zhang, Y.L., Lau, Y.L., Arakawa, E., and Leung, K.Y. (2003). Detection and genetic analysis of group II capsules in *Aeromonas hydrophila*. Microbiology *149*, 1051–1060.

New Developments on the Virulence Mechanisms of *Aeromonas hydrophila*

7

Elena V. Kozlova, Duraisamy Ponnusamy and Ashok K. Chopra

Abstract

This chapter covers new virulence factors/mechanisms that are associated with infections caused by *Aeromonas* species. We focused primarily on the type 3 secretion system (T3SS) effector AexU, which has ADP-ribosyltransferase and GTPase-activating protein activities and operates by inhibiting NF-κB in the host cells. Emphasis was also placed on the T6SS and its effectors haemolysin coregulated protein (Hcp) and valine-glycine-repeat G (VgrG) family protein members in terms of their roles in the pathogenesis of *Aeromonas* infections. While Hcp inhibits bacterial phagocytosis, VgrG1 through its carboxyl-terminal vegetative insecticidal protein-2 (VIP-2) domain, leads to ADP ribosylation of host cell actin, thus preventing its polymerization and cytotoxicity in the eukaryotic cells. Finally, the role is discussed of three quorum-sensing systems (AI-1-AI-3) in modulating bacterial virulence through second messenger cyclic-diguanylate monophosphate (c-di-GMP).

Introduction

Various virulence mechanisms and factors, as well as their molecular characterization both at the genetic and phenotypic level from species (spp.) of *Aeromonas*, have been described (Chopra *et al.*, 2009; Chopra and Houston, 1999; Fadl *et al.*, 2006; Galindo *et al.*, 2006; Sierra *et al.*, 2010). This book chapter provides updated information on the new type 3 secretion system (T3SS) effector AexU of *A. hydrophila* SSU, a diarrhoeal isolate, the T6SS and its effectors, and the quorum sensing (QS) mechanisms found in this isolate and how they dictate bacterial virulence.

T3SS

In *Aeromonas* spp., the presence of a plasmid encoding T3SS was first reported in *A. salmonicida*, a fish pathogen, and four T3SS effector proteins (AexT, AopP, AopH, and AopO) were identified (Braun *et al.*, 2002; Dacanay *et al.*, 2006; Fehr *et al.*, 2006). AexT is a homologue of ExoT/S, a bifunctional toxin of *Pseudomonas aeruginosa*, with a GTPase-activating protein (GAP) NH_2-terminal and an ADP-ribosyltransferase (ADPRT) COOH-terminal domain (Braun *et al.*, 2002; Pederson *et al.*, 1999; Sundin *et al.*, 2004). AopP is a member of the YopJ family, a group of T3SS effector proteins that interfere with the mitogen-activated protein kinase (MAPK) and/or the nuclear factor kappa B (NF-κB) signalling pathway.

However, unlike YopJ, AopP only inhibits the NF-κB pathway downstream of IκB kinase (IKK) activation, but does not inhibit the MAPK signalling pathway (Fehr *et al.*, 2006). The biological functions of AopO and AopH have not been fully elucidated yet; however, they are homologues of YopH and YopO in *Yersinia enterocolitica*. Interestingly, AexT is encoded on the chromosome, while the AopP, AopO, and AopH are all present on the plasmid in *A. salmonicida* (Braun *et al.*, 2002; Dacanay *et al.*, 2006; Fehr *et al.*, 2006).

While the T3SS gene clusters were identified in other spp. of *Aeromonas* (Dacanay *et al.*, 2006), we have focused here on a new T3SS effector (designated as AexU) found in the diarrheal isolate SSU of *A. hydrophila* (Sha *et al.*, 2005). This chromosomally located *aexU* gene coded for a 512 amino acid (aa)-long protein, with its NH_2-terminus (aa residues 1–231) exhibiting a 67% homology with the NH_2-terminus of AexT from *A. salmonicida*. Importantly, its COOH-terminus (aa residues 232–512) had no homology with any known functional proteins in the NCBI database; however, the full-length AexU retained ADPRT activity, associated with the carboxyl-terminal domains of other known ADP-ribosylating toxins, such as AexT and ExoT/S (Braun *et al.*, 2002; Sundin *et al.*, 2004).

The *ΔaexU* mutant of *A. hydrophila* SSU was attenuated in a septicaemic mouse model of infection, and we detected specific anti-AexU antibodies in the sera of mice that survived challenge by the wild-type (WT) bacterium (Sha *et al.*, 2007). Interestingly, both the full-length and the NH_2- and COOH- terminal domains of AexU exhibited ADPRT activity (Sha *et al.*, 2007; Sierra *et al.*, 2007). When the *aexU* gene was expressed and produced in the HeLa Tet-Off cell system, the host cells underwent actin reorganization and cell rounding, followed by their apoptosis (Sha *et al.*, 2007; Sierra *et al.*, 2007). These effects of AexU were more significantly attributed to its associated GAP activity rather than to the ADPRT activity (Sierra *et al.*, 2007, 2010). Based on cell signalling studies, AexU operated by inhibiting NF-κB and inactivating Rho GTPases, which could down-regulate IL-6 and IL-8 secretion from HeLa cells (Sierra *et al.*, 2010). Interesting, however, were our findings that when the Δ*aexU* null mutant of *A. hydrophila* SSU was complemented with the *aexU* gene devoid of both ADPRT and GAP activities, it became more virulent in a mouse model compared to the WT bacterium having the native *aexU* gene, with a concomitant increase in the production of pro-inflammatory cytokines (Sierra *et al.*, 2010). Thus, ADPRT and GAP activities of AexU seemed to negatively modulate host cytokine responses by influencing an as yet unknown function of AexU and needs further investigation (Sierra *et al.*, 2010).

T6SS

Currently, the majority of the information on this new secretion system is available through our studies on *A. hydrophila* SSU (Suarez *et al.*, 2008, 2010a,b). The T6SS is organized in a cluster of 20 different genes which contribute to either structural or functional elements of the system (Pukatzki *et al.*, 2009; Suarez *et al.*, 2008). The genes in the T6SS cluster of SSU code for some of the known protein homologues, such as haemolysin coregulated protein (Hcp), valine-glycine-repeat (Vgr)-G2 and -G3 proteins, virulence-associated secretion (Vas) proteins -A, -F, -H and -K, ATPase associated with diverse cellular activities (AAA+) protein ClpB, and many hypothetical proteins (Suarez *et al.*, 2008). The *vgrG1* gene is located outside of the T6SS gene cluster on the chromosome (Suarez *et al.*, 2010a). Proteins VasF and VasK are otherwise commonly referred to as IcmH- and IcmF-like proteins, respectively, which are members of IcmF-associated, homologous proteins (IAHP) and present in a variety of bacterial pathogens, such as *Yersinia pestis* CO92 and KIM, *Salmonella* Typhimurium, *Escherichia coli* O157:H7, and *P. aeruginosa* (Cascales, 2008; Das and Chaudhuri, 2003).

Protein sequences of VasF and VasH have led to the suggestion that they could be part of the T6SS structural components forming a conduit between the cytoplasm and the periplasm of the bacterium (Cascales, 2008). However, VasH could also be a transcriptional regulator controlling expression of the T6SS effectors (Suarez *et al.*, 2008). Deletion of the *vasK* gene from *A. hydrophila* SSU blocked Hcp secretion to the extracellular media, but the mutation did not interfere with either the translocation of Hcp to the eukaryotic cytoplasm or its expression in the bacterium (Suarez *et al.*, 2008), suggesting architectural conformation of VasK in T6SS that allows Hcp to pass freely into the extracellular environment in the absence of eukaryotic contact. Further, the *vasK* mutant was readily phagocytized by macrophages and exerted less cytotoxicity when compared with the results in the WT bacterium. In a septicaemic mouse model of infection, this mutant was highly attenuated in its virulence (Suarez *et al.*, 2008). These findings indicated that macrophage-mediated elimination of *A. hydrophila* could be the deciding factor for the subsequent disease outcomes associated with this pathogen.

It is likely that the anti-phagocytic property of VasK is related either to the obstruction of the macrophage phagocytic processes by VasK alone or in combination with other proteins localized on the surface of the bacterium. It is also possible that VasK directs Hcp to the extracellular environment to impair macrophage functions. VasH also exhibits pathogenic characteristics in the mouse model similar to those with VasK; the *A. hydrophila* SSU Δ*vasH* mutant showed increased phagocytic uptake by macrophages and was highly attenuated in a mouse model of infection. However, in contrast to the Δ*vasK* mutant, deletion of the *vasH* gene from *A. hydrophila* SSU led to the total absence of Hcp and VgrG2 and -3 in bacteria (Suarez *et al.*, 2008, 2010a), indicating that VasH could play a role in the transcriptional activation of genes *hcp* and *vgrG2* and *vgrG3*. In agreement with this speculation, the amino acid sequence of VasH reveals the presence of RNA polymerase binding sigma-54 domain and DNA binding helix–turn–helix motif (Morett and Segovia, 1993; Suarez *et al.*, 2008).

Although no proven model of T6SS exists to explain the mechanistic background of the system, a hypothetical model can be recreated from the available experimental findings (Fig. 7.1). Proteins VgrG and Hcp are believed to play a dual role in *A. hydrophila* SSU, both as the structural constituents and the effector proteins (Cascales, 2008; Morett and Segovia, 1993; Silverman *et al.*, 2012; Suarez *et al.*, 2010a,b). VgrG and Hcp have no known signal peptide or cleavage sites on their amino acid sequences, although VgrG and Hcp are detected inside the host cell cytoplasm when T6SS comes in contact with the eukaryotic cells or in the extracellular milieu in the absence of host cell contact (Rego *et al.*, 2010; Suarez *et al.*, 2008, 2010a,b). Hcp is possibly transported across the cytoplasmic membrane into the periplasm as a multiprotein complex in combination with VgrGs (Cascales, 2008). AAA+ family protein ClpB and transmembrane proteins VasF and VasH are believed to be involved in this transport process; by using ATP hydrolytic energy, ClpB could unfold VgrG and Hcp and push them across the cytoplasmic membrane into the periplasm through VasF and VasH (Cascales, 2008; Galan and Wolf-Watz, 2006; Pukatzki *et al.*, 2006; Silverman *et al.*, 2012; Yeo and Waksman, 2004).

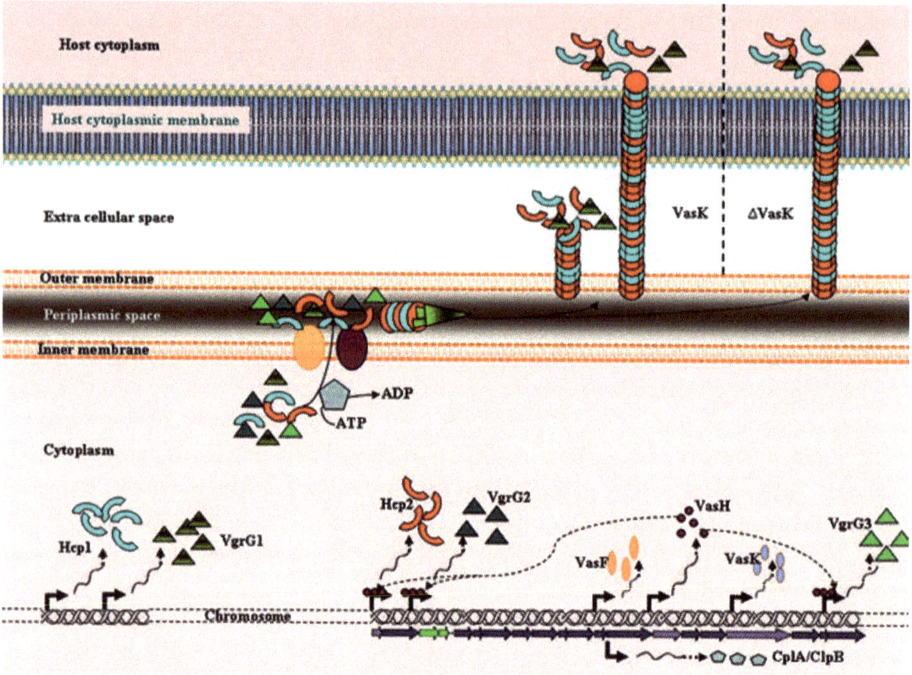

Figure 7.1 Schematic representation of T6SS in *A. hydrophila* SSU. This structural construct of *A. hydrophila* SSU T6SS is created based on the experimental evidence presented in references (Cascales, 2008; Morett and Segovia, 1993; Pukatzki *et al.*, 2006; Rego *et al.*, 2010; Silverman *et al.*, 2012; Suarez *et al.*, 2008, 2010a,b). Mechanistic aspects of the system have been explained in the text. ↱, possible promoter sequence.

In the periplasm, Hcp assembles into multiple hexameric rings that are subsequently stacked to form the tubular structure of T6SS. At the same time, VgrG forms a homotrimeric complex which is placed on the top of the growing tubular shaft. VgrG homotrimers may have a wider internal diameter on the side where they interact with tubular Hcp, but the diameter gradually becomes narrower towards the distal end of the complex, taking on a needle-like conformation (Cascales, 2008). This tubular structure is expected to grow from the periplasmic compartment through the outer bacterial membrane, and once the tip of it recognizes the eukaryotic cell membrane, both the tip and portion of the tubule are inserted and released into the cytoplasm of the host cells, where the VgrG and Hcp exert their effector roles (Fig. 7.1).

Interestingly, Hcp alone was unable to puncture the eukaryotic cell membrane, and neither the recombinant Hcp nor that purified from *A. hydrophila* SSU was able to translocate across the cytoplasmic membrane, although membrane binding was detected when mixed with RAW264.7 murine macrophages (Suarez *et al.*, 2010b). This finding clearly suggests that the needle-like structure of VgrG on the tip of the Hcp tubule is necessary to successfully puncture the host cell cytoplasmic membrane. In agreement with this, the structural organization of VgrG shows a striking homology to that of lambda phage tail tube protein gpV and T4 phage protein gp19, which aid in injection of phage DNA into the bacterial cytoplasm (Bonemann *et al.*, 2010; Leiman *et al.*, 2009; Pell *et al.*, 2009; Pukatzki *et al.*, 2007). Further, secretion of VgrG is mutually dependent on the presence of protein Hcp in the bacterium. In *A. hydrophila* SSU, mutational inactivation of the *vasH* gene blocked both the expression and the secretion of Hcp, but not the expression of the gene encoding VgrG1 located outside of the T6SS gene cluster. However, the secretion of VgrG1 did not occur in the absence of Hcp, indicating the interdependency of these proteins for secretion (Pukatzki *et al.*, 2006; Suarez *et al.*, 2010b).

Overall, these findings tempt us to hypothesize that VgrG and Hcp are translocated across the cytoplasmic membrane of *A. hydrophila* in an ATP-dependent manner. Later from the periplasm, the effector protein VgrG forms the cap on the developing Hcp tubule, and the whole structure grows until it reaches the eukaryotic cytoplasm or extracellular media without the help of being transported by another secretion system. This assumption is further strengthened by the findings that the shutdown of T3SS and/or flagellar apparatus did not affect either the extracellular secretion or the host cytoplasmic delivery of Hcp in *A. hydrophila* SSU (Suarez *et al.*, 2008).

Hcp from *A. hydrophila* SSU induces caspase-3 dependent apoptotic changes at the lower dose; on the other hand, at the higher concentration, Hcp hexamers formed intracellularly may embed themselves into the host cell membrane and create membrane pores, leading to necrotic cell death (Cascales, 2008; Mougous *et al.*, 2006; Suarez *et al.*, 2010b). Hcp preferably binds host macrophages to interfere with their phagocytic process; but to a lesser extent, Hcp also binds with Gr-1 positive granulocytes. Hcp did not show any role in granulocyte recruitment to the infection site, even though the experimental evidence supports the possibility that granulocyte activation could be hampered by Hcp (Suarez *et al.*, 2010b). The cytokine and chemokine profiles in mice treated with recombinant Hcp revealed the trend of macrophage deactivation, namely the down and up regulation of factors associated with macrophage activation and deactivation, respectively, such as IFN-γ, IL-1α, IL-10, macrophage colony-stimulating factor (M-CSF), granulocyte colony-stimulating factor (G-CSF), macrophage inflammatory protein (MIP)-1β and -2, transforming growth factor (TGF)-β, and keratinocyte-derived chemokine (KC) (Gordon, 2003; Hornef *et al.*, 2002; Suarez *et al.*, 2010b).

Among VgrG isomers in *A. hydrophila* SSU, VgrG1 is a relatively high molecular mass protein of 103 kDa, and VgrG2 and VgrG3 have molecular weights of approximately 77 kDa. Similar to Hcp translocation, the *vasH* mutant of *A. hydrophila* SSU was unable to deliver VgrG across the eukaryotic cytoplasmic membrane (Suarez *et al.*, 2010a). The NH_2-terminal VgrG domain is present in all of the isoforms of VgrG in *A. hydrophila* SSU and has a homology to the T4 phage tail spike proteins gp27 and gp5, which form the tip complex of T6SS (Bonemann *et al.*, 2010; Cascales, 2008; Suarez *et al.*, 2010a). The COOH-terminal of VgrG1 has homology with vegetative insecticidal protein (VIP)–2, which carries an ADPRT catalytic domain and is believed to freely float in the cytoplasm of the host cells upon insertion of the tip (Cascales, 2008; Suarez *et al.*, 2010a). The ADPRT activity disturbs actin polymerization, resulting in changes in the normal architectural features of epithelial cells into round phenotypes. Further, the COOH-terminal domain of VgrG1 showed the induction of caspase-9 dependent apoptotic changes when the protein was expressed in epithelial cells (Suarez *et al.*, 2010a), which may mean that VgrG1 has a possible virulence role during *A. hydrophila* SSU infection.

Quorum sensing-dependent virulence regulation in *Aeromonas hydrophila*

Quorum sensing (QS) is a mechanism of controlling gene expression in response to an expanding bacterial population (Fuqua *et al.*, 1994). Swift *et al.* (1997) found the genes for the QS signal generator and response regulator in *A. hydrophila* and *A. salmonicida* and termed them *ahyRI* and *asaRI*, respectively. They encode proteins belonging to the LuxRI family. The major signal molecule synthesized by AhyI and AsaI was identified as *N*-(butanoyl)-L-homoserine lactone (BHL) (Swift *et al.*, 1997) and represented the auto-inducer-1 (AI-1) QS system (Mok *et al.*, 2003). Later investigation showed that the *luxRI* QS gene homologues are universally present in the genus *Aeromonas* independent of their origin (Jangid *et al.*, 2007). Our recent study demonstrated that, in addition to *ahyR* and *ahyI*, two more *ahyR* homologues (*ahyR2* and *ahyR3*) exist in the genome of *A. hydrophila* SSU (Kozlova *et al.*, 2011).

In general, AHLs consist of fatty acids, varying in length and substitution, and are linked via a peptide bond to a homoserine lactone moiety. AHLs are commonly synthesized by members of the LuxI family of proteins and sensed by members of the LuxR family of transcriptional regulators (Fuqua *et al.*, 1996). After AHLs reach a threshold level (correlating to a certain bacterial cell density), a complex of AHL with its cognate LuxR transcriptional regulator is formed, enabling its binding to DNA, and, thereby, altering expression of multiple virulence genes (Jimenez *et al.*, 2012).

Several virulence-associated phenotypes in bacterial pathogens have been shown to be controlled by their QS systems (Winzer and Williams, 2001) and include biofilm formation (Croxatto *et al.*, 2002), the production of virulence factors such as proteases (Swift *et al.*, 1997, 1999; Croxatto *et al.*, 2002), a haemolysin (Kim *et al.*, 2003), the T3SS (Henke and Bassler, 2004), an extracellular toxin (toxin T1) (Manefield *et al.*, 2000), and siderophores (Lilley and Bassler, 2000).

In *A. hydrophila*, the AhyR/C4-HSL-dependent QS system regulates both extracellular protease production and biofilm development (Swift *et al.*, 1999; Lynch *et al.*, 2002). Mutation in either *ahyI* or *ahyR* abolishes both serine- and metallo-protease activities, which could be restored in the *ahyI* mutant by the addition of exogenous C4-HSL (Swift *et al.*, 1999). During the biofilm mode of growth, an *ahyI* mutant fails to form large 'mushroom-like' microcolonies, characteristic of the differentiated biofilms made by the parental *A. hydrophila* strain. In contrast, mutation in the *ahyR* gene results in the formation of a biofilm which occupies a far greater proportion of the available surface area than does the parental strain (Lynch *et al.*, 2002). Similar to these results is the significant decrease in biofilm formation exhibited by the *A. hydrophila* YJ-1 *ahyI* mutant (Chu *et al.*, 2011). Importantly, AhyR acts as both a negative and a positive regulator of *ahyI* and hence of C4-HSL production in a growth phase-dependent manner (Kirke *et al.*, 2004).

In our study of *A. hydrophila* SSU, we noted protease production and solid surface-associated biofilm formation to be significantly reduced in the *ahyRI* mutant compared to that in the wild-type (WT) strain. Scanning electron microscopic observations revealed that the *ahyRI* mutant was also defective in the formation of structured biofilm, as it was less filamentous and produced a distinct exopolysaccharide on its surface when compared with the structured biofilm produced by the WT strain (Khajanchi *et al.*, 2009).

Recently, studies have shown that *Aeromonas* flagella facilitate adherence to (and possibly invasion) of the epithelial cell line HEp-2 as well as biofilm formation on plastic (micro-titre plate assay), which may mean that these structures could also be virulence determinants for enteropathogenic strains (Gavin *et al.*, 2002, 2003). *Aeromonas* species possess polar flagella for swimming motility and lateral flagella for swarming motility (Kirov, 2003; Kirov *et al.*, 2002, 2004). Both flagella and pili of *A. caviae* have been linked to adherence to epithelial cells *in vitro* (Kirov *et al.*, 1998; Rabaan *et al.*, 2001), but lipopolysaccharide and membrane proteins have also been implicated in this property in other *Aeromonas* species (Francki and Chang 1994; Quinn *et al.*, 1994; Merino *et al.*, 1996). However, while the *ahyRI* mutant of *A. hydrophila* SSU had less filamentous structure and significantly reduced biofilm formation potential, the swimming and swarming motility of this mutant was unaffected (Khajanchi *et al.*, 2009).

While AHL-mediated QS in *A. hydrophila* SSU had no effect on the expression and translocation of the T3SS effector AexU in host cells (Khajanchi *et al.*, 2009), the T3SS up-regulation was demonstrated in the *A. hydrophila* AH-3 strain (Vilches *et al.*, 2005).

Interestingly, AHL mediated the QS-regulated secretion of the T6SS effectors Hcp2 and VgrG family of proteins in *A. hydrophila* SSU (Khajanchi *et al.*, 2009). Recently, we demonstrated the role of T6SS in biofilm formation and motility in *A. hydrophila* SSU. It should be noted that there are two copies of the *hcp* gene in the chromosome of SSU and three copies of the *vgrG* gene (Suarez *et al.*, 2010a,b). Finally, the *ahyRI* mutant of SSU demonstrated less virulence in a mouse model of infection as a result of a possible decrease in metalloprotease production, prevention of T6SS effector secretion, and attenuation in the biofilm formation (Khajanchi *et al.*, 2009).

Several studies show a close interaction between AHLs and the human immune system (Davis *et al.*, 2010; Wu *et al.*, 2005). Special attention should be given to a report showing that interferon-gamma is sensed by *P. aeruginosa* (PA) through binding to the outer membrane protein OprF, which results in the expression of PA-I lectin, a QS-dependent virulence determinant leading to an elevated pathogenicity of the bacterium (Wu *et al.*, 2005). Conversely, several groups have shown that 3-O-C12-HSL is able to modulate the immune system by inhibiting lymphocyte proliferation, T-cell differentiation, and production of IL-2 and cytokines (Telford *et al.*, 1998; Ritchie *et al.*, 2005). We investigated the immunomodulatory role of three AHLs, namely C4-HSL, C6-HSL and 3-oxo-C12-HSL, on the innate immune response to *A. hydrophila* infection in a septicaemic mouse model. Our study was the first to show that pretreatment of mice with AHLs prevented clinical sequelae to enhance their survivability after *A. hydrophila* infection, and we noted that the AHL-mediated QS system had no role in the regulation of the T3SS effector AexU in *A. hydrophila* SSU.

These AHLs seem to trigger an innate immune response in mice to clear bacteria from the blood and different mouse tissues (Khajanchi *et al.*, 2011). Our results also showed that the number of neutrophils was increased in the blood of infected mice treated with C4- and C6-HSLs, but not with 3-oxo-C12-HSL, although a reduction in the number of bacteria was seen with all of the AHLs. We believe that the protective effect of HSLs in mice is a result of a combination of several factors that culminated in reduced numbers of bacteria and of cytokines and chemokines levels in different tissues. Since, by using the neutropenic mouse model, we observed that neutrophils played an important role in the host defence against *A. hydrophila* infection, recruitment of these cells to the blood by C4- and C6-HSLs could be one of several mechanisms as to how mice were clearing bacteria from the blood during the course of infection (Khajanchi *et al.*, 2011).

The *luxS* QS system is present in approximately one-half of all sequenced bacterial genomes (Waters and Bonnie, 2005) and was first characterized as the regulator of bioluminescence in *Vibrio harveyi* (Suretteet *et al.*, 1999). Many bacterial phenotypes have been attributed to cell signalling *via* the LuxS-based AI-2 system (Kendall and Sperandio, 2007) in Gram-positive and Gram-negative bacteria that possibly means an autoinducer (AI)-2 is a universal language for interspecies communication (Xavier and Bassler, 2003). In addition, LuxS is also an integral component of the activated methyl cycle, which provides an alternative explanation for its widespread conservation (Winzer *et al.*, 2003). The synthesis of AI-2 is dependent on the enzyme coded by the *luxS* gene. The *luxS*-encoding enzyme converts S-ribosyl-homocysteine (SRH) into homocysteine and 4,5-dihydroxy-2,3-pentanedione (DPD), the latter of which is very unstable and reacts with water to form several different furanones (Schauder *et al.*, 2001; Sperandio *et al.*, 2003; Winzer *et al.*, 2002).

As reviewed (Schauder *et al.*, 2001), at high cell density, AI-2 binds LuxP, initiating a dephosphorylation cascade which results in the dephosphorylation and inactivation of the response regulator LuxO. Inactivation of LuxO and positive activation by LuxR allow expression of the bioluminescence operon in *V. harveyi*. Indeed, the structure of AI-2 bound to LuxP has been resolved and found to be a furanosyl borate diester (Chen *et al.*, 2002). However, AI-2 activity is not always associated with boron, as *S.* Typhimurium AI-2-binding protein LsrB (LuxS-regulated protein B), interacts with (2R,4S)-2-methyl-2,3,3,4-tetrahydroxytetrahydrofuran (*R*-THMF), which lacks boron (Miller *et al.*, 2004). In *V. cholerae* at low cell density, LuxQ acts as a kinase, transferring phosphate to the response regulator LuxO via the phosphotransfer protein LuxU. LuxO-P activates the transcription of genes encoding four small regulatory RNAs (sRNAs) called Qrr1–4 (quorum regulatory RNAs). The Qrr sRNAs, along with the sRNA chaperone, Hfq, base-pair with and destabilize the mRNA encoding *hapR*, and so no HapR is produced. Under this condition, HapR-repressed genes are expressed, whereas HapR-activated genes are not (Hammer, 2007; Bassler, 2007). Interestingly, the recently annotated genome sequence of *A. hydrophila* ATCC 7966 (Seshadri *et al.*, 2006) and our recent study of the *A. hydrophila* SSU genome indicated the presence of signal transduction cascade gene (*luxS, luxU, luxO, luxR*) homologues, but it does not appear to harbour homologues of *luxP* and *lsr* genes (Kozlova *et al.*, 2008, 2011), which encode the known receptors essential for AI-2 QS and LuxP or Lsr ABC-transporter, further suggesting to us that LuxS has primarily a metabolic role in this bacterium.

Recently, we noted the presence of a *litR* gene in *A. hydrophila* SSU (Kozlova *et al.*, 2011), which is a homologue of *hapR* gene in *V. cholerae* (Hammer, 2003; Bassler, 2003). HapR in *V. cholerae* regulates cyclic diguanosine monophosphate (c-di-GMP) metabolizing genes that lead to a net reduction in the level of cellular c-di-GMP, and, thereby, to the negative control of biofilm formation (Waters *et al.*, 2008). By directly down-regulating the expression of the *vpsT* gene (Waters *et al.*, 2008), HapR is also shown to control biofilm formation in a c-di-GMP-independent manner. VpsT is a transcriptional activator which, together with VpsR, is required for the expression of the *V. cholerae* polysaccharide biosynthesis operon (Casper-Lindley, 2001; Yildiz *et al.*, 2001; Yildiz, 2009; Visick, 2009). In addition to the *vpsT* homologues, we also found the presence of FleQ and FleN in *A. hydrophila* SSU (Kozlova *et al.*, 2011). FleQ, the first example of a c-di-GMP effector involved in transcription control, is a master regulator of flagellar gene expression in *P. aeruginosa* (Hickman, 2008; Harwood,

2008). FleN, an anti-activator of FleQ, down-regulates FleQ activity through direct interactions and thereby plays a crucial role in maintaining a single flagellum in *P. aeruginosa* (Dasgupta *et al.*, 2000, 2001).

The global regulators CsrA and HapR both largely contribute to the control of c-di-GMP levels by modulating selected panels of GGDEF and EAL domain-encoding genes in response to internal signals (Waters *et al.*, 2008; Lenz *et al.*, 2005) (Fig. 7.2). In addition to QS-dependent HapR regulating cascade, we also identified the VarS/VarA (LetS/LetA)-CsrA regulatory system in the *A. hydrophila* SSU genome (unpublished data) by examining gene homologues that exist in a variety of Gram-negative bacteria (Lenz *et al.*, 2005). However, subtle differences were noted, for example, between the HapR-dependent QS systems of *V. cholerae* and *A. hydrophila* SSU. Further, how these two regulatory systems co-function in *A. hydrophila* is not known yet.

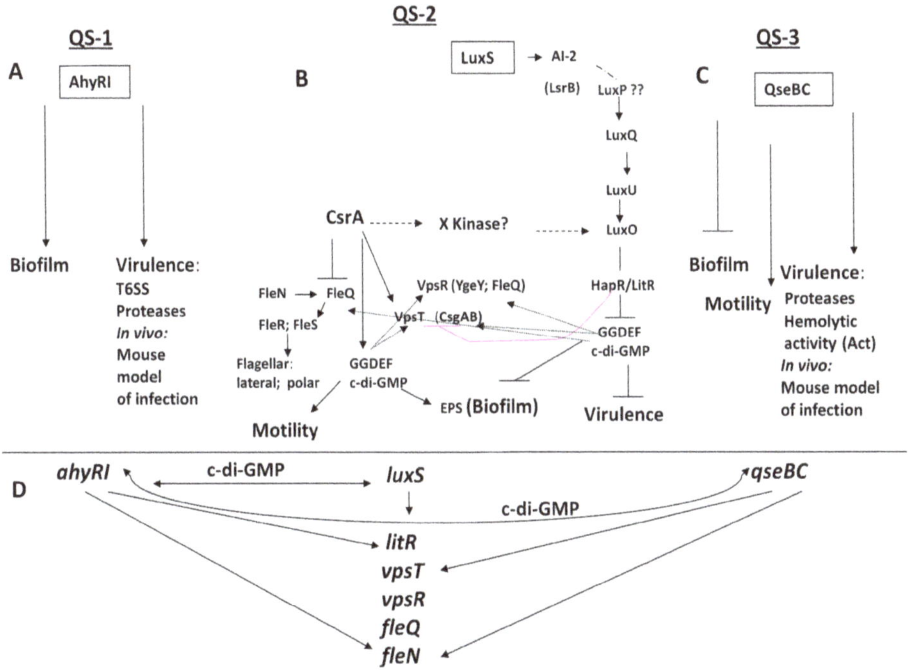

Figure 7.2 QS-dependent regulatory network in *A. hydrophila* SSU. Three modified autoinducer (AI) systems, which are known to date in other bacteria, exist in *A. hydrophila.* (A) While the *ahyRI*-based (AI-1) QS system positively regulates bacterial virulence (Khajanchi *et al.*, 2009), the LuxS-based QS system (AI-2) negatively regulates *A. hydrophila* virulence (Kozlova *et al.*, 2008) (B) The global regulators CsrA and HapR both contribute to the control of c-di-GMP levels by GGDEF and EAL domain-encoding genes expressing modulation in response to internal signals (Waters *et al.*, 2008; Lenz *et al.*, 2005). (C) QseBC has both positive and negative regulation of various virulence factors/mechanisms of *A. hydrophila* (Khajanchi *et al.*, 2012) that play an important role in fine tuning the expression of virulence genes (Kozlova *et al.*, 2012). (D) Cross-talk between AI-1- and QseBC- systems was identified (Kozlova *et al.*, 2012) in addition to interplay between the AI-1and AI-2 systems (Kozlova *et al.*, 2008). An alteration in the transcription of major genes involved in biofilm formation and motility of the WT *A. hydrophila* is dependent on the existence of AI-1, AI-2, and QseBC systems, and they are co-regulated via c-di-GMP.

Our study showed that the Δ*luxS* mutant of *A. hydrophila* SSU formed denser biofilm and exhibited decreased motility and enhanced virulence in the septicaemic mouse model; however, the *csgAB* (*vpsT*) transcript was down-regulated, while the *fleQ* (*vpsR*) was increased. Importantly, VpsR in *A. hydrophila* SSU was more closely related to FleQ in *P. aeruginosa* (Hickman, Harwood, 2008), thus making the LitR-VpsT-FleQ system of *A. hydrophila* SSU unique (Kozlova *et al.*, 2011).

Recently, we demonstrated the role of QseBC/AI-3 QS system in controlling the virulence of *A. hydrophila* SSU. The QseC, a sensor histidine kinase, and QseB, a response regulator (Khajanch *et al.*, 2012) that responds to QS signal, was first discovered in enterohaemorrhagic *Escherichia coli* (EHEC) (Sperandio *et al.*, 2002). Later, this system was also found in other pathogens, such as *Salmonella* Typhimurium (Bearson and Bearson, 2008), *Edwardsiella tarda* (Wang *et al.*, 2011), and uropathogenic *E. coli* (UPEC) (Kostakioti *et al.*, 2009). In addition to responding to AI-3, QseC sensor kinase has been shown in several reports to respond to eukaryotic hormones typified by adrenaline and/or noradrenaline (Clarke *et al.*, 2006; Waldor and Sperandio, 2007; Walters and Sperandio, 2006). Our data provided evidence that deletion of the *qseB* gene attenuated bacterial virulence in a septicaemic mouse model of infection, diminished swimming and swarming motility, and the mutant bacteria formed denser biofilms when compared with those from the parental strain of *A. hydrophila* SSU. The decrease in the virulence of *A. hydrophila* Δ*qseB* mutant correlated with the reduced production of protease and the cytotoxic enterotoxin, which has associated haemolytic activity. The QseBC system did not regulate the T3- and -6 secretion systems of *A. hydrophila* (Khajanch *et al.*, 2012). We demonstrated that the QseBC system, in addition to c-di-GMP-dependent AI-1 and AI-2 system regulation, is involved in the modulation of expression of two major regulators of biofilm and motility, such as VpsT and FleN (Kozlova *et al.*, 2012).

Cyclic-di-GMP, the bacterial intracellular second messenger, has also been implicated in the regulation of cell surface properties of several bacterial species (Hengge, 2009). Overexpression of the GGDEF domain protein in *A. veronii* increased c-di-GMP concentration and biofilm formation, while motility was reduced (Rahman *et al.*, 2007). We demonstrated that the effect of c-di-GMP overproduction on biofilm formation, motility and protease activity was dependent on the existence and functionality each of the three QS systems in *A. hydrophila* SSU. Indeed, a balance of the transcriptional levels of gene expression between *luxS* and *ahyR* was regulated by modulation of c-di-GMP levels (Kozlova *et al.*, 2011). Cross talk between AI-1- and QseBC- systems exists in *A. hydrophila* SSU, and c-di-GMP modulation on QseBC- system is dependent on the expression of the AI-1 system (Kozlova *et al.*, 2012).

Overall, *ahyRI* (AI-1) and LuxS-based (AI-2) QS systems were positive and negative regulators of virulence, respectively, while the QseBC two-component signal transduction system acted as both a negative and positive virulence regulator in a diarrheal isolate SSU of *A. hydrophila*. Based on our study of c-di-GMP over-expression in the QS mutants, we demonstrated c-di-GMP-dependent cross-talk between three QS systems which exist in *A. hydrophila* SSU and identified major components of a QS regulatory network.

Acknowledgements

We thank all the graduate students, postdoctoral fellows, research scientists, and research associates who contributed to the work presented in this chapter. The funding from the

National Institutes of Health, National Science Foundation, Environmental Protection Agency, and the American Water Works Association Research Foundation is greatly acknowledged.

References

Bearson, B.L., and Bearson, S.M. (2008). The role of the QseC quorum-sensing sensor kinase in colonization and norepinephrine-enhanced motility of *Salmonella enterica* serovar Typhimurium. Microb. Pathog. *44*, 271–278.

Bonemann, G., Pietrosiuk, A., and Mogk, A. (2010). Tubules and donuts: a type VI secretion story. Mol. Microbiol. *76*, 815–821.

Braun, M., Stuber, K., Schlatter, Y., Wahli, T., Kuhnert, P., and Frey, J. (2002). Characterization of an ADP-ribosyltransferase toxin (AexT) from *Aeromonas salmonicida subsp. salmonicida*. J. Bacteriol. *184*, 1851–1858.

Cascales, E. (2008). The type VI secretion toolkit. EMBO Reports 9, 735–741.

Casper-Lindley, C., and Yildiz, F.H. (2004). VpsT is a transcriptional regulator required for expression of vps biosynthesis genes and the development of rugose colonial morphology in *Vibrio cholerae* O1 El Tor. J. Bacteriol. *186*, 1574–1578.

Chen, X., Schauder, S., Potier, N., Van Dorsselaer, A., Pelczer, I., Bassler, B.L., and Hughson, F.M. (2002). Structural identification of a bacterial quorum-sensing signal containing boron. Nature *415*, 545–549.

Chopra, A.K., and Houston, C.W. (1999). Enterotoxins in *Aeromonas*-associated gastroenteritis. Microbes Infect. *1*, 1129–1137.

Chopra, A.K., Graf, J., Horneman, A.J., and Johnson, J.A. (2009). Virulence factor-activity relationships (VFAR) with specific emphasis on *Aeromonas species* (spp.). J. Water Health 7 *(Suppl. 1)*, S29–54.

Clarke, M.B., Hughes, D.T., Zhu, C., Boedeker, E.C., and Sperandio, V. (2006). The QseC sensor kinase: a bacterial adrenergic receptor. Proc. Natl. Acad. Sci. U.S.A. *103*, 10420–10425.

Croxatto, A., Chalker, V.J., Lauritz, J., Jass, J., Hardman, A., Williams, P., Camara, M., and Milton, D.L. (2002). VanT, a homologue of *Vibrio harveyi* LuxR, regulates serine, metalloprotease, pigment, and biofilm production in *Vibrio anguillarum*. J. Bacteriol. *184*, 1617–1629.

Dacanay, A., Knickle, L., Solanky, K.S., Boyd, J.M., Walter, J.A., Brown, L.L., Johnson, S.C., and Reith, M. (2006). Contribution of the type III secretion system (TTSS) to virulence of *Aeromonas salmonicida subsp. salmonicida*. Microbiology *152*, 1847–1856.

Das, S., and Chaudhuri, K. (2003). Identification of a unique IAHP (IcmF associated homologous proteins) cluster in *Vibrio cholerae* and other proteobacteria through *in silico* analysis. In Silico Biol. 3, 287–300.

Dasgupta, N., and Ramphal, R. (2001). Interaction of the antiactivator FleN with the transcriptional activator FleQ regulates flagellar number in *Pseudomonas aeruginosa*. J. Bacteriol. *183*, 6636–6644.

Dasgupta, N., Arora, S.K., and Ramphal, R. (2000). *fleN*, a gene that regulates flagellar number in *Pseudomonas aeruginosa*. J. Bacteriol. *182(2)*, 357–364.

Davis, B.M., Jensen, R., Williams, P., and O'Shea, P. (2010). The interaction of N-acylhomoserine lactone quorum sensing signaling molecules with biological membranes: implications for inter-kingdom signaling. PLoS One 5:e13522.

Fadl, A.A., Galindo, C.L., Sha, J., Erova, T.E., Houston, C.W., Olano, J.P., and Chopra, A.K. (2006). Deletion of the genes encoding the type III secretion system and cytotoxic enterotoxin alters host responses to *Aeromonas hydrophila* infection. Microb. Pathog. *40*, 198–210.

Fehr, D., Casanova, C., Liverman, A., Blazkova, H., Orth, K., Dobbelaere, D., Frey, J., and Burr, S.E. (2006). AopP, a type III effector protein of *Aeromonas salmonicida*, inhibits the NF-kappaB signalling pathway. Microbiology *152*, 2809–2818.

Francki, K.T., and Chang, B.J. (1994). Variable expression of O-antigen and the role of lipopolysaccharide as an adhesin in *Aeromonas sobria*. FEMS Microbiol. Lett. *122*, 97–101.

Fuqua, W.C., Winans, S.C., and Greenberg, E.P. (1994). Quorum sensing in bacteria-the LuxR-LuxI family of cell density-responsive transcriptional regulators. J. Bacteriol. *176*, 269–275.

Fuqua, C., Winans, S.C., and Greenberg, E.P. (1996). Census and consensus in bacterial ecosystems: the LuxR-LuxI family of quorum-sensing transcriptional regulators. Annu. Rev. Microbiol. *50*, 727–751.

Galan, J.E., and Wolf-Watz, H. (2006). Protein delivery into eukaryotic cells by type III secretion machines. Nature *444*, 567–573.

Galindo, C.L., Gutierrez, C., Jr., and Chopra, A.K. (2006). Potential involvement of galectin-3 and SNAP23 in *Aeromonas hydrophila* cytotoxic enterotoxin-induced host cell apoptosis. Microb. Pathog. *40*, 56–68.

Gavín, R., Rabaan, A.A., Merino, S., Toma´s, J.M., and Gryllos, I. (2002). Lateral flagella of *Aeromonas species* are essential for epithelial cell adherence and biofilm formation. Mol. Microbiol. *43*, 383–397.

Gavín, R., Merino, S., Altarriba, M., Canals, R., Shaw, J.G., and Tomás, J.M. (2003). Lateral flagella are required for increased cell adherence, invasion and biofilm formation by *Aeromonas* spp. FEMS Microbiol. Lett. *224*, 77–83.

Gordon, S. (2003). Alternative activation of macrophages. Nat. Rev. Immunol. *3*, 23–35.

Hengge R. (2009). Principles of c-di-GMP signalling in bacteria. Nat. Rev. Microbiol. *7*, 263–273.

Hammer, B.K., and Bassler, B.L. (2003). Quorum sensing controls biofilm formation in *Vibrio cholerae*. Mol. Microbiol. *50*, 101–104.

Hammer, B.K., and Bassler, B.L. (2007). Regulatory small RNAs circumvent the conventional quorum sensing pathway in pandemic *Vibrio cholerae*. Proc. Natl. Acad. Sci. U.S.A. *104*, 11145–11149.

Henke, J.M., and Bassler, B.L. (2004). Quorum sensing regulates type III secretion in *Vibrio harveyi* and *Vibrio parahaemolyticus*. J. Bacteriol. *186*, 3794–3805.

Hickman, J.W., and Harwood, C.S. (2008). Identification of FleQ from *Pseudomonas aeruginosa* as a c-di-GMP-responsive transcription factor. Mol. Microbiol. *69*, 376–389.

Hornef, M.W., Wick, M.J., Rhen, M., and Normark, S. (2002). Bacterial strategies for overcoming host innate and adaptive immune responses. Nat. Immunol. *3*, 1033–1040.

Jangid, K., Kong, R., Patole, M.S., and Shouche, Y.S. (2007). luxRI homologs are universally present in the genus *Aeromonas*. BMC Microbiol. *7*, 93.

Jimenez, P.N., Koch, G., Thompson, J.A., Xavier, K.B., Cool, R.H., and Quax, W.J. (2012). The multiple signaling systems regulating virulence in Pseudomonas *aeruginosa*. Microbiol. Mol. Biol. Rev. *76*, 46–65.

Kendall, M.M., and Sperandio, V. (2007). Curr. Opin. Gastroenterol. *23*, 10–15.

Khajanchi, B.K., Sha, J., Kozlova, E.V., Erova, T.E., Suarez, G., Sierra, J.C., Popov, V.L., Horneman, A.J., and Chopra, A.K. (2009). N-acylhomoserine lactones involved in quorum sensing control the type VI secretion system, biofilm formation, protease production, and *in vivo* virulence in a clinical isolate of *Aeromonas hydrophila*. Microbiology *155*, 3518–3531.

Khajanchi, B.K., Kirtley, M.L., Brackman, S.M., and Chopra, A.K. (2011). Immunomodulatory and protective roles of quorum-sensing signaling molecules N-acyl homoserine lactones during infection of mice with *Aeromonas hydrophila*. Infect. Immun. *79*, 2646–2657.

Khajanchi, B.K., Kozlova, E.V., Sha, J., Popov, V.L., and Chopra, A.K. (2012). The two-component QseBC signaling system regulates *in vitro* and *in vivo* virulence of *Aeromonas hydrophila*. Microbiology. *158*, 259–271.

Kim, S.Y., Lee, S.E., Kim, Y.R., Kim, C.H., Ryu, P.Y., Choy, H.E., Chung, S.S., and Rhee, J.H. (2003). Regulation of *Vibrio vulnificus* virulence by the LuxS quorum-sensing system. Mol. Microbiol. *48*, 1647–1664.

Kirke, D.F., Swift, S., Lynch, M.J., and Williams, P. (2004). The *Aeromonas hydrophila* LuxR homologue AhyR regulates the N-acyl homoserine lactone synthase, AhyI positively and negatively in a growth phase-dependent manner. FEMS Microbiol. Lett. *241*, 109.

Kirov, S.M. (2003). Bacteria that express lateral flagella enable dissection of the multifunctional roles of flagella in pathogenesis. FEMS Microbiol. Lett. *224*, 151–159.

Kirov, S.M., Sanderson, K., and Dickson, T.C. (1998). Characterisation of a type IV pilus produced by *Aeromonas caviae*. J. Med. Microbiol. *47*, 527–531.

Kirov, S.M., Tassell, B.C., Semmler, A.B., O'Donovan, L.A., Rabaan, A.A., and Shaw, J.G. (2002). Lateral flagella and swarming motility in *Aeromonas* species. J. Bacteriol. *184*, 547–555.

Kirov, S.M., Castrisios, M., and Shaw, J.G. (2004). *Aeromonas* flagella (polar and lateral) are enterocyte adhesins that contribute to biofilm formation on surfaces. Infect. Immun. 72, 1939–1945.

Kostakioti, M., Hadjifrangiskou, M., Pinkner, J.S., and Hultgren, S.J. (2009). QseC-mediated dephosphorylation of QseB is required for expression of genes associated with virulence in uropathogenic *Escherichia coli*. Mol. Microbiol. 73, 1020–1031.

Kozlova, E.V., Popov, V.L., Sha, J., Foltz, S.M., Erova, T.E., Agars, S.L., Horneman, J., and Chopra, A.K. (2008). Mutation in the S-Ribosylhomocysteinase (*luxS*) gene involved in quorum sensing affects biofilm formation and virulence in a clinical isolate of *Aeromonas hydrophila*. Microb. Pathog. *45*, 343–354.

Kozlova, E.V., Khajanchi, B.K., Sha, J., and Chopra, A.K. (2011). Quorum sensing and c-di-GMP-dependent alterations in gene transcripts and virulence-associated phenotypes in a clinical isolate of *Aeromonas hydrophila*. Microb. Pathog. *50*, 213–223.

Kozlova E.V., Khajanchi, B.K., Popov, V.L., Wen, J., and Chopra, A.K. (2012). Impact of QseBC system in c-di-GMP-dependent quorum sensing regulatory network in a clinical isolate SSU of *Aeromonas hydrophila*. Microb. Pathog. 53, 115–124.

Leiman, P.G., Basler, M., Ramagopal, U.A., Bonanno, J.B., Sauder, J.M., Pukatzki, S., Burley, S.K., Almo, S.C., and Mekalanos, J.J. (2009). Type VI secretion apparatus and phage tail-associated protein complexes share a common evolutionary origin. Proc. Natl. Acad. Sci. U.S.A. *106*, 4154–4159.

Lenz, D.H., Miller, M.B., Zhu, J., Kulkarni, R.V., and Bassler, B.L. (2005). CsrA and three redundant small RNAs regulate quorum sensing in *Vibrio cholerae*. Mol. Microbiol. *58*, 1186–1202.

Lilley, B.N., and Bassler, B.L. (2000). Regulation of quorum sensing in *Vibrio harveyi* by LuxO and sigma-54. Mol. Microbiol. *36*, 940–954.

Lynch, M.J., Swift, S., Kirke, D.F., Keevil, C.W., Dodd, C.E., and Williams, P. (2002). The regulation of biofilm development by quorum sensing in *Aeromonas hydrophila*. Environ. Microbiol. *4*, 18–28.

Manefield, M., Harris, L., Rice, S.A., de Nys, R., and Kjelleberg, S. (2000). Inhibition of luminescence and virulence in the black tiger prawn (Penaeus monodon) pathogen *Vibrio harveyi* by intercellular signal antagonists. Appl. Environ. Microbiol. *66*, 2079–2084.

Merino, S., Rubires, X., Aguilar, A., and Tomás, J.M. (1996). The O:34-antigen lipopolysaccharide as an adhesin in *Aeromonas hydrophila*. FEMS Microbiol. Lett. *139*, 97–101.

Miller, S.T., Xavier, K.B., Campagna, S.R., Taga, M.E., Semmelhack, M.F., Bassler, B.L., and Hughson, F.M. (2004). *Salmonella typhimurium* recognizes a chemically distinct form of the bacterial quorum-sensing signal AI-2. Mol. Cell. *15*, 677–687.

Mok, K.C., Wingreen, N.S., and Bassler, B.L.(2003). *Vibrio harveyi* quorum sensing: a coincidence detector for two autoinducers controls gene expression. EMBO J. *22*, 870–881.

Morett, E., and Segovia, L. (1993). The sigma 54 bacterial enhancer-binding protein family: mechanism of action and phylogenetic relationship of their functional domains. J. Bacteriol. *175*, 6067–6074.

Mougous, J.D., Cuff, M.E., Raunser, S., Shen, A., Zhou, M., Gifford, C.A., Goodman, A.L., Joachimiak, G., Ordonez, C.L., Lory, S., *et al.* (2006). A virulence locus of *Pseudomonas aeruginosa* encodes a protein secretion apparatus. Science *312*, 1526–1530.

Pederson, K.J., Vallis, A.J., Aktories, K., Frank, D.W., and Barbieri, J.T. (1999). The amino-terminal domain of *Pseudomonas aeruginosa* ExoS disrupts actin filaments via small-molecular-weight GTP-binding proteins. Mol. Microbiol. *32*, 393–401.

Pell, L.G., Kanelis, V., Donaldson, L.W., Howell, P.L., and Davidson, A.R. (2009). The phage lambda major tail protein structure reveals a common evolution for long-tailed phages and the type VI bacterial secretion system. Proc. Natl. Acad. Sci. U.S.A. *106*, 4160–4165.

Pukatzki, S., Ma, A.T., Sturtevant, D., Krastins, B., Sarracino, D., Nelson, W.C., Heidelberg, J.F., and Mekalanos, J.J. (2006). Identification of a conserved bacterial protein secretion system in *Vibrio cholerae* using the Dictyostelium host model system. Proc. Natl. Acad. Sci. U.S.A. *103*, 1528–1533.

Pukatzki, S., Ma, A.T., Revel, A.T., Sturtevant, D., and Mekalanos, J.J. (2007). Type VI secretion system translocates a phage tail spike-like protein into target cells where it cross-links actin. Proc. Natl. Acad. Sci. U.S.A. *104*, 15508–15513.

Pukatzki, S., McAuley, S.B., and Miyata, S.T. (2009). The type VI secretion system: translocation of effectors and effector-domains. Curr. Opin. Microbiol. *12*, 11–17.

Quinn, D.M., Atkinson, H.M., Bretag, A.H., Tester, M., Trust, T.J., Wong, C.Y.F., and Flower, R.F.L.P. (1994). Carbohydrate-reactive, pore-forming outer membrane proteins of *Aeromonas hydrophila*. Infect. Immun. *62*, 4054–4058.

Rabaan, A.A., Gryllos, I., Toma´s, J.M., and Shaw, J.G. (2001). Motility and the polar flagellum are required for *Aeromonas caviae* adherence to HEp-2 cells. Infect. Immun. *69*, 4257–4267.

Rahman, M., Simm, R., Kader, A., Basseres, E., Römling, U., and Möllby, R. (2007). The role of c-di-GMP signaling in an *Aeromonas veronii* biovar *sobria* strain. FEMS Microbiol. Lett. *273*, 172–179.

Rego, A.T., Chandran, V., and Waksman, G. (2010). Two-step and one-step secretion mechanisms in Gram-negative bacteria: contrasting the type IV secretion system and the chaperone-usher pathway of pilus biogenesis. Biochem. J. *425*, 475–488.

Ritchie, A.J., Jansson, A., Stallberg, J., Nilsson, P., Lysaght, P., and Cooley, M.A. (2005). The *Pseudomonasaeruginosa* quorum-sensing molecule *N*-3-(oxododecanoyl)-L-homoserine lactone inhibits T-cell differentiation and cytokine production by a mechanism involving an early step in T-cell activation. Infect. Immun. *73*, 1648–1655.

Schauder, S., Shokat, K., Surette, M.G., and Bassler, B.L. (2001). The LuxS family of bacterial autoinducers: biosynthesis of a novel quorum-sensing signal molecule. Mol. Microbiol. *41*, 463–476.

Seshadri, R., Joseph, S.W., Chopra, A.K., Sha, J., Shaw, J., Graf, J., Haft, D., Wu, M., Ren, Q., Rosovitz, M., *et al.* (2006). Genome sequence of *Aeromonas hydrophila* ATCC 7966T: jack of all trades. J. Bacteriol. *188*, 8272–8282.

Sha, J., Pillai, L., Fadl, A.A., Galindo, C.L., Erova, T.E., and Chopra, A.K. (2005). The type III secretion system and cytotoxic enterotoxin alter the virulence of *Aeromonas hydrophila.* Infect. Immun. *73*, 6446–6457.

Sha, J., Wang, S.F., Suarez, G., Sierra, J.C., Fadl, A.A., Erova, T.E., Foltz, S.M., Khajanchi, B.K., Silver, A., Graf, J., *et al.* (2007). Further characterization of a type III secretion system (T3SS) and of a new effector protein from a clinical isolate of *Aeromonas hydrophila*--part I. Microb. Pathog. *43*, 127–146.

Sierra, J.C., Suarez, G., Sha, J., Foltz, S.M., Popov, V.L., Galindo, C.L., Garner, H.R., and Chopra, A.K. (2007). Biological characterization of a new type III secretion system effector from a clinical isolate of *Aeromonas hydrophila*-part II. Microb. Pathog. *43*, 147–160.

Sierra, J.C., Suarez, G., Sha, J., Baze, W.B., Foltz, S.M., and Chopra, A.K. (2010). Unraveling the mechanism of action of a new type III secretion system effector AexU from *Aeromonas hydrophila.* Microb. Pathog. *49*, 122–134.

Silverman, J.M., Brunet, Y.R., Cascales, E., and Mougous, J.D. (2012). Structure and regulation of the type VI secretion system. Ann. Rev. Microbiol. *66*, 453–472.

Sperandio, V., Torres, A.G., and Kaper, J.B. (2002). Quorum sensing *Escherichia coli* regulators B and C (QseBC): a novel two-component regulatory system involved in the regulation of flagella and motility by quorum sensing in *E. coli.* Mol. Microbiol. *43*, 809–821.

Sperandio, V., Torres, A.G., Jarvis, B., Nataro, J.P., and Kaper, J.B. (2003). Bacteria–host communication: the language of hormones. Proc. Natl. Acad. Sci. U.S.A. *100*, 8951–8956.

Suarez, G., Sierra, J.C., Sha, J., Wang, S., Erova, T.E., Fadl, A.A., Foltz, S.M., Horneman, A.J., and Chopra, A.K. (2008). Molecular characterization of a functional type VI secretion system from a clinical isolate of *Aeromonas hydrophila.* Microb. Pathog. *44*, 344–361.

Suarez, G., Sierra, J.C., Erova, T.E., Sha, J., Horneman, A.J., and Chopra, A.K. (2010a). A type VI secretion system effector protein, VgrG1, from *Aeromonas hydrophila* that induces host cell toxicity by ADP ribosylation of actin. J. Bacteriol. *192*, 155–168.

Suarez, G., Sierra, J.C., Kirtley, M.L., and Chopra, A.K. (2010b). Role of Hcp, a type 6 secretion system effector, of *Aeromonas hydrophila* in modulating activation of host immune cells. Microbiology *156*, 3678–3688.

Sundin, C., Hallberg, B., and Forsberg, A. (2004). ADP-ribosylation by exoenzyme T of *Pseudomonas aeruginosa* induces an irreversible effect on the host cell cytoskeleton *in vivo.* FEMS microbiol. Lett. *234*, 87–91.

Surette, M.G., Miller, M.B., and Bassler, B.L. (1999). Quorum sensing *in Escherichia coli, Salmonella* Typhimurium, and *Vibrio harveyi*: a new family of genes responsible for autoinducer production. Proc. Natl. Acad. Sci. U.S.A. *96*, 1639–1644.

Swift, S., Karlyshev, A.V., Fish, L., Durant, E.L., Winson, M.K., Chhabra, S.R., Williams, P., Macintyre, S., and Stewart, G.S. (1997). Quorum sensing in *Aeromonas hydrophila* and *Aeromonas salmonicida*: identification of the LuxRI homologs AhyRI and AsaRI and their cognate N-acylhomoserine lactone signal molecules. J. Bacteriol. *179*, 5271–5281.

Swift, S., Lynch, M.J., Fish, L., Kirke, D.F., Tomás, J.M., Stewart, G.S., and Williams, P. (1999). Quorum sensing-dependent regulation and blockade of exoprotease production in *Aeromonas hydrophila.* Infect. Immun. *67*, 5192–5199.

Telford, G., Wheeler, D., Williams, P., Tomkins, P.T., Appleby, P., Sewell, H., Stewart, G.S., Bycroft, B.W., and Pritchard, D.I. (1998). The *Pseudomonas aeruginosa* quorum-sensing signal molecule N-(3-oxododecanoyl)-L-homoserine lactone has immunomodulatory activity. Infect. Immun. *66*, 36–42.

Vilches, S., Jimenez, N., Tomás, J.M., and Merino, S. (2005). *Aeromonas hydrophila* AH-3 type III secretion system expression and regulatory network. Appl Environ Microbiol. *75*, 6382–6392.

Waldor, M.K., and Sperandio, V. (2007). Adrenergic regulation of bacterial virulence. J. Infect. Dis. *195*, 1248–1249.

Walters, M., and Sperandio, V. (2006). Autoinducer 3 and epinephrine signaling in the kinetics of locus of enterocyte effacement gene expression in enterohemorrhagic *Escherichia coli.* Infect. Immun. *74*, 5445–5455.

Waters, C.M., and Bonnie, B.L. (2005). Quorum sensing: cell-to-cell communication in bacteria. Annu. Rev. Cell Dev. Biol. *21*, 319–346.

Waters, C.M., Lu, W., Rabinowitz, J.D., and Bassler, B.L. (2008). Quorum sensing controls biofilm formation in *Vibrio cholerae* through modulation of cyclic di-GMP levels and repression of *vpsT*. J. Bacteriol. *90*, 2527–2536.

Wang, X., Wang, Q., Yang, M., Xiao, J., Liu, Q., Wu, H., and Zhang, Y. (2011). QseBC controls flagellar motility, fimbrial hemagglutination and intracellular virulence in fish pathogen *Edwardsiella tarda*. Fish Shellfish Immunol. *30*, 944–953.

Weihua Chu, W., Yan Jiang, Y., Yongwang, L., and Zhu, W. (2011). Role of the quorum-sensing system in biofilm formation and virulence of *Aeromonas hydrophila*. Afr. J. Microbiol. Res. *5*, 5819–5825.

Winzer, K., and Williams, P. (2001). Quorum sensing and the regulation of virulence gene expression in pathogenic bacteria. Int. J. Med. Microbiol. *291*, 131–143.

Winzer, K., Hardie, K.R., and Williams, P. (2002). Bacterial cell-to-cell communication: sorry, cannot talk now – gone to lunch! Curr. Opin. Microbiol. *5*, 216–222.

Winzer, K., Hardie, K.R., and Williams, P. (2003). LuxS and autoinducer-2: their contribution to quorum sensing and metabolism in bacteria. Adv. Appl. Microbiol. *53*, 291–396.

Wu, L., Estrada, O., Zaborina, O., Bains, M., Shen, L., Kohler, J.E., Patel, N., Musch, M.W., Chang, E.B., Fu, Y.X., *et al.* (2005). Recognition of host immune activation by *Pseudomonas aeruginosa*. Science *309*, 774–777.

Xavier, K.B., and Bassler, B.L. (2003). LuxS quorum sensing: more than just a numbers game. Curr. Opin. Microbiol. *6*, 191–197.

Yeo, H.J., and Waksman, G. (2004). Unveiling molecular scaffolds of the type IV secretion system. J. Bacteriol. *186*, 1919–1926.

Yildiz, F.H., and Visick, K.L. (2009). Vibrio biofilms: so much the same yet so different. Trends Microbiol. *17*, 109–118.

Yildiz, F.H., Dolganov, N.A., and Schoolnik, G.K. (2001). VpsR, a member of the response regulators of the two-component regulatory systems, is required for expression of *vps* biosynthesis genes and EPS (ETr)-associated phenotypes in *Vibrio cholerae* O1 El Tor. J. Bacteriol. *183*, 1716–1726.

Molecular Diagnostics by Genetic Methods

8

Antonio Martínez-Murcia and Brigitte Lamy

Abstract

This chapter deals with most, if not all, genetic methods used for classification and molecular diagnostics of *Aeromonas* species. These methods include currently used strategies designed for identifying strains at the species level from pure cultures, for detection of species on complex samples, e.g. waters or food, and for genotyping strains. The genus *Aeromonas* represents one of the best examples to illustrate applications of molecular biology techniques in microbiology, as it early took the benefits from cutting-edge molecular biology developments to be deeply renovated and, at the same time, represents a particularly difficult example of application for molecular diagnostic microbial methods due to its complex historical taxonomy, recent changes, increasing number of new species descriptions, and their extremely close phylogenetic relationships. We have reviewed past, current and new genetic methods, aiming to give an overview and opinions about advantages and disadvantages, performance and limitations. After a brief description of the principles of the procedures, each method is illustrated by applications either in clinical diagnosis, microbial ecological issues, epidemiological surveys, population diversity studies, or in taxonomy rearrangements and new taxa descriptions. Present trends and future perspectives of new revolutionary technology, expected at short and medium term, are also discussed.

Introduction

In recent decades the experimental approaches of microbiology have undergone enormous changes. The shift has been on such a scale that today most relevant studies to determine microbial diversity, or the fastest and precise techniques for pathogen detection and diagnosis of infections, do not necessarily require the use of historically indispensable tools in microbiology: the Petri dish and microscope. The development of molecular biology has provided to microbiologists a variety of strategies to easily reach all genetic diversity, or detect any specific microorganism, in any kind of sample. Most relevant molecular biology methods to ascertain genetic information were the so-called 'Sanger Sequencing' developed by Frederick Sanger and colleagues in 1977 and the polymerase chain reaction (PCR) first envisioned by Kary Mullis (Nobel Prize in Chemistry in 1993) 30 years ago, the bases for latter development of quantitative PCR in the mid-1990s. Actually, the development of high-throughput sequencing (termed next-generation sequencing) may provide thousands or millions of sequences from a sample.

A reflection on the reasons for these changes on this experimental discipline seems appropriate preamble to facilitate an entry into the Molecular Diagnosis by Genetic Methods, those used in the classification, identification, and detection of bacteria of the genus *Aeromonas*. And most probably, the genus *Aeromonas* represents one of the best examples to illustrate the recent history of change in systematics. This group of ubiquitous microorganisms have been systematically studied, from the early 80s to the present day, by employing each genetic technique that chronologically appeared as a result of advances in molecular biology. A look to the relevant chapter of *Bergey's Manual of Systematic Bacteriology* (Popoff, 1984) is enough to appreciate the influence of this technological revolution in the classification and identification of *Aeromonas* species as, at that time, the author could only contribute with four described and validated species in the genus. In addition to this scarce knowledge of diversity, a long sea of taxonomic controversies were added in the coming years, mainly due to the complexity of the genus and the difficulties in establishing phenotypic clusters and clear-cut splitting species. Thirty years later, we may contribute to this chapter with a huge number of new species, almost 30, having most taxonomic problems clearly solved. Moreover, the recently provided Multilocus Phylogenetic Analyses (MLPA) have been evaluated as representing 'one of the first deep studies empirically confirming the great usefulness of MLPA for bacterial systematic, even for difficult groups such as *Aeromonas*'.

Classification frames and approaches for identification of microorganisms, the purposes of systematics on prokaryotes, has been deeply renovated as a consequence of genetic technology development, mostly due to a great change of our basic concepts after the recent discovery of real microbial diversity and evolution modes hypothesized. The increasingly rapid capacity to obtain genetic information has undoubtedly produced in microbiology the biggest impact ever. Until the advent of molecular sequencing, phylogeny based on bacterial evolution could not be empirically approached. Early taxonomies were generated on data relevant to the role bacteria plays in the environment, largely based on overall resemblance of phenotypic properties (biochemical and physiological), the only technologies available in most twentieth century. It is now generally recognized that, because the great heterogeneity of bacteria and their capacities to change phenotype as a response to selective pressures, numerical taxonomy has not been always reliable for constructing phylogenetic (natural) classifications.

Genomic relatedness, however, reflects to some extent the natural change occurred during evolution of species from their immediate ancestors giving raise possible, for first time in the history of microbiology, the provision of phylogenetic classifications. Genetic information is considered superior in two main ways: first, it is more reliable and precisely interpreted (objective), and second, innately more informative about natural relationships. All this together was served with the hypothetical concept of the molecular chronometer of evolution: the evolutionary history of microorganisms may be recorded in the primary chemical structure of some molecules. Although very first attempts to ascertain bacterial phylogenies used protein sequences (Zuckerkandl and Pauling, 1965), this revolution in microbial systematics grown exponentially with the work initiated by Carl Woese (1928–2012) by using the nucleotide sequences of ribosomal RNA. In 1967, he broached the notion of a prebiotic 'RNA world' with his book *The Genetic Code*, but his most ground-breaking achievement was to report in 1977, together with George Fox, the first scientifically based phylogenetic tree that revealed the long term course of evolution. Not only relevant to microbiology and evolutionary biology, but he forever changed our understanding of the biosphere and its evolution.

As a consequence of the new genetic approaches used for classification and identification, today we count with a more stable classification frame which can accommodate newly described diversity without dramatic changes, and very fast, reliable, sensitive, and precise methods for detection and/or identification of bacterial taxa. Genetic methods may be applied directly to samples without the need of previous microbial isolations, neither enrichments nor the use of selective media. This is probably the greatest advantage from a practical point of view, making the control of microbiota at a real-time fashion approachable. The independence of the culture media has allowed us to describe a vast diversity, most may not be adaptable to laboratory conditions. Incidentally, this hitherto unknown microbiota represents almost all life present in the environment. Obviously the impact on diagnosis of infectious diseases is considerable as specific pathogen detection may be performed in less than one hour with extreme level of sensitivity and specificity.

In the present chapter, most, if not all, the genetic methods applied to the genus *Aeromonas* are reported. Applications included species identification from pure cultures, species and/or specific gene detection from samples, and strain genotyping methods used in a variety of manners: for taxonomic studies and genetic relatedness, for the analysis of full microbial populations in complex matrices, and to ascertain strain relationships between clinical and environmental isolates belonging to the same species.

Species identification of pure cultures

Aeromonas isolation has largely been based on different kind of rich media, somehow bypassed by the use of selective antibiotics. Once *Aeromonas* strain has been isolated, it must be identified to the species level, which constitutes a particular difficulty considering the number of taxa, current taxonomic controversies, and the wide phenotypic and genetic diversity encountered, all of which reflect the complexity of structure within this genus. Genetic methods are today being recognized by far as the most accurate for identification to the species level. Historically, a first improvement in identification was provided by 16S rRNA gene sequencing, followed by 16S rDNA PCR-amplified restriction fragment length polymorphism, single housekeeping gene sequencing and at last multilocus sequence analysis. All these methods, thanks to their intrinsic chronometric properties and usefulness as phylogenetic markers, easiness and rapidness, have revolutionized in their time both species delineation, appraisal of *Aeromonas* genetic diversity, and provided robust framework for identification to the species level.

Sequencing and phylogenetic analysis of the 16S rRNA gene

The advent of ribosomal RNA sequence-base phylogenetic analysis, used during the past two decades, has provided robust classification frames, which facilitates objective and simple identification of new isolates, also identifying bacteria independently of growth conditions and media used. The rRNA molecules are not subjected to a degenerative code, rather their primary structures are highly conserved as they are compromised by their own interactions to form harping loops and these with ribosomal proteins to maintain structure/function of the ribosome. As a consequence, the very few mutations allowed in the rRNA sequences, which appear related with evolutionary events, are comprised in signature regions showing a gradient from high mutations rates to extremely conserved sequences. Mosaics of these signatures are relatively encompassed with the different taxonomic ranks, which makes this molecule the most powerful so far known to establish phylogenies above the genus level. These features make the alignment of homologous rRNA positions a straightforward step prior to ascertain the number and location of mutations events when comparing rRNA sequences. For identification and detection purposes, these taxa-specific signature regions can serve as targets to design oligonucleotide primers and probes for specific amplification by PCR. Highly conserved regions, however, have been used as the binding sites of 'universal' primers enabling researches to reach the rRNA sequences from virtually all living microorganisms, whether or not they have been cultured.

First phylogenetic analysis of the genus *Aeromonas* based on the 16S ribosomal RNA gene sequences was published 20 years ago (Martínez-Murcia *et al.*, 1992). This pioneer work helped to clarify the taxonomy of the genus, although some lack of congruence with previous species delineation was early elucidated. In other words, most findings (separated phylogenetic lineages) were generally in agreement to accepted DNA-hybridization grouping and previously established species. However, in the genus *Aeromonas*, rRNA sequences are extremely conserved and, in particular, a few closely related species show almost identical 16S rRNA with only one or two nucleotide differences (Martínez-Murcia *et al.*, 2005). Nevertheless, some signatures of the molecule are highly specific to the genus level which gave a good basis to develop qPCR tests for *Aeromonas* spp. (see 'Quantitative real-time PCR' in this chapter), but also specific of Aeromonadaceae (Martínez-Murcia *et al.*, 1992). For instance, phylogenetic tree of Fig. 8.1 clearly shows the inter-genera relationships of mentioned family and evidenced the proposed species *A. sharmana* does not belong to the genus *Aeromonas* (Martínez-Murcia *et al.*, 2007)

The 16S rRNA-based phylogenetic tree of *Aeromonas* lacks ultrametricity (shows very uneven tips) suggesting chronometric distortions such as inconstant evolution rates and parallel or convergent evolutionary changes (Martinez-Murcia, 1993; Sneath, 1993). Genetic crossing-over of *rrn* from different individuals has been suggested to explain that evolution of different parts (domains) of the molecule were not concordant (Sneath, 1993). However, part of this problem may derive from the organization of rRNA genes (rDNA) as a multigene family, often as *rrn* operons that, in *Aeromonas*, vary from 8 to 11. The members of a multigene family are subjected to a homogenization process by molecular mechanisms such as gene conversion, and rDNA sequences tend to evolve in concert. If this process is not uniform along the length of the molecule (or unequally directed), homogenization of ancestral regions may be combined with evolved parts resulting in hybrids showing homoplasy (Martinez-Murcia *et al.*, 1999). Consequently, non-uniform concerted evolution may also induce the chronometric distortions responsible for a loss of ultrametricity in phylogenetic trees, particularly of very closely related micro-organisms.

An analysis of almost-complete 16S rRNA gene sequences (1503 nt) of 28 type and reference strains show that only *A. simiae*, and now *A. cavernicola*, could be distinguished from the other species, according to Stackebrandt and Ebers standards, based on a sequence similarity lower than 98.7% (Stackebrandt and Ebers 2006). Considering all the other species, sequences varied by 10 or less nucleotides with the sequences of one or more other species, of which half (12 out of the 24) species show fewer than five varying positions. Informative nucleotide positions of the sequences, mainly located on region V3 and in a lesser extent in V1 and V2, are shown in Table 8.1.

Sequencing allows obtaining the complete information of the small-subunit ribosomal RNA primary structure, an advantage over alternative 16S rRNA derived methods such as

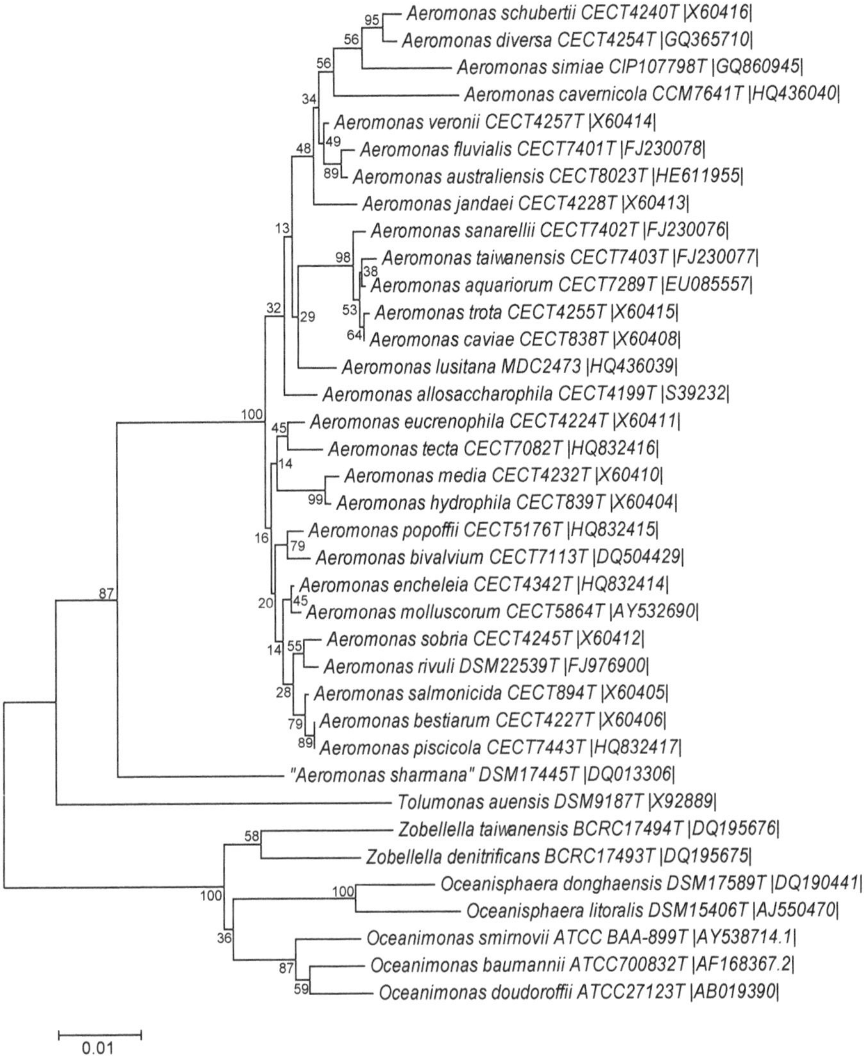

Figure 8.1 Phylogenetic NJ-tree derived from the 16S rRNA gene sequences (1297 nucleotides) showing relationships of all described *Aeromonas* species and other genera of family *Aeromonadaceae*. Numbers at nodes indicate bootstrap values (percentage of 1000 replicates). Bar, 0.001 estimated nucleotide substitutions per site. EMBL Accession numbers of selected reference 16S rRNA gene sequences are in parenthesis.

those targeting only very few signature nucleotides. After these two decades, the method has been universalized, automated, and may be relatively accessible to most laboratories. Phylogeny based on 16S rRNA gene sequences has played an important role in the identification of *Aeromonas* species, providing relevant support for new species descriptions which undoubtedly has enrich our knowledge about diversity of this genus and is widely used for the diagnosis of *Aeromonas* infections.

The presence of 'microheterogeneities', nucleotide diversity at homologous positions of the rRNA multicopy gene family, interferes with proper identification, being more likely

Table 8.1 Nucleotide composition at variable 16S rRNA gene sequence sites found in type strains representing all *Aeromonas* species known to date. Numbers, position in the sequence. Dots, identical composition to this of *A. veronii*

	A.veronii	A.fluvialis	A.trota	A.caviae	A.aquariorum	A.taiwanensis	A.sanarellii	A.media	A.hydrophila	A.eucrenophila	A.tecta	A.encheleia	A.molluscorum	A.popoffii	A.bivalvium	A.sobria	A.rivuli	A.bestiarum	A.piscicola	A.salmonicida	A.lusitana	A.allosaccharophila	A.jandaei	A.schubertii	A.diversa	A.cavernicola	A.australiensis	A.simiae
77	A	·	·	·	·	·	·	·	·	·	·	·	·	·	·	·	·	·	·	·	·	·	·	G	–	·	·	G
94	T	·	·	·	·	·	·	·	·	·	·	·	·	·	·	·	·	·	·	·	·	·	·	·	–	·	·	·
129	G	·	A	A	A	A	A	A	A	·	·	·	·	·	·	·	·	·	·	·	·	·	·	·	·	A	·	A
131	T	·	A	A	A	A	A	A	A	·	·	·	·	·	·	·	·	·	·	·	·	·	·	·	·	A	·	A
132	C	·	T	T	T	T	T	T	T	·	·	·	·	·	·	·	·	·	·	·	·	·	·	·	·	T	·	T
138	A	·	·	·	·	·	·	·	·	·	·	·	·	·	·	·	·	·	·	·	·	·	·	·	·	T	·	·
140	T	·	·	·	·	·	·	·	·	·	·	·	·	·	·	·	·	·	·	·	·	·	·	·	·	A	·	·
141	C	·	·	·	·	·	·	·	·	·	·	·	·	·	·	·	·	·	·	·	·	·	·	·	·	T	·	T
143	A	·	·	·	·	·	·	·	·	·	·	·	·	·	·	·	·	·	·	·	·	·	·	·	·	T	·	T
154	T	·	A	A	A	A	A	A	A	A	·	A	A	A	A	A	A	A	A	A	A	A	·	·	·	·	·	·
155	A	·	G	G	G	G	G	G	G	G	·	G	G	G	G	G	G	G	G	G	G	G	·	·	·	·	·	·
156	C	·	T	T	T	T	T	T	T	T	·	T	T	T	T	T	T	T	T	T	T	T	·	·	·	·	·	·
165	G	·	A	A	A	A	A	A	A	.	·	A	A	A	A	A	A	A	A	A	·	A	·	·	·	·	·	·
166	T	·	C	C	C	C	C	C	C	C	·	C	C	C	C	C	C	C	C	C	C	C	·	·	·	·	·	·
167	A	·	T	T	T	T	T	T	T	T	·	T	T	T	T	T	T	T	T	T	T	T	·	·	·	·	·	G
181	A	·	·	·	·	·	·	·	·	·	·	·	·	·	·	·	·	·	·	·	·	·	·	·	·	G	·	·
182	C	·	·	·	·	·	·	·	·	·	·	·	·	·	·	·	·	·	·	·	·	·	·	·	·	A	·	·
183	G	·	·	·	·	·	·	·	·	·	·	·	·	·	·	·	·	·	·	·	·	·	·	·	·	T	·	·
184	C	·	·	·	·	·	·	·	·	·	·	·	·	·	·	·	·	·	·	·	·	·	·	·	·	G	·	·
193	G	·	·	·	·	·	·	·	·	·	·	·	·	·	·	·	·	·	·	·	·	·	·	·	·	C	·	·
194	G	·	·	·	·	·	·	·	·	·	·	·	·	·	·	·	·	·	·	·	·	·	·	·	·	C	·	·
199	C	·	·	·	·	·	·	·	·	·	G	G	G	G	G	G	G	G	G	G	·	·	·	·	·	A	·	·
200	A	·	·	·	·	·	·	·	·	·	·	·	·	·	·	·	·	·	·	·	·	·	·	·	·	G	·	·
217	T	·	·	·	·	·	·	·	·	·	·	·	·	·	·	·	·	·	·	·	·	·	·	·	·	C	·	·
218	G	·	·	·	·	·	·	·	·	·	T	T	T	T	T	T	T	T	T	T	·	·	·	·	·	T	·	·
223	A	·	·	·	·	·	·	·	·	·	·	·	·	·	·	·	·	·	·	·	·	·	·	·	·	T	·	·
224	T	·	·	·	·	·	·	·	·	·	·	·	·	·	·	·	·	·	·	·	·	·	·	·	·	C	·	·
225	T	·	·	·	·	·	·	·	·	·	·	·	·	·	·	·	·	·	·	·	·	·	·	·	·	A	·	·
230	G	·	A	A	A	A	A	A	A	·	·	·	·	·	·	·	·	·	·	·	·	·	·	·	·	A	·	A
231	A	·	T	T	T	T	T	T	T	·	·	·	·	·	·	·	·	·	·	·	·	·	·	·	·	T	·	T
232	A	·	G	G	G	G	G	G	G	·	·	·	·	·	·	·	·	·	·	·	·	·	·	·	·	G	·	G
250	A	·	·	·	·	·	·	T	·	·	·	·	·	·	·	·	·	·	·	·	·	·	·	T	T	·	·	T
258	A	·	·	·	·	·	·	·	·	G	G	G	·	G	G	G	·	G	G	G	G	·	·	·	·	·	·	·
264	T	·	·	·	·	·	·	·	·	·	·	·	·	·	·	·	·	·	·	·	·	·	C	C	C	·	·	·
285	C	·	·	·	·	·	·	·	·	·	·	·	·	·	·	·	·	·	·	·	·	·	·	·	·	·	·	T
320	A	·	·	·	·	·	·	·	·	·	·	·	·	·	·	·	·	·	·	·	·	·	·	·	·	·	·	G
333	T	·	·	·	·	·	·	·	·	·	·	·	·	·	·	·	·	·	·	·	·	·	·	·	·	·	·	C
456	T	·	·	·	·	·	·	·	·	·	·	·	·	·	A	·	·	·	·	·	·	·	·	·	·	·	·	·
457	T	·	C	C	C	C	C	·	·	·	·	·	·	·	·	·	·	·	·	·	C	·	C	·	·	·	·	·
458	G	·	A	A	A	A	A	·	·	·	·	·	·	·	·	·	·	·	·	·	A	·	A	·	·	·	·	·
459	G	·	·	·	·	·	·	A	A	A	A	A	A	T	C	·	·	·	·	·	·	·	·	·	·	·	·	·
460	T	·	·	·	·	·	·	·	·	·	·	·	·	·	·	C	C	C	C	C	·	·	·	·	·	·	·	·
461	A	·	·	·	·	·	·	G	G	G	G	G	G	G	G	·	·	G	G	G	G	·	·	G	·	·	·	·
462	G	C	·	·	·	·	·	C	C	C	C	C	C	·	·	·	·	C	C	C	·	·	·	·	·	·	·	·
463	C	·	·	·	·	·	·	·	·	·	·	·	·	·	·	·	·	·	·	·	·	·	·	T	·	·	·	·
464	T	G	·	·	·	·	·	·	·	·	·	·	·	·	·	·	·	·	·	·	·	G	·	·	·	G	G	·
469	A	·	T	T	T	T	C	C	C	C	C	C	C	C	T	T	T	C	C	C	·	·	T	C	·	·	·	·
470	C	·	·	·	·	·	·	G	G	G	G	G	G	·	·	·	·	G	G	G	·	·	·	·	·	·	·	·
471	T	·	·	·	·	·	·	C	·	C	C	·	·	C	C	·	·	·	·	·	·	·	·	·	·	·	·	·
472	G	·	·	·	·	·	·	A	A	A	A	A	A	A	A	·	·	·	·	·	A	·	·	·	·	·	·	·
473	C	·	·	·	·	·	·	T	T	T	T	T	T	G	G	T	T	T	T	T	·	·	·	·	·	·	·	·
474	C	·	T	T	T	T	T	·	·	·	·	·	·	·	·	·	·	·	·	·	T	·	T	·	·	·	·	·
475	A	·	G	G	G	G	G	·	·	·	·	·	·	·	·	·	·	·	·	·	G	·	G	·	·	·	·	·
476	G	·	·	·	A	A	·	·	A	·	·	A	A	A	T	·	A	A	A	A	·	·	·	·	·	·	·	·
594	T	·	·	·	·	·	·	·	·	·	·	·	·	·	·	·	·	·	·	·	·	·	·	·	·	·	·	C
632	T	·	·	·	·	·	·	·	·	·	·	·	·	·	·	·	·	·	·	·	·	·	·	·	·	·	·	A
647	C	·	·	·	·	·	·	·	·	·	·	·	·	·	·	T	·	·	·	·	·	·	·	·	·	·	·	·
649	A	G	G	·	·	·	·	·	·	·	·	·	·	·	·	·	·	·	·	·	·	·	·	G	G	·	·	G
650	G	·	·	·	·	·	·	·	·	·	·	·	·	·	·	·	·	·	·	·	·	·	·	A	A	·	·	·
707	C	·	·	·	·	·	·	·	·	G	·	·	·	·	·	·	·	·	·	·	·	·	·	·	·	·	·	·
834	C	·	·	·	·	·	·	·	·	·	·	·	·	·	·	·	·	·	·	·	·	·	·	T	T	·	·	T
839	T	·	·	·	·	·	·	·	·	·	·	·	·	·	·	·	·	·	·	·	·	·	·	C	C	·	·	C
847	A	·	·	·	·	·	·	·	·	·	·	·	·	·	·	·	·	·	·	·	·	·	·	G	G	·	·	G
1006	T	·	·	·	·	·	·	·	·	·	·	·	·	·	·	·	·	·	·	·	·	·	·	·	·	·	·	C
1008	C	·	·	·	·	·	·	·	·	·	·	·	·	·	·	·	·	·	·	·	·	·	·	·	·	·	·	T
1009	T	C	·	·	·	·	·	·	·	·	·	·	·	·	·	·	·	·	·	·	·	C	·	·	·	·	C	·
1010	G	·	·	·	·	·	·	·	·	·	·	·	·	·	·	·	·	·	·	·	·	T	·	·	·	·	T	·
1011	T	G	C	C	C	C	C	C	C	·	·	·	·	C	·	C	·	C	C	·	·	A	C	·	C	·	A	·
1018	A	T	G	G	G	G	·	G	G	·	·	·	·	G	·	·	·	G	G	·	·	T	G	·	G	·	T	·
1019	C	G	·	·	·	·	·	·	·	·	·	·	·	·	·	·	·	·	·	·	·	G	·	·	·	·	G	·
1021	G	·	·	·	·	·	·	·	·	·	·	·	·	·	·	·	·	·	·	·	·	·	·	·	·	·	·	A
1023	A	·	·	·	·	·	·	·	·	·	·	·	·	·	·	·	·	·	·	·	·	·	·	·	·	·	·	G
1153	G	·	·	·	·	·	·	·	·	·	·	·	·	·	·	·	·	·	·	·	·	·	·	A	A	·	·	A
1251	G	·	·	·	·	A	·	·	·	·	·	·	·	·	·	·	·	·	·	·	A	·	·	·	·	·	·	·
1285	A	·	·	·	·	·	·	·	·	·	·	·	·	·	·	·	·	·	·	·	·	·	·	T	T	·	·	T
1286	A	·	·	·	·	·	·	·	·	·	·	·	·	·	·	·	·	·	·	·	G	·	·	·	·	·	·	·
1308	C	·	T	T	T	T	T	·	·	·	·	·	·	·	·	·	·	·	·	·	·	·	·	T	T	·	·	T
1329	G	·	A	A	A	A	A	·	·	·	·	·	·	·	·	·	·	·	·	·	·	·	·	A	A	·	·	A
1355	A	·	·	·	·	G	G	·	·	·	·	·	G	·	·	·	G	G	G	G	G	·	·	·	·	·	·	·
1367	T	·	·	·	·	C	C	·	·	·	·	·	C	·	·	·	C	C	C	C	C	·	·	·	·	·	·	·

the finding of some species with identical sequence. Consequently, the 16S rRNA gene sequence analysis has thus been criticized and withdrawn for housekeeping gene sequence analysis. The rate of intragenomic heterogeneities, formerly estimated to 8.1% and 21% of the strains (Alperi *et al.*, 2008; Morandi *et al.*, 2005), has been revised, by using PCR-temporal temperature gel electrophoresis (PCR-TTGE), to 41.5% of the strains, but still under estimated since the authors characterized *rrs* intragenomic heterogeneities only in the V3 *rrs* region (Roger *et al.*, 2012b). Performing standard 16S rRNA gene amplification (i.e. without discriminating the sequences before sequencing), undoubtedly results in losing information, probably the most valuable.

The number of aeromonad 16S rRNA sequences (or their corresponding genes) deposited in the public databases increased from about 1200 entries (as annotated in 2011) to exactly 5662 *Aeromonas* 16S rRNA sequences registered at the EMBL (summer 2013), which represents about a fivefold increase. Of these, only 1494 entries (ca. 26%) were sequences ranging 1300–1500 bp, and 3354 sequences (ca. 60%) below 1000 bp long, indicating most are due to a single-primer sequencing run for fast identification adding low information to the database. Curiously, 451 sequences (ca. 8%) were > 1500 bp, counting 216 and 92 of them were recently deposited by emerging scientific groups from China and India respectively. However, mistakes in taxonomy labelling and/or very poor sequencing quality are of concern for which caution is deeply advised.

Sequencing of single coding-protein housekeeping genes

Earlier studies reported that some genes coding for proteins are suitable phylogenetic markers for bacterial systematics (Yamamoto and Harayama, 1995, 1996, 1998; Huang, 1996; Stackebrandt *et al.*, 2002), at least, to short term phylogenetic relationships. In fact, it is well known that, unlike rRNA, these molecular chronometers showed a much better intra-genus resolution, even to split closely related species. Housekeeping genes coding for proteins are those clearly essential to maintain relevant functions in the cell life, which represents a selective pressure to changes. Only protein coding genes strictly essential, specifically those involved in DNA processing including transcription, and translation, show optimal chronometric capacity to resolve close bacterial relationships. This aspect is relevant when selection of housekeeping candidates is made.

The amino acid sequence is not fully free to change at random, but the corresponding coding gene is subjected to a degenerative code: different codons may translate for the same amino acid to protein, having the coding-gene more freedom to mutate in the third position of triplets, also some change is allowed in the second position. Consequently, the function may be maintained (protein sequence conserved) by nucleotide sequences differing considerably. A relevant consequence affects to their evolution mode, in contrast to concrete rRNA variation located at signature regions, mutations on these sequences may spread along the molecule. This makes the coding-protein genes very powerful to discriminate highly related strains, although the finding of conserved binding sites to design primers is sometimes extremely difficult. Definitely, the design of universal primers for sequencing protein-coding genes is not possible, even primers useful for a family or a group of several genera. In our experience, the primers designed in several genes in the genus *Aeromonas*, and in others, occasionally not worked for some of the more distant species, being necessary to generate an alignment only to venture other design possibilities.

Another obvious consequence of this mode of evolution subjected to degenerative code shown by the coding-protein housekeeping genes sequences is that, contrarily to a popular claim among some researches, phylogenies derived from the corresponding amino acid sequences do not behave as good molecular chronometers to split species, of course the trees are not congruent to those obtained from gene sequences. Closely related species may show highly similar or 'identical' protein sequence, although considerable nucleotide changes; consequently, protein sequences are not good resolving closely related taxa. In our experience, trees from amino acid sequences illustrate very well this statement (unpublished). Apart of this, as a standard control for quality of nucleotide sequences, we usually recommend to translate determined gene sequences to protein sequences in order to check open reading frames (detect possible stop codons), consistency of insertions/deletions, and to analyse possible mistakes during the alignments.

The first housekeeping coding-protein gene used for a phylogenetic analysis including all *Aeromonas* species described at that time was the *gyrB* gene, which encodes for the subunit B of the DNA gyrase (Yáñez *et al.*, 2003). The use of this gene as a phylogenetic marker for identifying aeromonads to the species level revealed to be valuable thanks a higher diversity enabling better capacity to differentiate between species. Another finding indicated that inter-species relationship distances (branch lengths between species clusters) were not the same for all species because, as it is generally recognized, they all do not evolve at exactly the same rate. The use of different algorithms does not alter the taxonomic conclusions (species clustering).

A further work, compared the *gyrB* phylogeny with the one obtained by using the *rpoD* gene, which encodes the σ70 factor of the RNA polymerase. Clustering of the *gyrB*- and *rpoD*-derived trees were highly similar (Soler *et al.*, 2004), indicating that both phylogenetic markers are synchronized to that point to split species. Other housekeeping genes used in *Aeromonas* were the *rpoB*, which encodes for the RNA polymerase β-subunit (Küpfer *et al.*, 2006), the *dnaJ*, coding a thermal shock protein (Nhung *et al.*, 2007), the *recA*, which encodes for a DNA repair protein (Sepe *et al.*, 2008) and the gene coding for the Cpn60 chaperone type I, *cpn60*, involved in protein assembly (Miñana-Galbis *et al.*, 2009). The analysis of a 1.1 kb *gyrB* sequence confirmed that this gene has high resolving power, with maximal interspecies divergence of 15.2%. Similar results were obtained by partial sequencing of the *rpoB* (517 bp), which showed maximal interspecies divergence of 13%. The topologies of the *gyrB*- and *rpoB*-derived trees were similar and confirmed the close relationship of species within the genus *Aeromonas* (Kupfer *et al.*, 2006). Similarly, the sequences of *cnp60* (555 bp) showed inter-species divergence rates between 3.7% and 16.9% and intra-species divergence rates ≤3.5% and displayed similar discriminatory power to *gyrB* and *rpoD* sequences (Miñana-Galbis *et al.*, 2009). Based on 50 strains isolated from environmental (lake water and wastewater) and animal sources, *rpoB* was validated as a valuable identification tool for the aeromonads (Lamy *et al.*, 2010).

Sequencing of these gene fragments is relatively easy and cheap, consequently the approach has been recommended for fast and reliable identification of presumptive *Aeromonas* isolates. In fact, this molecular approach has been determinative in rapidly increasing the number of strains belonging to new *Aeromonas* species that were recently described on the basis of a single strain (i.e. *A. schubertii*, *A. simiae*, *A. fluvialis*, *A. taiwanensis*, *A. sanarellii*, *A. cavernicola*; some data unpublished). However, for accurate identification, a single-gene phylogeny should be contrasted with other housekeeping-gene trees to avoid possible misclassifications as this shown in Fig. 8.2. Owing to the close *gyrB*-relationship of the single *A. sanarelli* strain to *A. caviae* (represented by 33 strains), the branch of the former may cluster at a border-line position within *A. caviae*, giving rise to an erroneous conclusion.

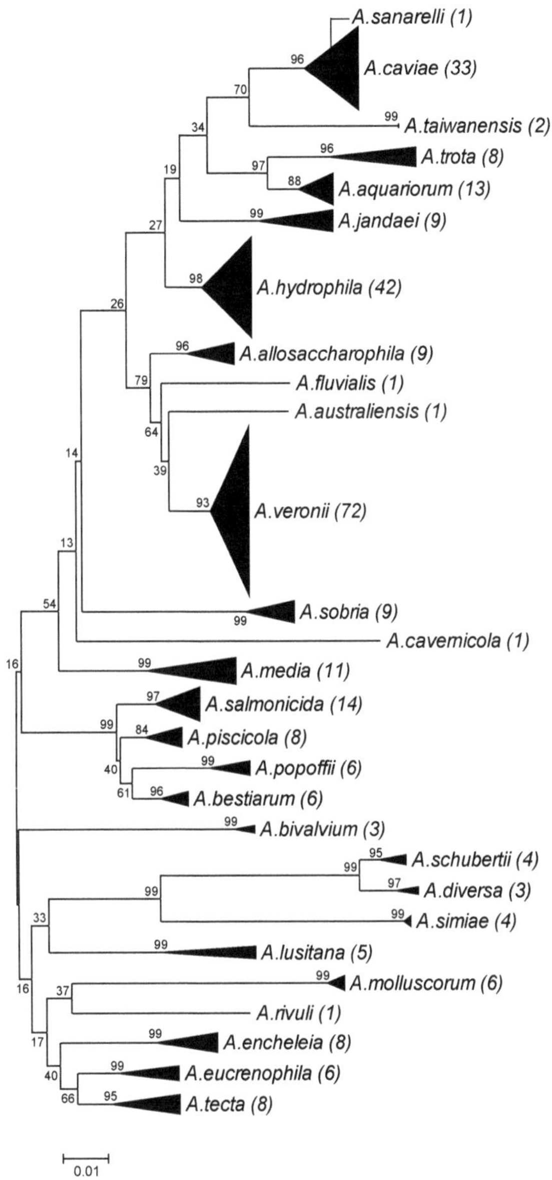

Figure 8.2 Phylogenetic NJ-tree derived from a 759 bp fragment of *gyrB* sequences belonging to 294 *Aeromonas* strains available at the EMBL public database, including type and reference strains of all know described species and food strains (Martinez-Murcia and co-workers), clinical and environmental isolates (Lamy and co-workers). Numbers at nodes indicate bootstrap values (percentage of 1000 replicates). Bar, 0.01 estimated nucleotide substitutions per site. *A. sanarelli* branch has been manually added to highlight the borderline position within *A. caviae* cluster.

Multilocus phylogenetic analysis of protein encoding housekeeping gene fragments

Multilocus phylogenetic analysis (MLPA), also known as multilocus sequence analysis (MLSA), is the approach on which a concatenated sequence, comprising data from several loci, is subjected to phylogenetic inference. The strategy may be confused with the multilocus sequence typing (MLST), a highly discriminative method for characterizing microbial isolates (Maiden *et al.*, 1998). MLST was designed firstly for global epidemiology and surveillance, and has been successfully applied to study bacterial populations of a single species (intra-species level) in the epidemiological characterization of clinically relevant pathogens. MLST is based on the same principles as multilocus enzyme electrophoresis (MLEE) but relies directly on allelic differences derived from DNA sequences of internal fragments of protein encoding housekeeping genes. According to Vauterin and Vauterin (2006), better use of the term multilocus sequencing analysis (MLSA) when strains are clustered based upon multiple housekeeping gene sequences, whereby the analysis is not necessarily based on allele numbering but on the calculation of total sequence similarity. Since the DNA sequence data are subsequently subjected to phylogenetic analysis, the term multilocus phylogenetic analysis (MLPA) has recently been considered to be more appropriate (Martínez-Murcia *et al.*, 2011).

MLPA of representative strains of a genus offers the opportunity to incorporate concatenated multigene phylogenies into bacterial systematics, as regularly recommended (Stackebrandt *et al.*, 2002; Schleifer, 2009; Tindall *et al.*, 2010). The strategy deals with the hypothesis of that the relationships obtained reflect those contained in the complete genomes, i.e. are synchronized with the evolution of the entire genetic content. The sequencing of multiple loci to assess phylogenetic relationships is a very powerful method because it (i) cumulates a large number of nucleotides from different and unrelated genes, therefore increasing the amount of information representative of the whole genome and evaluating levels of congruence from each selected gene, (ii) allows a more accurate analysis of species delineation, (iii) enables to quantify genetic diversity, (iv) enables a better resolution in the identification, including the intra-species level, (v) limits consequences of erroneous identification due to horizontal gene transfer which can be detected by comparing single-gene trees.

Several authors reported *Aeromonas* phylogenies based on several genes with the aim to improve species discrimination, identification accuracy and robustness. The housekeeping sequencing works reported contribute to date with up to 17 distinct genes of which partial-sequence are available: *atpD, cnp60, dnaJ, dnaK, dnaX, gltA, groL, gyrA, gyrB, metG, ppsA, radA, recA, rpoB, rpoD, tsf,* and *zipA* (Yañez *et al.*, 2003; Soler *et al.*, 2004; Küpfer *et al.*, 2006; Nhung *et al.*, 2007; Sepe *et al.*, 2008; Nováková *et al.*, 2009; Miñana-Galbis *et al.*, 2009; Martinez-Murcia *et al.*, 2011; Martino *et al.*, 2011; Roger *et al.*, 2012a), currently in all described species of the genus *Aeromonas*. This probably makes the *Aeromonas* group one of the best genus ever been phylogenetically investigated (Fig. 8.3).

To date, three MLPA schemes for the genus *Aeromonas* have been published, with different objectives. The first, based on seven genes (4705 bp), provided a phylogenetic frame with taxonomic purposes (Martinez-Murcia *et al.*, 2011). A second study, using six genes (3084 bp), was developed for achieving precise genotyping of outbreak traceability of strains isolated from fish, crustaceans and molluscs (Martino *et al.*, 2011). The third MLPA so far published, based on seven genes (3993 bp), was developed for studying population

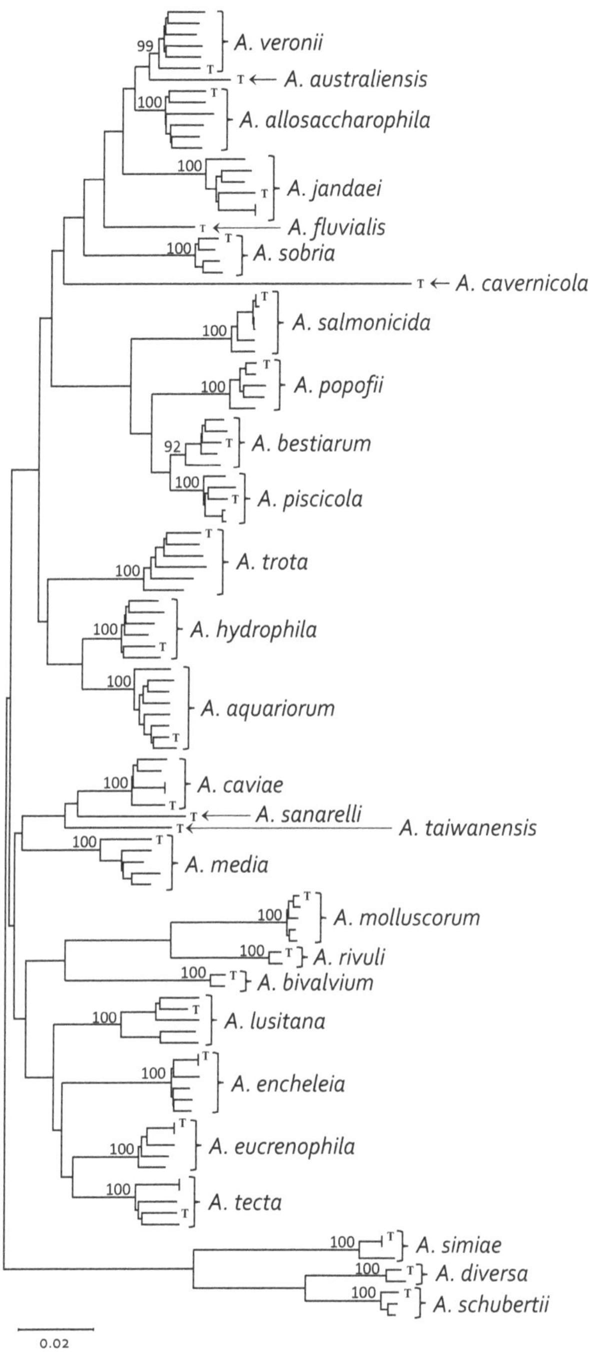

Figure 8.3 Unrooted neighbour-joining phylogenetic tree derived from the MLPA of concatenated sequences of six housekeeping genes (*gyrB*, *rpoD*, *recA*, *dnaJ*, *gyrA* and *dnaX*; total 4204 bp) showing the relationships of all known *Aeromonas* species. Numbers indicate bootstrap values. Sequences provided by Martínez-Murcia and co-workers available at the EMBL. Bar, 0.02 estimated substitutions per site. T, branch of type strain from corresponding species.

structure and phylotaxonomic issues (Roger *et al.*, 2012a). Despite these studies used different genes in their analyses, all yielded congruent phylogenetic trees, and demonstrate the ability of this approach to solve classification issues.

The work from Martinez-Murcia *et al* (2011) was based on the *gyrB, rpoD, recA, dnaJ, gyrA, dnaX* and *atpD* genes, the two latest encoding for elongation factor of DNA polymerase III and the F1 sector of the membrane-bound ATP synthase β subunit, respectively. For all the single genes but one (*atpD*), there was two different levels of clearly delimited genetic divergence, corresponding to the intra-species and to the inter-species levels (Fig. 1 in Martínez-Murcia *et al.*, 2011). The topologies of the single-gene phylogenetic trees, at the deep branching nodes, were obviously not exactly the same in all the phylogenies obtained, because mutation rates are not the same in all genes (according to function relevancy) and coding-protein genes do not offer a good resolution at long-term relationships. The relationships at the ancestry of the genus, at time of speciation, are uncertain and deep branching in the tree can be only regarded as a mere hypothesis. However, what is relevant, single-gene analyses showed that strains analysed grouped in tight clusters, with high bootstrap values (100%), which clearly separated all the species. The robustness of the species clustering shown by the concatenated tree was the greatest, enabling distinction even between closely related species. No overlapping between intra- and inter-species ranges was persisting with the concatenated MLPA. The few chronometric distortions encountered in single genes, whether or not corresponding to possible horizontal gene transfer and recombination, or simply the low resolution promptly found in some genes (i.e. *atpD*) probably due to a relative housekeeping condition, were absorbed by the concatenated analysis and disappeared in the final MLPA tree. Although for further new species descriptions the analysis of several housekeeping genes (at least five) is recommended, preliminary screening of isolates may be undertaken by using only two selected by the researcher. Another important aspect to consider is the number of reference strains from each species used in the MLPA. If not balanced, misleading interpretations with consequences for their taxonomy may occur.

The multilocus scheme developed by Martino *et al.* (2011) achieved precise strain sequencing typing (MLST), and the phylogenetic analysis of concatenated sequences delineated the relationship among the taxa belonging to the genus *Aeromonas*, providing a powerful tool for outbreak traceability, host range diffusion, and ecological studies. It enabled species identification although one can regret that only 15 type strains were included in the analysis. However, the concatenated sequence analysis provided was a mere consequence of the data obtained for their MLST analysis, main purpose of this study. The scheme was based on *groL* which encodes for chaperonin groEL, *ppsA* which encodes for phosphoenolpyruvate synthase, *metG* which encodes for methionyl-tRNA synthetase, *gyrB, gltA*, and *recA*. A Web-based MLST sequence database (http://pubmlst.org/*aeromonas*) specific for the *Aeromonas* genus was developed and implemented with all the results.

Roger *et al.* (2012a) based their MLPA scheme on seven genes encoding housekeeping proteins, including *gyrB, rpoB, radA* encoding for DNA repair protein RadA, *gltA* encoding for the type II citrate synthase, *dnaK, encoding for* heat shock 70 kDa protein (HSP70), *tsf*, encoding elongation factor Ts, and *zipA* which encodes for cell division protein ZipA. The population structure was inferred from MLPA following reconstruction of the distance and ML trees from the concatenated sequences. This work was based on a 195 strain collection (58% from clinical origins). The authors estimated a high genetic diversity, recombination rates and horizontal gene transfer in *Aeromonas* species (5.8% in the studied population). Altogether, this highlighted the mode of evolution in this group as complexes of species, with a possible speciation in progressing locally, giving raise some closely related species which may explains the usual difficulties in bacterial identification. The analysis showed small clusters of strains associated with human infection and provided phylotaxonomic data that confirmed previous results (Martínez-Murcia *et al.*, 2011) in helping to clarify some confusing taxonomy issues of the genus *Aeromonas*, largely reported in the literature. In agreement with results from Martinez-Murcia *et al.* (2011), the MLPA tree reconstructed from the concatenated sequences did not show overlapping between intra- and inter-species ranges (Fig. 8.4). Using MLPA, Roger *et al.* (2012a), but also Martino *et al.* (2011), performed a more appropriate identification of some strains which probably were deposited in strain collections under erroneous identification or incomplete nomenclature.

Restriction fragment length polymorphism (RFLP) of PCR fragments

16S rDNA – RFLP

Because most of the very few informative 16S rRNA nucleotide positions found to split species are grouped in concrete regions, the use of some endonucleases which cleaves on sites showing these target signatures may yield on patterns differing by the length of resulting fragments (Fig. 8.5). Early, Borrell *et al.* (1997) took advantage of that and developed a species identification scheme based on the RFLP analysis of the 16S rRNA gene of the different species, also known as ARDRA (amplified ribosomal DNA restriction analysis). The 16S rRNA gene (1503 bp) was amplified by PCR before being subjected to restriction analysis by the enzyme pair *Alu*I and *Mbo*I. When 76 previously characterized clinical isolates were compared with RFLP analysis, patterns were in agreement with biochemical identification except for few cases: one strain identified as *A. veronii* by biochemical tests showed the RFLP pattern of *A. hydrophila* while, conversely, three strains of *A. hydrophila* showed RFLP patterns of *A. veronii*. This added more evidences of that some biochemical key tests may lead to incorrect species determination. Used endonucleases were not enough to distinguish patterns of a few species, for instance *A. salmonicida* from *A. encheleia*. Thus, additional enzyme treatment was needed to split these species, involving digestion with other enzymes (*Nar*I and *Hae*III), (Borrell *et al.* 1997). The schemes were progressively improved by increasing the list of enzymes used, which today includes: *Alu*I, *Mbo*I, *Nar*I, *Hae*III, *Alw*NI, *Pst*I, *Sfa*NI, *Cfo*I, *Mnl*I (Borrell *et al.*, 1997; Graf *et al.*, 1999; Figueras *et al.*, 2000).

ARDRA has also been used to describe new RFLP patterns corresponding to non-described species, and detect the presence of microheterogeneities (Martínez-Murcia *et al.*, 2005; Alperi *et al.*, 2008; Figueras *et al.*, 2011b). Description of 11 additional new *Aeromonas* species since the year 2000 has open a limitation to the usefulness of this method because some RFLP patterns from new species were found to be identical to these from other species. Ciapini *et al.* (2002) noted discrepancies between Biolog and ARDRA results, and between chromosomal DNA restriction and ARDRA profiles.

Some limitations of this approach include (i) relatively laborious method to perform and expensive, (ii) insufficient discriminatory power intrinsic to the extremely conserved sequence of the targeted molecule. Non-described species may be undetected. The method was very helpful at that time (late 1990s to early 2000s) particularly for laboratories showing limitations accessing to facilities for sequencing. At present, 16S rRNA based techniques for

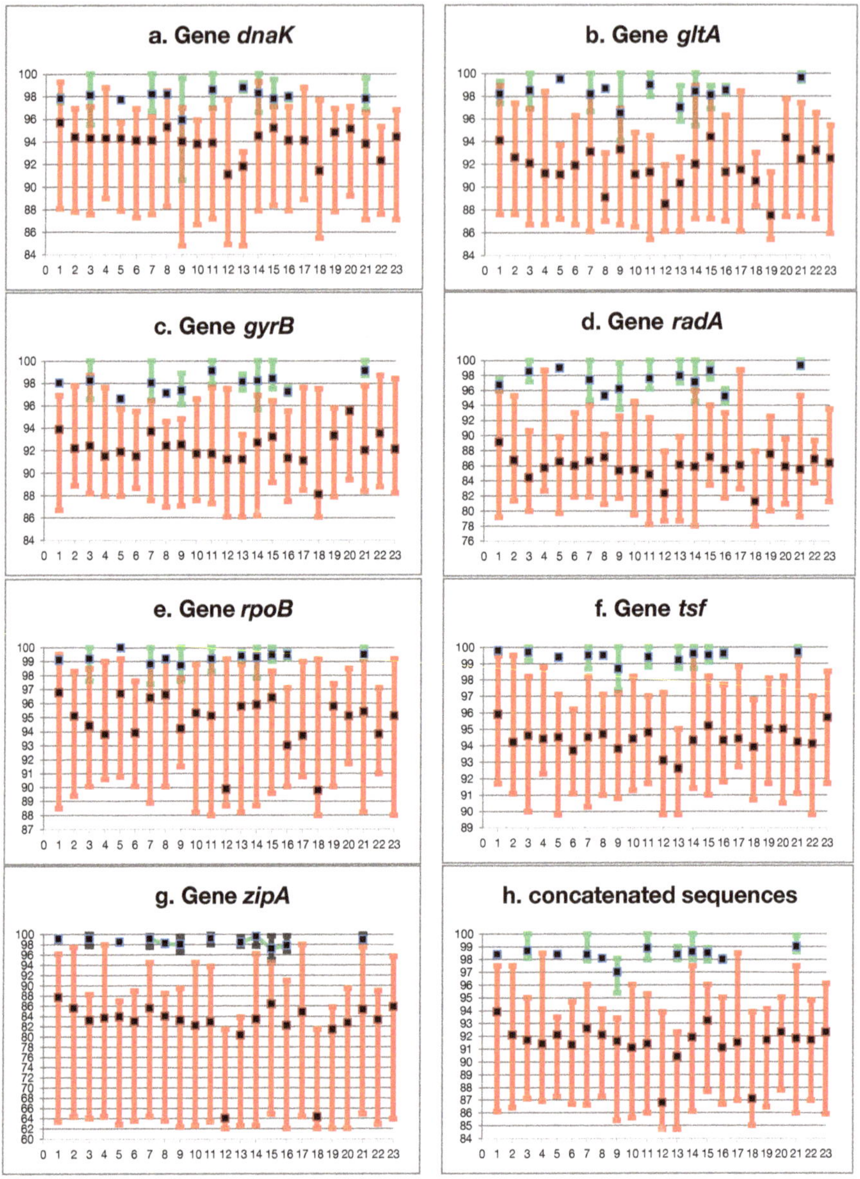

Figure 8.4 Ranges of intra- and inter-species sequence similarities between 191 strains of aeromonad collection for seven genes (a. to g.) and for concatenated sequences (h.) (*dnaK-gltA-gyrB-radA-rpoB-tsf-zipA*). Difference in sequence similarity: intra-specific (green); inter-specific (red); mean group (black square). Absciss, Numbers 1 to 23 correspond to taxa (with *n* = number of strains): 1. *A. allosaccharophila* (*n* = 3); 2. *A. bestiarum* (*n* = 1); 3. *A. caviae* (*n* = 34); 4. *A. encheleia* (*n* = 1); 5. *A. enteropelogenes* (*n* = 2); 6. *A. eucrenophila* (*n* = 1); 7. *A. hydrophila* (*n* = 35); 8. *A. jandaei* (*n* = 2); 9. *A. media* (*n* = 6); 10. *A. popoffi* (*n* = 1); 11. *A. salmonicida* (*n* = 8); 12. *A. schubertii* (*n* = 1); 13. *A. sobria* (*n* = 5); 14. *A. veronii* (*n* = 71); 15. *A. aquariorum* (*n* = 8); 16. *A. tecta* (*n* = 3); 17. Hybridation Group HG11 (*n* = 1); 18. *A. diversa* (*n* = 1); 19. *A. fluvialis* (*n* = 1); 20. *A. taiwanensis* (*n* = 1); 21. *A. piscicola* (*n* = 3); 22. *A. sanarellii* (*n* = 1); 23. Souche *A. bestiarum* CCM 1271 (*n* = 1).

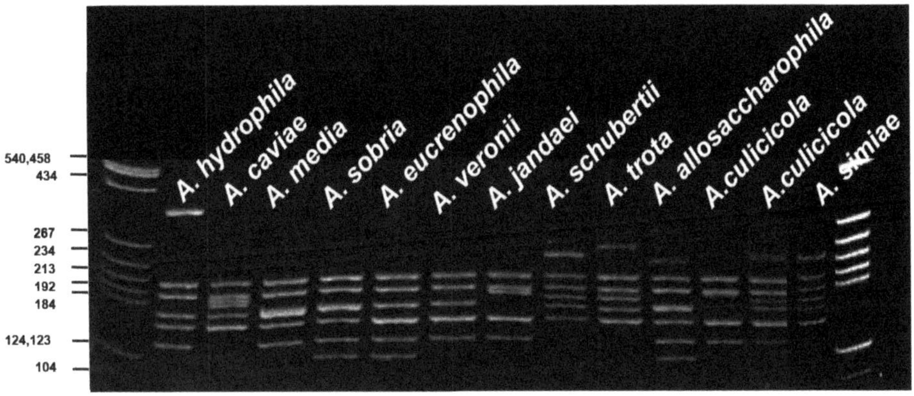

Figure 8.5 16S rDNA-RFLP patterns obtained from type strains of some *Aeromonas* species, following the protocol from Borrel *et al.* (1997) and Figueras *et al.* (2000).

identification of *Aeromonas* species have been superseded by protein-coding housekeeping gene sequencing analysis.

RFLP of PCR-amplified toxin-coding genes

Similarly, the strategy to characterize strains by using PCR-amplified restriction fragment length polymorphism may be applied to genetic sequences coding for virulent factors (such as toxins) described to be intrinsic to one or more *Aeromonas* species. Kingombe *et al.* (1999) used the cytolytic enterotoxin (ahcytoen) from *A. hydrophila* and the homologous aerolysin and haemolysin genes from different *Aeromonas* spp. to develop PCR primers that were used in RFLP and/or PCR amplicon sequence analysis (PSA) to characterize the distribution of these genes in various environmental, food, clinical and reference strains ($n = 350$). A later study to evaluate the utility of this methodology included a total of 201 strains obtained from 78 raw and 123 processed and ready-to-eat retail food samples (Kingombe *et al.* 1999). In this work, a variant of this method included further characterization of toxin-gene by amplicon sequencing analysis (PCR-ASA). By using the conventional cultivation method based on the Ampicillin-Dextrin Agar (ADA) medium, authors found presumptive *Aeromonas* colonies in 65 samples out of 201 (32.3%), and revealed the presence the targeted virulence genes in only 51/201 (25.4%) of the tested samples.

The approach intuitively may help to address the prevalence of *Aeromonas* harbouring some genes coding for virulence factors, by screening previously isolated strains or directly on the samples. May be worth to remind that non-described homologous genes may contain variation at the targets of designed primers and toxicity may be underestimated when drawing conclusions. On the opposite, it is not clear how well ahcytoen is accurate (or sensitive, or relevant) to detect virulent aeromonads.

Temporal temperature gel electrophoresis (TTGE) of 16S rDNA variable regions

One of the problems encountered when 16S rRNA gene PCR-products are directly subjected to sequencing is derived from the existence of multiple copies of *rrn* operon (organized as a multigene family) showing discrete sequence nucleotide polymorphisms (so-called 'microheterogeneities') at some variable regions. PCR of this multigene family yields a heterogeneous template giving raise to sequence ambiguities at positions where composition is polymorphic. Above is particularly relevant when closely related microorganisms are under analysis, precisely the case for the *Aeromonas* species. Temporal temperature gel electrophoresis (TTGE) makes possible to separate such a mix of 16S rDNA PCR products, on the basis of their GC content. By regularly increasing the temperature during electrophoresis, separation of DNA fragments of the same length containing different nucleotide sequences may occur. Rather a limitation, the 16S rRNA gene polymorphisms are thus exploited to provide strain-specific patterns. The method enables to evaluate quickly and easily a large number of strains, although TTGE separation is limited to only short fragments (Fig. 8.6).

This method is relevant for *Aeromonas*, a genus which strains contain a high number of *rrn* operons (from 8 to 11) showing a high frequency of microheterogeneities (Morandi *et al.*, 2005; Alperi *et al.*, 2008; Roger *et al.*, 2012b). Roger *et al.* (2012b) used TTGE after PCR-amplification of a 199 bp fragment which overlaps the 16S rDNA variable region V3. They identified 38 bands from a 195 strain collection that included all the described species (Roger *et al.*, 2012b). Specific bands or, more frequently, combination of two or three bands were informative for identifying most aeromonads at the level of species. An illustrative example is given by the case of *A. caviae* and *A. trota*: TTGE allows easy and robust split of this pair of taxa while they are distinguishable by only a two nucleotide difference. The number of bands observed on a pattern was also found to be informative of the species: only five species showed three or four bands, while other species showed one or two bands, although this high intragenomic heterogeneity did not concern all the members of a species. *A. caviae* and *A. media* are to be mentioned because they were associated with the highest rate of strains with three or four bands (44% and 30%, respectively), compared with *A. hydrophila* (3%). Roger *et al.* (2012b) accurately identified 152 (78%) strains out of 195 evaluated, being only the 199 bp (V3 region) analysed. Discrepancies included five erroneous identification results and 37 unresolved cases. TTGE, currently based on the 199 bp

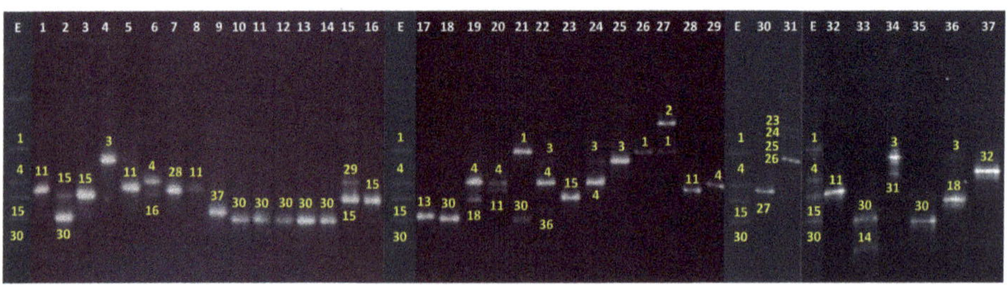

Figure 8.6 TTGE-PCR patterns obtained from the type strains of species and subspecies within the genus *Aeromonas*. 1: *A. diversa*; 2: HG11; 3: *A.* tecta; 4: *A. aquariorum*; 5: *A. veronii* bv. *veronii*; 6: *A. trota*; 7: *A. sobria*; 8: *A. simiae*; 9: *A. schubertii*; 10: *A. salmonicida* subsp. *smithia*; 11: *A. salmonicida* subsp. *salmonicid*; 12: *A. salmonicida* subsp. *pectinolytica*; 13: *A. salmonicida* subsp. *masoucida*; 14: *A. salmonicida* subsp. *achromogenes*; 15: *A.* popoffii; 16: *A. media*; 17: *A. allosaccharophila*; 18: *A. bestiarum*; 19: *A. caviae*; 20: *A. culicicola*; 21: *A. encheleia*; 22: *A. enteropelogenes*; 23: *A. eucrenophila*; 24: *A. hydrophila* subsp. *anaerogenes*; 25: *A. hydrophila* subsp. *dhakensis*; 26: *A. hydrophila* subsp. *hydrophila*; 27: *A. hydrophila* subsp. *ranae*; 28: *A. ichthiosmia*; 29: *A. jandaei*; 30: *A. bivalvium*; 31: *A. molluscorum*; 32: *A. veronii* bv. *sobria*; 33: *A. fluvialis*; 34: *A. taiwanensis*; 35: *A. piscicola*; 36: *A. sanarellii*; 37: *A. rivuli*.

fragment, could not distinguish the species *A. bestiarum* and *A. piscicola* from *A. salmonicida*, *A. jandaei* from *A. caviae*, *A. simiae* and *A. diversa* from *A. veronii*, and *A. eucrenophila* from *A. tecta*. This study included only a limited number of strains belonging to species other than *A. caviae*, *A. hydrophila*, *A. veronii*, *A. aquariorum* and *A. media*. It is likely that the method is improvable by amplifying complementary regions, e.g. V1 and V2 regions for some taxa.

Species detection on samples

Direct detection of *Aeromonas* DNA markers by using genetic methods, mostly polymerase chain reaction (PCR) together with the use of genetic probes, has rapidly advanced in the past decade. Genetic methods have been developed for detection of *Aeromonas* spp. in a variety of matrices, focusing one or multiple targets, and using several methodological strategies. Genetic methods are more sensitive and faster than culture, not only for complex matrices with dense background flora that interferes with culture detection, but of course for organisms for which culture methods are inadequate or unavailable. The use of selective agents, particularly antibiotics (i.e. ampicillin), in culture media may inhibit some strains of the target bacteria (i.e. *A. trota*). The independence of culturing of the nucleic acid analysis makes this new genetic approach for detection very fast, reliable, sensitive, and precise.

PCR analysis

Direct detection of aeromonads by polymerase chain reaction (PCR), and their variants, has been by far the most popular technique used in recent methodology developments for *Aeromonas* detection. Selected targets were often the rRNA genes and their intergenic spacers, but of course, also using primers designed from housekeeping protein-coding genes and factors considered specific moieties of relevant *Aeromonas* species.

A pioneer work developing a PCR protocol for identification of *Aeromonas sobria* was based on the 16S rRNA sequence data (Ash *et al.*, 1993a). PCR targeting species-specific 16S rRNA gene signatures were also developed to identify rarely isolated species *A. schubertii* and *A. jandaei* (Ash *et al.*, 1993b). Further 16S rDNA-based PCR method were developed to identify *A. hydrophila* and *A. veronii*, from environmental *Aeromonas* isolates obtained from sewage effluents and surface waters around Sydney Dorsch *et al.* (1994), or to detect *A. caviae* and *A. trota*, with a limit between 50 and 100 cells per gram of crab meat (Khan and Cerniglia, 1997). Besides, a couple of genus-specific *Aeromonas* PCR primers targeting the 16S–23S rDNA intergenic spacer region were developed by Kong *et al.*, (1999). The potential role of houseflies in the epidemiology of *A. caviae* was investigated by examining the prevalence of *A. caviae* in houseflies collected from two South Carolina farms and one restaurant, by means of PCR amplification of 16S rRNA *Aeromonas* genus-specific signature followed by partial sequencing (Nayduch *et al.*, 2001). *A. caviae* was detected in 78% dairy farm flies, 55% pig farm flies, 39% restaurant flies, and faeces from cows and pigs (58% and 100%, respectively).

Some authors have reported PCR methods using 16S rRNA universal primers combined with other techniques on which resides the specificity of the test. Peng *et al.* (2002) described the use of universal primer PCR for the detection of 16S ribosomal RNA (rRNA) genes from *A. hydrophila* captured by anti-*A. hydrophila* antibody coupled to a microplate. The approach combining immuno-capture with 'universal' PCR provided a quick, sensitive (10 CFU), and reproducible way for the detection of *A. hydrophila* cells. Universal primer

PCR combined with denaturing gradient gel electrophoresis (DGGE) was evaluated by Ji *et al.* (2004) as a method permitting the rapid detection causative bacterial pathogens (*Pseudomonas fluorescens, Vibrio anguillarum, Aeromonas hydrophila, Vibrio fluvialis, Providencia rettgeri* and *Aeromonas sobria*) rapidly. Since then, this method has been increasingly used to study the bacterial communities including *Aeromonas* and their dynamics under various conditions, e.g. the changes of the contaminating bacteria in food (Lafarge *et al.*, 2004; Li *et al.*, 2006), the intestinal microbial communities of the rainbow trout under various treatments (Dimitroglou *et al.*, 2009), the phylogenetic diversity of bacterioneuston (Azevedo *et al.*, 2012), the bacterial diversity in intestinal tracts of earthworm (Hong *et al.*, 2011a,b), or the detection of *Aeromonas salmonicida* and other bacteria in skin mucus from wild salmon and cultured trout (Quinn and Stevenson, 2012). PCR-DGGE developed by Tacão *et al.* (2005b) used *gyrB* sequences to study the diversity of *Aeromonas* in wastewater treatment plant finding it was a powerful tool to monitor these species in complex water samples. Calhau *et al.* (2010) further developed a PCR-DGGE based on three genes (*gyrB, rpoD, sodB*) with *Aeromonas* specific-primers and applied to *Aeromonas* communities in estuarine water. The combined use of more than one set of primers was advisable.

Several PCR based detection schemes use virulence factors as genetic markers to differentiate between the species *A. hydrophila*, as considered to be the most likely species to cause disease in humans. A PCR assay targeting the *lip* gene which encodes a thermostable extracellular lipase of *A. hydrophila*. With *A. hydrophila* cells, the sensitivity of the PCR assay was <10 CFU (Cascon *et al.*, 1996). Baloda *et al.* (1995) developed a PCR method used to assay the presence of the aerolysin gene (encodes the β-subunit of aerolysin) in a total of 89 *A. hydrophila* and *A. sobria* strains isolated from drinking water, fish and foods. PCR products were amplified only from A. *hydrophila* strains with a haemolytic phenotype but not from haemolytic *A. sobria* strains. The possibility of false negative results was not investigated in deep. Identification of *A. trota* (hybridization group 13) by amplification of the aerolysin gene was achieved by Khan *et al.* (1999) and direct detection of diarrhoeagenic *Aeromonas* from faeces by targeting aerolysin toxin gene by Kannan *et al.* (2001). Out of 602 faecal samples collected from patients with acute diarrhoea admitted at the hospital, 68 (11.29%) were found to be possessing *Aer* gene. Tests mentioned above detected *A. trota* strains, currently missed using culture media containing ampicillin. Kong *et al.* (2002) also used the *aerA* gene in a multiplex PCR that allowed the simultaneous detection of six commonly encountered waterborne pathogens. Detection limit of 2 $\log_{10}$ CFU was reported for the marine samples processed. A duplex real-time SYBR green PCR assays for detection of 17 species of food- or waterborne pathogens in stools used the *aerA* gene as target for *A. hydrophila* detection (Fukushima *et al.*, 2003).

Nucleic acid amplification PCR methods targeting virulence genes have been used to study the distribution of this features among environmental samples, but also to differentiate pathogenic from non-pathogenic strains (Kingombe *et al.*, 1999, 2004; Chacon *et al.*, 2003; Sen and Rodgers 2004). Gonzalez-Rodriguez, *et al.* (2002) developed a protocol that involves an overnight selective enrichment step in tryptic soy broth yeast extract containing ampicillin followed by PCR amplification of two haemolysin genes, *hlyA* and *aerA*, that contribute to the virulence of *A. hydrophila*. Detection limit was 1–10 cfu/g within 24 hours. Wang *et al.* (2003) developed a multiplex PCR method for detection of haemolysin and aerolysin genes in *A. hydrophila* and *A. sobria*, and used it to screen 121 clinical and 7 reference strains for *ahh1*, *asa1*, and *aerA* genes. Phenotypic expression of toxins was evaluated in Vero cell culture cytotoxicity assay and the presence and number of haemolysin genes correlated with cytotoxin titres.

Other authors took the option to target several targets in the same PCR reaction with purposes of species identification. Ozbas *et al.* (2000) developed a multiplex and semi-nested PCR method for detection of *A. hydrophila* in raw milk. The detection limit was 100 CFU/g and the detection rate was 23% for PCR and 14% for culture. Chu and Lu (2005) developed a multiplex PCR (mPCR) method that targeted a gene associated to virulence and a region of the 16S rDNA that is species-specific. However, the possibility of that the two target genes, if present, could arise from different strains, was not discussed. Another multiplex for detection of *Aeromonas* isolates targeted three genes lipase, elastase, and DNA gyraseB (Sen, 2005). Three separated multiplex reactions were required due to the large number of primers involved. For rapid detection of six virulence factors of *Aeromonas* isolates, a hexaplex PCR assay was developed by Nam and Joh (2007). The virulence factors detected were aerolysin (*aer*), GCAT (*gcat*), serine protease (*ser*), nuclease (*nuc*) lipase (*lip*) and lateral flagella (*laf*). The authors relate the dominance of *Aeromonas sobria* isolation with detection of virulence genes *aer* and *nuc*. However, sensitivity of PCR targeting virulence genes has never been deeply evaluated.

Finally, the loop-mediated isothermal amplification strategy, a method that not need run PCR in thermocyclers, has been used to target the *A. caviae* 16S-23S intergenic spacer regions (internal transcribed spacer). Eighteen reference strains and 109 clinical samples were analysed founding this a reliable method (Wei *et al.*, 2008).

Quantitative real time PCR (qPCR)

Developed in the mid-1990s for the analysis and quantification of nucleic acids, real-time PCR (qPCR) has become the method of choice for molecular biological testing, gaining rapidly in popularity, to the point where it is now accepted in many laboratories as the standard method for detecting microbial pathogens. The advancement provided by the real-time version of PCR is due to its unique ability to monitor the complete DNA amplification process, while PCR reaction occurs. This is achieved by using fluorescent DNA binding dyes resulting in PCR-kinetics. As the amount of detected fluorescence is proportional to the amplified copy number, a kinetic parameter (Ct) allows us to ascertain the original quantity of template. Real-time PCR has catalysed wider acceptance of PCR because it is more rapid, sensitive and reproducible, making the implementation of real-time PCR attractive to many.

Shannon *et al.* (2007) developed 13 probe and primer sets to detect a number of pathogens in wastewater: *Aeromonas* hydrophila, *Bacillus cereus, Clostridium perfringens, Enterococcus faecalis, Escherichia coli, E. coli* O157:H7, *Helicobacter pylori, Klebsiella pneumoniae, Legionella pneumophila, Listeria monocytogenes, Pseudomonas aeruginosa, Salmonella* sp., *and Staphylococcus aureus.* The adhesin gene (*ahaI*) was used as target by quantitative TaqMan(r) PCR (qPCR) to detect *A. hydrophila.* The sensitivity of the assay was 100 fg of genomic DNA (= 22 gene copies), based on a standard curve generated using *A. hydrophila* purified DNA. This qPCR method was effective for the purpose and confirmed the risk of exposure to pathogens in the wastewater discharge, as established in the Environment Canada guidelines.

qPCR has been developed by Trakhna *et al.* (2009) for identifying reference strains, clinical, and environmental isolates of *A. hydrophila.* The duplex TaqMan-probe PCR developed targeted both, the 16S rRNA and the *aerA* enterotoxin genes. They showed a correlation with biochemical analysis, whatever the origin of the isolate (France, Lebanon, Tunisia). LabChip preparation time was 20 minutes and the PCR products were revealed by using a micro-channel fluidic instrument in 5 minutes.

During present year, a qPCR test based on a genus-specific 16s rRNA target to detect all known species of *Aeromonas* has been developed by Martinez-Murcia and co-workers (unpublished), and is commercially available (Genetic PCR Solutions, Spain). Thermal regime is a fast two-step protocol of 45 cycles on which annealing of primers and extension by polymerization occurs simultaneously at stringent temperature of 60°C in 45 seconds. Full qPCR reaction takes less than one and half hours. The qPCR method has been calibrated by using a synthetic DNA fragment containing the amplified 119 bp sequence and inserted in plasmid pEX-, a vector backbone of 2450 bp, as standard DNA template. Tenfold dilutions containing 10^6 to 10^1 copies of this standard DNA were subjected to qPCR, using the reagents supplied in the kit (AerSpp qPCR-test; GPS), including a negative control with no sample (Fig. 8.7). Validation following international norm for laboratory assays and calibration guidelines (ISO 17025) has been performed, including specificity (inclusivity and exclusivity), reproducibility, repeatability, and determination of the limits for detection and quantification. The test has been conceived with some practical advantages for routine laboratory screening: (i) all reagents (primers/probe, dNTPs, Taq polymerase, and buffers) are together in the same tube; (ii) they are presented at a ready-to-use 'mono-dose' format, avoiding manipulation of stocks, damage of dye, and cross contamination; and (iii) the components are dehydrated, enabling transportation and manipulation at room Tª without continuous frosting–defrosting of reagents, known to degrade oligonucleotides.

Problems due to possible contamination of *Aeromonas* DNA are one of concern when using so much sensible as the qPCR. Liu *et al.* (2012), using a multiplex PCR-Luminex for a pathogen panel that includes *Aeromonas,* reported that traces of *Aeromonas* DNA may exist in certain components of commercially available kit for DNA extraction (QIAamp DNA stool minikit) which may lead to false-positive results. The aerolysin amplicons obtained were sequenced and found to be 100% identical to this of *Aeromonas hydrophila* strain ZN1, only 92% identity to this of spiked *A. hydrophila* strain 28SA. In addendum, authors indicated that further batch of the contaminated extraction reagent was found to be negative for the assay. Using the AerSpp qPCR-test mentioned above, Martínez-Murcias and co-workers experienced false positives due to *Aeromonas* DNA traces of reagents used for PCR set-up. Depending of the ready-to-use master-mixes employed, the quantity (as shown by Ct values) of targeted DNA found in the negative control assays was relatively high, even reached Ct estimated to correspond > 10^4 genomic copies. However, some master-mixes (i.e. MixStable qPCR.5X, Genetic PCR Solutions, Spain) showed no signal at all in the negative controls. Technicians using qPCR should be aware of any reagent as a source of *Aeromonas* DNA, including buffers and ultra-pure water which are expected to be DNA-free.

Probe-based blot hybridization assays

Probe-based blotting combines the transfer of electrophoresis-separated DNA fragments to a filter membrane and subsequent fragment detection by probe hybridization. The probe DNA is labelled so that it can be detected, usually by either incorporating radioactivity or tagging the molecule with a fluorescent or chromogenic dye.

Digoxigenin (DIG)-labelled DNA probe based on a 237 bp of the glycerophospholipid-cholesterol acyltransferase gene (GCAT) was developed and was tested in a colony

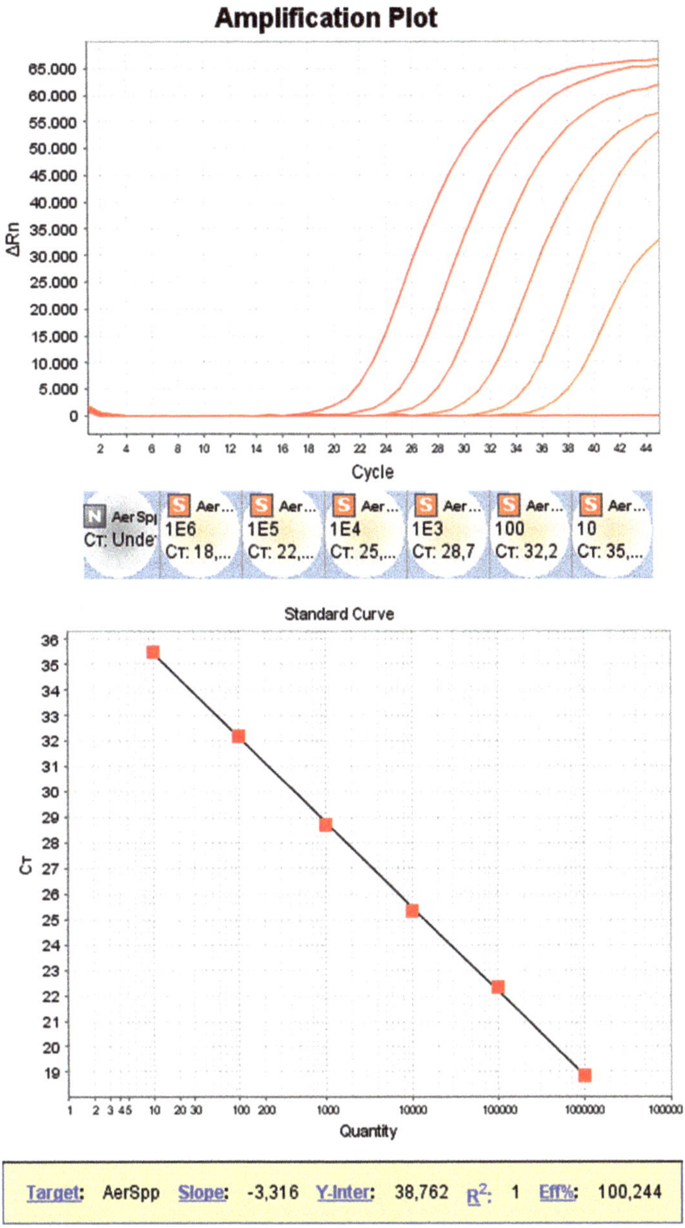

Figure 8.7 Kinetics and calibration curve obtained by using the AerSpp *qPCR-test* kit over tenfold dilution series (10^6 to 10 copies) of standard template (specific synthetic target sequence cloned on a plasmid).

hybridization assay from primary isolation media (Chacon *et al.*, 2002). The probe hybridized with all *Aeromonas* species tested ($n = 16$) but not with strains of other enteropathogenic bacteria ($n = 20$). Although design of primers and probe was only based on a couple of sequences available (from *A. hydrohila* and *A. salmonicida*), the assay was successfully tested with available species. However, validation only included a single strain (type or reference strain) of each *Aeromonas* species. Aguilera-Arreola *et al.* (2005) developed and used DIG-labelled probes to aerolysin-haemolysin, *ast* and *lafA* genes to determine by dot blot the presence and distribution of these genes coding for putative virulence factors in 109 *Aeromonas* clinical and environmental isolates. Another DIG-labelled DNA probe based on an outer membrane protein target, was developed for the specific detection of *Aeromonas* sp. from food, environmental, and clinical samples (Khusiramani *et al.*, 2009). Dot blot reaction was positive for all the 40 *Aeromonas* isolates tested and negative for *Escherichia coli, Pseudomonas* sp., *Klebsiella* sp., *Vibrio parahaemolyticus, V. harveyi, V. alginolyticus, V. vulnificus, Edwardsiella tarda* and *Staphylococcus* sp. This protein was considered conserved in various *Aeromonas* species, but assay was tested in only six *Aeromonas* species.

Khajanchi *et al.* (2010) developed DNA probes directed to 11 virulence factor genes *act, alt, ast, ascV, aexU, gidA*, enolase gene, *dam, tagA, hlyA*, and *ahyRI* to determine the virulence profile of 227 clinical and environmental *Aeromonas* isolates from different geographical locations in the United States. Based on colony blot DNA hybridization, the heat-labile cytotonic enterotoxin gene and the DNA adenosine methyltransferase gene were more prevalent in clinical isolates than in water isolates for all three biochemically identified *Aeromonas* groups, named as the *A. hydrophila* group, the *A. caviae-A. media* group, and the *A. veronii-A. sobria* group. Isolates showing the same virulence signature, were undistinguishable by pulsed-field gel electrophoresis (PFGE) fingerprinting. Authors considered these findings provide the first suggestive evidence of successful colonization and infection by particular strains of certain *Aeromonas* species after transmission from water to humans.

Fluorescence *in situ* hybridization (FISH)

Probe technology is also applied to fluorescence *in situ* hybridization (FISH), a straightforward technique that consists in hybridizing a fluorescently labelled, specific oligonucleotide probe to its complementary sequence on whole-cell preparations previously fixed in a surface. Probes and targets are finally visualized *in situ* by microscopy analysis. These whole-cell hybridization techniques offer, not only detection of specific bacteria at the single-cell level, but also determination of their abundance in complex communities, allowing to improve knowledge of the structure and dynamics of natural microbial communities. As a combined molecular and cytological approach, the major advantage of this visually appealing technique resides in its unique ability to retain information at the single-cell level.

First methods in microbiology were based on rRNA-targeted oligonucleotide probes and found applications in microbial ecology for ecophysiological investigations of complex microbial communities. One drawback of rRNA-targeted FISH, however, is the limited detection of cells with low ribosome content. Another drawback of FISH is the low degree of resolution obtained with rRNA sequences to distinguish different species because of the extremely slow rate of evolution of rRNA. The application of FISH targeting housekeeping genes (e.g. *gyrB*) is hampered by the low copy number of target molecules. Therefore, several FISH-adapted methods have been developed, e.g. the cycling primed *in situ* amplification–fluorescence *in situ* hybridization (CPRINS-FISH), which corresponds to an *in situ* PCR with functional probes. It offers the possibility to address a specific gene in a bacterial cell *in situ*, regardless of copy number or metabolic activity. After cell permeabilization, CPRINS uses one primer and results in linear amplification of the target DNA, which has been used to detect low-copy-number, even single-copy, nucleic acids in cells. Extension products are several kilobases in length and are thus less likely to leak from the cells. PCR products are then detected by using multiply labelled probes (Kenzaka *et al.*, 2005).

Microbial ecology studies concerning *Aeromonas* have been developed with different types of applications: (i) localization of bacterioplankton (Kenzaka *et al.*, 1998, 2005), (ii) identification in biofilms (Bomo *et al.*, 2004) and (iii) microbial identification and quantification in animal intestinal tracts, mainly fish, protist and leech (Asfie *et al.*, 2003; Huber *et al.*, 2004; Jezbera *et al.*, 2005; Perkin *et al.*, 2005; Silver *et al.*, 2007; Kikuchi and Graf, 2007).

To analyse the bacterial community structure of planktonic bacteria in river water at two sites differing in organic carbon pollution level, Kenzaka *et al.* (1998) developed HNPP-FISH, using 2-hydroxy-3-naphthoic acid 2′-phenylanilide phosphate (HNPP) and Fast Red TR. The fluorescent signals enhanced eightfold compared to standard FISH with mono-FITC-labelled oligonucleotide probes. This method however could only distinguish bacteria from five bacterial groups (*Flavobacterium–Cytophaga*; *Burkholderia–Pseudomonas* (rRNA III)-authentic *Alcaligenes*; *Vibrio–Aeromonas*; *Pseudomonas* (rRNA I); the genus *Acinetobacter*) without differentiating Vibrio from *Aeromonas*. The approach was later improved by their authors by using CPRINS-FISH, targeting of the *rpoD* genes of *Aeromonas sobria* and *A. hydrophila.* The method could quantify loads of 10^2 cells/ml in the water samples, providing a high sensitivity for microscopic detection of bacteria carrying a specific gene in natural aquatic samples, and enabling monitoring of target bacteria at the species level for various microorganisms (Kenzaka *et al.*, 2005). Bomo *et al.* (2004) used FISH to examine *Aeromonas* spp. in biofilms and demonstrated that none of the isolates belonged to *A. caviae, A. hydrophila,* or *A. veronii.* A subpopulation of FISH-positive *A. hydrophila* was, however, able to establish in a pre-formed biofilm, thereby demonstrating their ability to incorporate and persist in those biofilms formed within distribution pipe systems. A similar observation was not made for culturable aeromonads, though the exact human health significance of this remains unknown, which questions the adequacy of culture-based techniques. Besides, several works were based on FISH for characterizing bacterial community structure of fish intestinal tract samples. Asfie *et al.* (2003) studied the faecal microflora of goldfish by using tetramethylrhodamine labelled rRNA targeted oligonucleotides as group specific probes (AER66 for *Aeromonas*). *Aeromonas* spp. made up 18.9% to 34.5% of total cell density in faecal samples in goldfish, at counts from 2.5×10^8 to 2.3×10^{10} cells/g. Huber *et al.* (2004) studied the dominant culturable and non-culturable microbiota of rainbow trout intestine by using a set of oligonucleotide probes to detect and enumerate the bacterial community structure of the gastrointestinal tract. Members of the gamma subclass of Proteobacteria (mainly *Aeromonas* and Enterobacteriaceae) dominated the bacterial population structure. Another FISH-adapted method application consisted in analysing bacterial prey composition in protistan food vacuoles in both laboratory and natural populations, enabling the authors to show that flagellate species preferably ingested *Aeromonas hydrophila* over *Pseudomonas fluorescens,* while ciliate *Cyclidium glaucoma* showed no clear preferences (Jezbera *et al.*, 2005). Finally, FISH was used to study cell ability to phagocytose *Aeromonas veronii* in the leech digestive tract (Silver *et al.*, 2007). *Aeromonas veronii* colonize to leech digestive tract and colonization defects in the leech of a mutant (*A. veronii* JG752) was related to an *ascU* homologue, encoding an essential component of type III secretion systems (T3SS). FISH enabled to show that at 42 hours 38% of JG752 cells were phagocytosed by leech macrophage-like cells compared with $< 0.1\%$ of the parental strain (Silver *et al.*, 2007). Kikuchi and Graf (2007) also investigated by FISH the spatial and temporal population dynamics of a microbial community inside the leech digestive tract. The method enabled to study spatial and temporal changes of the localization and microcolony structure of the native symbionts in the crop, after ingestion of a sterile blood meal and contributed to highlight a synergistic interaction of the symbionts since the sizes of the mixed microcolonies were consistently larger than the size of either single-species microcolony.

Biosensors

Taking benefit from nucleic acid hybridization for the detection and analysis of specific DNA sequences, DNA biosensors, also called 'genosensors', have been subjected to large development (Bakker and Qin, 2006). They are analytical devices that result from the integration of single-stranded DNA fragments (ssDNA probes) and a transducer, which translates the recognition event into a physically measurable signal. Highly selective base-pairing interaction between the probe and target DNA leads to the formation of specific double-stranded hybrid (dsDNA) in the biosensor detection layer. The hybridization reaction is then recognized usually by electrochemical, piezoelectric or optical transduction system (Lucarelli *et al.*, 2008). Taking into account its generally high sensitivity, simplicity, low cost, small dimensions and suitability to miniaturization, the electrochemical and piezoelectric methods are one of the most interesting solutions in evolving biosensing techniques (Bakker and Qin, 2006; Lucarelli *et al.*, 2008). The hybridization event is detected via direct measurements of electrical response of DNA strands or application of electroactive redox indicator that provides discrimination between ssDNA probes and newly formed dsDNA (Lucarelli *et al.*, 2008).

Works on biosensors for detecting *Aeromonas* are currently scarce. First work concerned a DNA piezoelectric biosensor coupled with PCR used to distinguish pathogenic *Aeromonas* from other bacteria present in water, vegetable or clinical samples (Tombelli *et al.*, 2000). The *aer* gene of *Aeromonas* was amplified by PCR and the PCR products were then detected by the biosensor which contained a biotinylated capture probe immobilized on the streptavidin coated surface of a quartz crystal. The quartz crystal microbalance technique enabled efficient detection of specific DNA fragments. The detection procedure was simple, but the preparation of biosensor detection interface was quite time- and labour-consuming.

Recently, Tichoniuk *et al.* (2010) developed an electrochemical DNA biosensor for the detection of *A. hydrophila* in food via identification of nucleic acid fragments typical of the aerolysin gene. The biosensor had a recognition layer consisting of thiolated single-stranded DNA of the aerolysin gene. Response was proportional to the concentration of target DNA fragments in the solution and it was not interfered with by the presence of non-complementary nucleic acid fragments when *A. hydrophila* DNA was present at the concentration of $2.5\,\mu g/cm^3$ among other microbial DNA. The optimal hybridization reaction time was 15 minutes.

Microarrays

Unlike other bacterial characterization methods that use a small number of gene or DNA-based functions, microarrays are comprehensive analytical tools that enable to examine thousands of genes. Among the many different applications, microarrays has been used to (i) detect many different types of bacteria in a given ecosystem, (ii) compare genome content between strains and (iii) evaluate global transcriptional responses in various conditions or distinct strains. These three types of applications have concerned *Aeromonas*.

One application of microarrays concerns multiple bacterial species detection. Stine *et al.* (2003) constructed microarrays of DNA probes to study the population dynamics of microbial communities, the interactions of marine bacteria in coastal waters where aeromonads make up a significant percentage of the microbial flora. Microarrays also provides a powerful tool for studying bacterial populations in biofilms. Another microarray was developed with short oligonucleotide probes targeting 16S rRNA sequences in order to detect bacterial pathogens in wastewater (Lee, D.Y. *et al.*, 2008). The microarray (62 probes designed against 38 species, four genera, one family of pathogens, including *Aeromonas*) successfully detected multiple wastewater pathogen species at different stages of the disinfection process. This constitutes an interesting alternative for environmental pathogen monitoring, although with limits given that was based on 16S rRNA gene. Zhou *et al.* (2011) have improved the approach with an oligonucleotide-based microarray using the 16S-23S rDNA internal transcribed spacer regions (ITS) and *gyrB* gene, for the most prevalent pathogenic bacteria in drinking water samples. The 26 specific probes can simultaneously detect *Aeromonas hydrophila, Klebsiella pneumoniae, Legionella pneumophila, Pseudomonas aeruginosa, Salmonella* spp., *Shigella* spp., *Staphylococcus aureus, Vibrio cholerae, Vibrio parahaemolyticus, Yersinia enterocolitica* and *Leptospira interrogans*. This technology may also be applied to environmental pathogen monitoring and food safety. A second application concerns gene content based on comparative genomic hybridizations between strains. Galindo *et al.* (2003) and Galindo *et al.* (2004) first used microarrays to identify *A. hydrophila* cytotoxic enterotoxin-inducing genes in macrophages, thus demonstrating the potential of microarrays in elucidating intracellular mechanisms of pathogenesis. In order to better understand virulence and pathogenicity in *A. salmonicida*, Nash *et al.* (2006) constructed a DNA microarray using a subset of 2024 genes from the draft genome sequence of *A. salmonicida* subsp. *salmonicida* strain A449. The microarray included genes encoding known virulence-associated factors in *A. salmonicida* and homologues of virulence genes of other pathogens. Using this tool, the carriage of virulence-associated genes *A. salmonicida* subspecies and other *Aeromonas* species from different hosts and geographic locations was evaluated, and an increased variation in gene content across sub-species and species boundaries was observed, the greatest variation being observed among genes associated with plasmids and transposons. Antimicrobial resistance genes in bacteria including aeromonads from water and ornamental fish have also been evaluated using microarrays (Verner-Jeffreys *et al.*, 2009).

Finally, a third application concerns the evaluation of global transcriptional responses. Microarrays and proteomics have been used to examine the effects of cytotoxic enterotoxin on human epithelial cells (Galindo *et al.*, 2005), to study the global transcription response of wild-type *Aeromonas hydrophila* and its virulence-deficient mutant in a murine model of infection (Fadl *et al.*, 2007) or to describe the gene expression profiles from immune tissues of fish along the infection process with *Aeromonas* (Millan *et al.*, 2011).

Strain genotyping

Genetic analysis provides a more stable basis for microbiological strain typing than phenotypic methods. DNA-typing methods generally refer to techniques which differentiate isolates, as belonging or not to the same strain (clonality), and may be useful for determining the intra-species relatedness. Several molecular techniques, including finger-printing by ribotyping, pulsed-field gel electrophoresis (PFGE), amplified fragment length polymorphism (AFLP), 16S-23S rRNA spacers-RFLP, and PCR-based strategies such as those using random primers (RAPD-PCR), or targeting one or more repetitive elements (ERIC, BOX or REP-PCR), have been used to study strain genomic relatedness in the genus *Aeromonas*. Some of them are easy and fast to perform but a recognized disadvantage is their poor reproducibility. Since the discriminatory power of typing methods varies among bacterial species, the genotyping technique employed should be validated to prove discrimination capacity for the strains under analysis. If no previous reference genotyping analysis has been published, several methods applied simultaneously are recommended.

DNA typing methods have been used in a variety of ways: (i) systematic analysis of genetic relatedness to propose taxonomic frames or describe new species; (ii) analysis of environmental microbial populations in complex matrices; and (iii) to ascertain genetic relationships between clinical and environmental strains of the same species for epidemiologic purposes.

Description of new bacterial species may be better, and is desirable, to be defined on the basis of several isolates, following recommendations of the 'ad hoc committee for the re-evaluation of the species definition in bacteriology' (Stackebrandt *et al.*, 2002). Genotyping avoids a species being defined on the basis of isolates that could belong to the same strain. This may be critical for isolates recovered from the same source and/or site, where different strains or multiple isolates of the same strains may co-exist. Consequently, we strongly encourage taxonomists to check for redundancy of isolates by genotyping and recommend to be mandatory for new species descriptions, as several cases of *Aeromonas* species definitions were based on genetically highly related, if not the same strain (Figueras *et al.*, 2006).

On the other hand, DNA-typing has opened a possibility to reach the genomic diversity of a single species on nature, and has been highly useful to determine epidemiological relationships between clinical isolates and wild environmental strains. Very often, researchers expect the finding of a close genetic clustering related to niches, or to pathogenicity. However, these expectations were rarely confirmed indicating that a general genotyping pattern specifically related to virulence seems more a desire than the real world. The actual scenario allow us to believe that any genotype may spread everywhere (ubiquity) and shows the 'plasticity' to adapt by expressing specific functions according to concrete niches colonized.

Ribotyping

This method, ribotyping, should not be confused with amplified ribosomal DNA restriction analysis (ARDRA), the later corresponds to restriction fragment length polymorphism (RFLP) of 16S rRNA gene after its amplification (see 'Restriction fragment length polymorphism (RFLP) of PCR fragments' in this chapter). Ribotyping exploits the fact that bacteria have highly conserved DNA which codes for their rRNA. It is based on the hybridization of digested chromosomal DNA with probes derived from specific rRNA genes. When bacterial chromosomes are digested with restriction enzymes (i.e. *Sma*I, *Pst*I), the ribosomal genes are distributed in the resulting fragments, producing strain-specific patterns after separation by electrophoresis and hybridization with a suitable rRNA-labelled probe. Either the entire *rrn*B operon of *Escherichia coli* (coding for 16S, 5S and 23S rRNA), or a mixture of few oligonucleotides designed from conserved regions of 16S-23S rRNA genes, are currently used as probes, making the method a universal approach for bacterial strain typing. Both, the quality of the DNA purification and the choice of the restriction enzyme are critical to obtain a optimal discriminatory power. This technique produces specific gel electrophoretic patterns which have been used to differentiate strains of a single species (Grimont and Grimont, 1986). Ribotyping became attractive for scientists working on *Aeromonas* as soon as it became available (early 1990s). The method was shown to be useful for establishing an epidemiological link as well as a taxonomic tool on the basis of a characteristic pattern for an *Aeromonas* strain or species (Martinetti and Altwegg, 1992). However, it has never seen large diagnostic applications owing to its long and fastidious implementation despite its promising character. Ribotyping was also applied on aeromonads for identification at the species/HG or subspecies levels and in polyphasic studies for typing purposes at the strain level. An example of this is shown in Fig. 8.8 (unpublished), which supported the species description of *A. cavernicola* (Martínez-Murcia *et al.*, 2013).

Typing aeromonads at the strain level was first evaluated on aeromonads by Altwegg Lüthy-Hottenstein J (1991) and Moyer *et al.* (1992) to demonstrate, by using *Sma*I derived ribopatterns, that clinical strains of *Aeromonas* had a different pattern than those recovered from a water distribution system, the associated treatment plant, and source water (Moyer *et al.*, 1992). The method was also used to investigate and exclude, on the basis of ribopatterns, epidemiological links between strains isolated from travellers' diarrhoea during the same trip to Morocco (Hänninen *et al.*, 1995b). Ribotyping was further used along a polyphasic study for typing strains and studying the routes of transmission of *A. salmonicida* subsp. *salmonicida* in Finnish fish farms and the spread of infection. On the basis of ribopatterns, derived from one or several enzymes, the authors identified a likely Swedish origin of the infecting strains. Ribotyping with four different enzymes (*Sma*I, *Bgl*I, *Pst*I and *Cla*I) proved to be the most sensitive method for studying genetically homogeneous *A. salmonicida* subsp. *salmonicida*. Elsewhere, the combination of *Cla*I/*Pst*I ribopatterns was used to improve the discriminatory power of the method, using numerical analysis and similarity matrix according to the unweighted pair-group method to produce dendrogram (Hänninen and Hirvelä-Koski, 1997). Ribotyping was regularly used as strain typing method in epidemiological studies (e.g. Pedersen *et al.*, 1996; Demarta *et al.*, 2000; Bauab *et al.*, 2003), although it was further shown that other typing methods were associated with a better discriminatory power, for instance random amplification polymorphic DNA (Hänninen *et al.*, 1995a), or the pulsed-field gel electrophoresis, (Hänninen and Hirvelä-Koski, 1999). The lack of discriminatory power of ribotyping is partly reflected in the fact that the number of bands was lower than number of *rrn*. Martinetti and Altwegg (1992) estimated the number of *rrn* to be between 6 and 8 for aeromonads, although it was further shown to be between 8 and 11. This was likely because most *rrn* in *Aeromonas* cluster in a rather small region of the chromosome (Seshadri *et al.*, 2006; Reith *et al.*, 2008) and ribotypes did not reflect the overall extent shown by PFGE or RFLP (Bingen *et al.*, 1994).

Typing aeromonads at the species or subspecies level has also been tried on the *Aeromonas*. By including the type strains of each species or HG in the analysis, it has been shown that strains could be assigned to a DNA–DNA hybridization group (HG), a species or a subspecies on the basis of ribopattern comparison (Martinetti and Altwegg, 1992; Hänninen and Hirvelä-Koski, 1997) and the method was further used for identifying purposes. Several enzymes have been tested, including *Sma*I, *Cla*I or *Pst*I, and ribopatterns were analysed alone or combined (Martinetti and Altwegg, 1992; Hanninen *et al.*, 1995a; Hänninen and Siitonen, 1995). *Sma*I ribopattern analysis was used to differentiate strains at

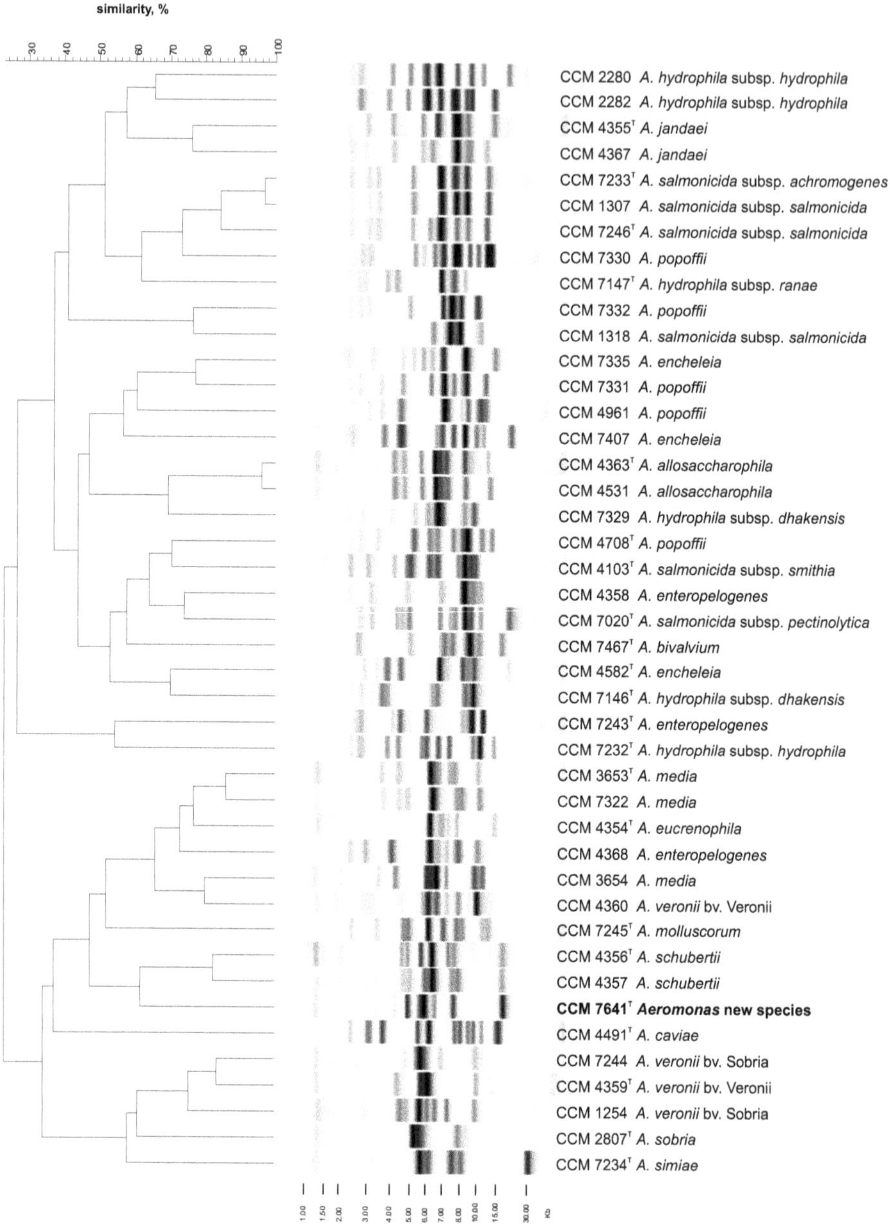

Figure 8.8 Ribotyping patterns and dendrogram showing relationships of some *Aeromonas* species to type strain CCM 7641 of *Aeromonas cavernicola*.

the HG level, and *Pst*I to differentiate *A. salmonicida* subsp. *salmonicida* from atypical *A. salmonicida* because *Sma*I could not digest the DNA of about 30% *A. salmonicida* strains; the combination of *Cla*I/*Pst*I ribopatterns improved the discriminatory power of the method and could identify a ribopattern specific of all typical *A. salmonicida* strains (Hänninen and Hirvelä-Koski, 1997; Austin *et al.*, 1998). The method has been also useful in distinguishing strains of HG2 from HG3 which were not easy to distinguish by phenotypic methods used at that time (Hänninen and Hirvelä-Koski, 1997; Altwegg, 1999). From this point of view, ribotyping was included in polyphasic studies designed to characterize strains belonging to a given taxon (Huys *et al.*, 1997; Austin *et al.*, 1998; Demarta *et al.*, 2004; Krejcí *et al.*, 2009; Nováková *et al.*, 2009; Sedlacek *et al.*, 2012). However, several limitations have been highlighted. The method, evaluated on 332 strains from various origins, could not identify the HG of environmental isolates with atypical biochemical characteristics and impact of intra-specific variability is likely to be noteworthy (Hänninen and Siitonen, 1995) although insufficiently evaluated. Patterns of *A. jandaei* (HG 9) and *A. veronii* (HG 8/10) cannot be distinguished (Altwegg, 1999). Therefore, although ribotyping results have once been recognized to have sufficient taxonomic value to permit assignment of strains to DNA–DNA HG (Martinetti and Altwegg, 1992, 1996), and despite an obvious intrinsic value for identifying strain to the species or subspecies level, the method has insufficient performance for identifying any strain. The conventional ribotyping is also time-consuming and is limited in its robustness and reproducibility, being investigator or laboratory dependent.

Ribotyping, at time to be described, represented a real improvement of available tools for studying aeromonads, although today, other methods have been further developed either for identification purposes (e.g. ARDRA, housekeeping gene sequencing) or strain typing purposes (e.g. PFGE, RAPD-PCR, ERIC-PCR), that surpass its performances, and are easier to use.

Total DNA restriction analysis by pulsed-field gel electrophoresis

The digestion of highly purified chromosomal DNA with restriction endonucleases generates a series of fragments which, subjected to agarose gel electrophoresis with a periodic change of the orientation of the electric field, enables separation of high molecular weight fragments. A clear advantage of the PFGE method is that it addresses a large portion of an investigated genome (>90%). The fingerprint consists of large fragments (10–800 kb) with a high discriminatory power that may be compared to establish epidemiological links between strains. By selecting appropriate restriction enzyme or a set of enzymes, the method is either species or strain significant and the choice of enzymes is therefore critical. Three main applications on aeromonads were developed and were directly related to the enzymes used: first, epidemiologic studies aimed at (i) seeking a link between strains from clinical and environmental origin with patterns strain-significant sought, which was mainly achieved by using *Xba*I, *Swa*I or *Spe*I; or (ii) typing strains recovered from food; second, genetic relatedness of 'typical' and 'atypical' *A. salmonicida* strains, using *Xba*I and *Spe*I; (iii) studying, mainly with I-*Ceu*I, the genetic and physical map of the chromosome of aeromonads, which holds information on identification at the species level. The widespread application of PFGE in microbiology laboratories lead to recognition that genomic DNA of aeromonads may be characterized by PFGE (Garcia *et al.*, 2000), and PFGE has been replacing ribotyping as the preferred method for genomic typing of aeromonads (Hänninen and Hirvelä-Koski, 1999; Bonadonna *et al.*, 2002). However, the analysis of PFGE results is prone to some subjectivity and the continuous quality control and portability of data are limited.

First work on aeromonad typing using PFGE has been conducted in 1996 by Talon *et al.* (1996) to characterize *Aeromonas hydrophila* strains isolated from a cluster of hospital-acquired infections. The authors evaluated the reproducibility and discriminatory power

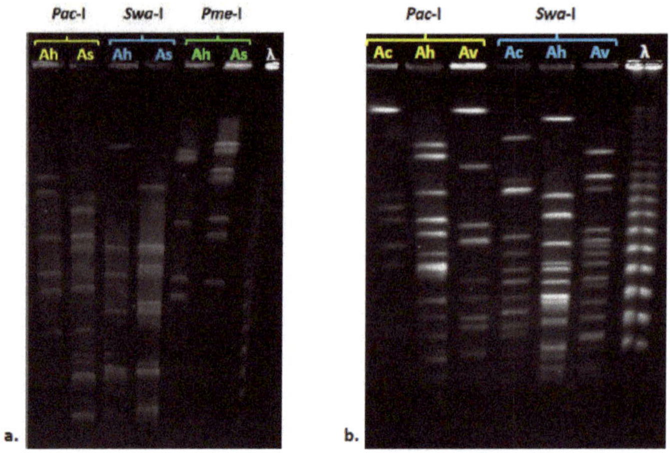

Figure 8.9 PFGE Pulse-types obtained after DNA digestion by restriction enzymes *Pac*I, and *Swa*I. Ac, *A. caviae;* Ah, *A. hydrophila;* As, *A. salmonicida*; Av, *A. veronii.* λ, phage λ concatemer.

of the method using *Xba*I, *Swa*I, *Spe*I and *Dra*I, and all but *Dra*I gave satisfactory patterns and could exclude the water supply as the source of infection. Fig. 8.9 shows an example of typing using different enzymes. Hänninen and Hirvelä-Koski (1999) confirmed that *Xba*I and *Spe*I digests produced concordant results on an 88 atypical *Aeromonas salmonicida* strain collection. Additionally, PFGE and ribotyping (*Cla*I/*Pst*I) methods correlated well suggesting genetic relatedness, but in most cases PFGE pattern analysis subdivided ribotypes into several PFGE genotypes and within a PFGE genotype into subtypes (Chomarat *et al.*, 1998; Hänninen and Hirvelä-Koski, 1999). In parallel, the PFGE method was found less simple, cheap and quick than RAPD-PCR for typing strains in a context of outbreak (Talon *et al.*, 1998). Finally, *Spe*I has arisen from several works that mainly focused on *A. salmonicida* as the most appropriate endonuclease out of about ten evaluated, including *Kpn*I, *Not*I, *Sac*I, *Sma*I, *Spe*I, *Xba*I, *Swa*I and *Dra*I (Talon *et al.*, 1996; Chomarat *et al.*, 1998; Hänninen and Hirvelä-Koski, 1999; Garcia *et al.*, 2000; Giraud *et al.*, 2004). Borchardt *et al.* (2003) compared *Aeromonas* isolates from diarrhoea patients with those obtained from their drinking water by PFGE and found them to be genetically unrelated, suggesting that gastrointestinal infection were not linked with groundwater exposure. Both Pablos *et al.* (2010) and Khajanchi *et al.* (2010) used PFGE patterns and virulence gene profile to demonstrate genetic relatedness between *Aeromonas* diarrheal isolates and water source isolates that had identical PFGE patterns and the same virulence gene profile, bringing evidence for waterborne infection. PFGE was used as a strain typing tool to ascertain the prevalence and genotypes of *Aeromonas* in ready-to-eat food in Italy, where a high proportion was found positive (14.4%, range 0–45%) with a high level of genetic heterogeneity (Villari *et al.*, 2000); in mineral water, where strains belonged to only two clones, which was related to a likely consequence of a localized development of a biofilm (Villari *et al.*, 2003); or in chicken carcasses in Libya (Abdullah *et al.*, 2003). PFGE was also used for clonal analysis in studies on antimicrobial resistance determinants present in the aeromonads of aquaculture origin (Giraud *et al.*, 2004; Ndi *et al.*, 2011).

Umello and Trust (1998) examined the genomic relationship between typical and atypical strains of *A. salmonicida* isolated from diverse origins. The results from the I-*Ceu*I genomic digestion fingerprints confirmed the homogeneity of 'typical' strains, whereas extensive polymorphism was observed in the number and size of the fragments obtained following digestion of the genome of the 'atypical' strains with I-*Ceu*I, which underlined that comparison of I-*Ceul* digestion fingerprints could be used as a powerful tool to subdivide the atypical strains (Umelo and Trust, 1998). The PFGE method became popular to study this group at the beginning of 2000s (Hänninen and Hirvelä-Koski, 1999; O'hlci *et al.*, 2000; Garcia *et al.*, 2000). For instance, O'hlci *et al.* (2000) showed on the basis of RAPD-PCR and PFGE patterns *A. salmonicida* constituted a genetically heterogeneous group of strains, encompassing within a homogeneous or clonal lineage comprised solely of typical strains and the *A. salmonicida* subsp. *salmonicida* type strain.

PFGE has also been used to provide a typing scheme in *Aeromonas* at the species level. PFGE enabled the rapid development of physical maps for bacterial genome analysis. Simultaneously, Dodd and Pemberton (1998) and Umello and Trust (1998) determined the physical and genetic map of the chromosome of *Aeromonas hydrophila* and *Aeromonas salmonicida*, respectively, by using the I-*Ceul* and *Pme*I PFGE, which has provided an initial insight into the genome with a gene organization within the genome, improved genomic knowledge on *Aeromonas* and supplied further information on the molecular evolution of the genus *Aeromonas.*

In fact, I-*Ceu*I is specific for cleavage in the *rrl* (23s) gene, so that number of *rrn* copies and their position on the genome can be deduced from the number and size of I-*Ceu*I-generated fragments (Dodd and Pemberton, 1998; Umelo and Trust, 1998). Taking advantage of this characteristic and enlarging Umelo and Trust's observations, Roger *et al.* (2012b) used PFGE to determine *rrn* operon number and distribution across the chromosome and the diversity of the chromosomal *rrn* copy organization on a 195-strain collection. Most *Aeromonas* species had a quite specific I-*Ceu*I fingerprint with the exception of *A. aquariorum*, that displayed a roughly similar pattern to the one observed for the species *A. hydrophila*. Another notable exception was *A. media* that displayed highly diverse *rrn* chromosomal repartition. Despite intra-species variability, the size of most genomic fragments enabled to distinguish a single species from each other. Results were reinforced by combining I-*Ceu*I PFGE pattern with ribosomal sequence diversity using TTGE in the V3 *rrs* region (Roger *et al.*, 2012b).

Amplified fragment length polymorphism (AFLP)

The AFLP method combines the principle of restriction fragment length polymorphism analysis with highly specific PCR amplification. Briefly, the genomic DNA is digested with restriction enzymes, and restriction half site-specific double-stranded oligonucleotide linkers (adaptors) are ligated to the population of restriction fragments obtained. These fragments are resolved by using high resolution denaturing poly-acrylamide gel or capillary electrophoresis. Enzymes used for *Aeromonas* are *Apa*I and *Taq*I (Janssen *et al.*, 1996; Huys *et al.*, 1996). Correlation with the DNA HG groups was acceptable (Huys *et al.* 1996). Strains belonging to the same HG clearly clustered together although, within a given HG, most strains could be easily differentiated from each other (Janssen *et al.*, 1996). Some of the differences were that *A. allosaccharophila* and *A. encheleia* did not represent separate AFLP cluster, but were genotypically related to HG8/10 and HG6 respectively. HG6 or

A. eucrenophila strains demonstrated significant heterogeneity and thus may comprise more than one species. The method was applied along studies conducted for *A. hydrophila* subspecies descriptions of subsp. *ranae* and subsp. *dhakensis* (Huys *et al.*, 2002, 2003) or *A. salmonicida* 'atypical' strain typing (Lund *et al.*, 2003). Despite promising performances, AFLP has been rarely used for aeromonad identification, with the exception of Rahman *et al.* (2002) work, probably superseded by housekeeping gene sequencing and MLSA. However, this is a highly reproducible method with a recognized potential to genotype strains at the intra-species level, as shown in Fig. 8.10.

16S-23S rRNA spacers – RFLP

The gene encoding 23S rRNA and the sequence of the intergenic spacer region, between the 16S and 23S rRNA genes (16S–23S ISR), was first determined in *Aeromonas hydrophila* by East and Collins (1993). Some authors early used RFLP of the 16S–23S ISR for genotyping *Aeromonas veronii* in investigations of water treatment plant colonization and foodborne outbreaks (Martinez-Murcia *et al.*, 2000). Because this intergenic regions are believed not to be subjected to the same functional compromise as their rRNA counterparts, their sequences accumulates considerable variation often showing insertions/deletions. A comparison of the RFLP of the 16S-23S ISR with other two genomic typing methods, i.e. enterobacterial repetitive intergenic consensus sequence PCR (ERIC-PCR), and repetitive extragenic palindromic (REP-PCR) on 26 isolates of *A. popoffii* from different geographical regions suggested that all three methods were required to obtain some geographical discrimination among aeromonads (Soler *et al.*, 2003b). Cluster analysis of the 16S–23S ISR-RFLP have been subsequently used for typing *A. culicicola* (Pidiyar *et al.*, 2003), several *Aeromonas* spp. (Łaganowska and Kaznowski, 2004), strains of *A. salmonicida* and *A. bestiarum* (Martinez-Murcia *et al.*, 2005), and presumptive virulent *Aeromonas hydrophila* strains isolated from several disease outbreaks (Pridgeon and Klesius, 2011; Mendes-Marques *et al.*, 2012; Pridgeon *et al.*, 2013).

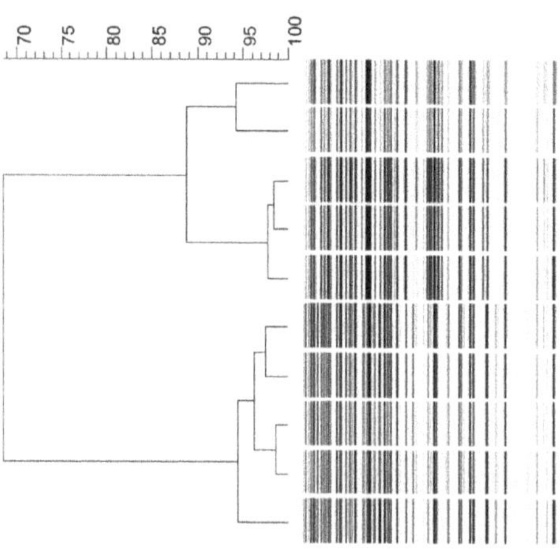

	Origin	MIC gentamicin
AEW-111	Fish	64
AEW-113	Duckweed	64
AEW-114	Sewage	64
AEW-115	Lagoon	64
AEW-112	Fish pond	64
AEW-117	Lagoon	8
AEW-118	Duckweed	8
AEW-116	Fish pond	8
AEW-120	Fish	8
AEW-119	Sewage	8

Figure 8.10 UPGMA/Pearson correlation clustering of normalized AFLP fingerprints of a selection of multiresistant *Aeromonas* isolates from different sampling points of a duckweed aquaculture-based hospital sewage water recycling system in Bangladesh. The fragments (roughly 50–800 bp) were electrophoresed using an ABI 3130 Genetic Analyzer.

Genetic typing by using PCR-based techniques

PCR-based typing methods enable differentiation at the strain level within the same species. These include Random amplification of polymorphic DNA (RAPD-PCR), enterobacterial repetitive intergenic consensus (ERIC-PCR), BOX element PCR, Repetitive extragenic palindromic PCR (REP-PCR), and data are analysed alone or in combination to form a composite fingerprint. They may be highly sensitive and reliable for typing the *Aeromonas* isolate from different sources, but reproducibility is sometimes a drawback (Szczuka and Kaznowski, 2004; Tacão *et al.*, 2005a).

RAPD-PCR is based on the amplification of randomly sized DNA fragments, while REP, ERIC- or BOX-PCR is based on the amplification short repetitive DNA sequences, widely distributed in the genome. The differences in band sizes in all these methods represent polymorphisms due to the different lengths of fragments amplified with the random primers or targeted repetitive elements. A consistent difference on the patterns should be considered as they coming from different strains. Hence, identical banding profiles can be taken as an unequivocal proof of clonality, or perhaps they belong to the same strain. Performances of RAPD-, ERIC- and REP-PCR have been compared. Both RAPD- and ERIC-PCR showed similar level of discriminatory power, between 0.94 and 0.96 (Szczuka and Kaznowksi, 2004; Aguilera-Arreola *et al.*, 2005; Maiti *et al.*, 2009), with good correlation, and provided the same utility for epidemiological investigations and population genetic analysis, whereas REP-PCR was less effective for differentiating the isolates of *Aeromonas* spp. (Szczuka and Kaznowksi, 2004; Maiti *et al.*, 2009). RAPD- has been used in combination with ERIC-PCR to improve discrimination between strain and typing performance (Aguilera-Arreola *et al.*, 2005; Davin-Regli *et al.*, 1998). To date, ERIC-PCR is the most frequently used for typing strains isolated from various origins.

Random amplification of polymorphic DNA (RAPD-PCR)

The RAPD method involves the use of short random sequence oligonucleotide primers, usually 9 to 10 nucleotides long, at a low-stringency primer annealing conditions to arbitrarily amplify DNA fragments from an extracted total genome used as template. The single primers should anneal on those sites of the genome where a complementary-enough sequence exists and, if two anti-sense priming sites are sufficiently close, PCR then amplifies the DNA fragment between them. Level of complementarities and distance in primer binding (both unknown and based in a hope) are limited and the threshold values of these two parameters are a source of possible lack of reproducibility. However, this may be solved by performing some experimental validations. After amplification the PCR products are electrophoresed on usual agarose gel in TBE buffer, and the gels visualized. The use of a molecular size marker is recommended to help in size identification (Martinez-Murcia *et al.*, 2008).

Primers used for RAPD for *Aeromonas* have been designed for other genera (Neilan, 1995; Martinez-Murcia *et al.*, 1995; Davin-Regli *et al.*, 1998; Maiti *et al.*, 2009) or specifically designed for aeromonads (Oakey *et al.*, 1996, 1998, 1999; Talon *et al.*, 1998) and these latest have been the most widely used in recent studies (e.g. Szczuka and Kaznowksi; 2004; Aguilera-Arreola *et al.*, 2005). Some primers were found to yield species-specific bands in all *A. hydrophila or A. salmonicida* strains tested and were used to develop probes and PCR primers for the respective species detection (Oakey *et al.*, 1998, 1999).

First applied on *Aeromonas* by Miyata *et al.* (1995), most studies have used RAPD-PCR for documenting relationships between strains of clinical and environmental origin for epidemiologic purposes. Davin-Regli *et al.* (1998) used RAPD combined with ERIC-PCR (described in the next bullet) to study a hospital-occurring outbreak involving seven diarrhoea cases suspected to be caused by *Aeromonas*. The methods enabled to clarify a complex situation, demonstrating a nosocomial transmission relationship between two of these cases, the other five being only sporadic. Alavandi *et al.* (2001) used RAPD-PCR and whole cell protein profiles by SDS-PAGE to examine the relationship between *Aeromonas* isolates from children with diarrhoea and their water sources. Water isolates clustered in two major groups while clinical isolates were scattered across the dendrogram. RAPD-PCR was found to be useful in distinguishing isolates but did not correspond with the phenospecies or genomospecies of *Aeromonas*. Similarly, Delamare *et al.* (2002) used RAPD and offered strain differentiation but was not useful in species identification. Szczuka and Kaznowksi (2004) analysed 120 isolated from water and human stools from patients with gastroenteritis by RAPD-PCR, REP-PCR, and ERIC-PCR. No determinant clone was associated with human diarrhoea. There was no genetic similarity between clinical and environmental strains isolated from the same or distinct geographical areas. Some clones colonized a specific ecosystem, e.g. drinking water. These techniques were used to show that the hospital environment was not the source of aeromonads in hospitalized children with diarrhoea since no clonal relationship could be demonstrated between stool and environmental isolates. RAPD PCR in combination with ERIC-PCR was also used to determine and quantify the genetic diversity of *A. hydrophila* strains obtained from clinical as well as environmental sources (Aguilera-Arreola *et al.*, 2005). Both methods consistently performed well in discriminating between strains. Finally, Maiti *et al.* (2009) used RAPD-PCR for typing of 42 *Aeromonas* isolates from clinical and environmental sources. No correlation in banding pattern was found between *Aeromonas* isolates from environmental and clinical sources.

On a distinct application, Martinez-Murcia *et al.* (2008) used RAPD-PCR for typing strains in order to describe the novel species *A. aquariorum*, on the basis of non-clonal isolates but also embracing clinical strains of *Aeromonas hydrophila* subsp. *dhakensis* (Martinez-Murcia *et al.*, 2009). They showed that 19 out of 23 isolates with identical *gyrB* partial sequences belonging to *A. aquariorum* clones showed 19 different RAPD patterns, indicating that, although highly similar, they are different individuals.

Methods based on amplification of a single type of repetitive element (ERIC, BOX or REP)

Enterobacterial repetitive intergenic consensus (ERIC)

First described by Versalovic *et al.* (1991), the method is based on specific primers designed to match inverted repeat consensus sequences commonly found in enterobacteria, is easy to perform, is powerful to differentiate strains, and is currently one of the most popular method. Although a full primer matching is not fully guaranteed, contrary to random-primers, the reproducibility of ERIC-profiles is expected to be higher, although not very high.

ERIC was, since the early 2000s, widely used for typing strains in epidemiologic studies involving isolates from animal, food, environmental and/or clinical sources. Fontes *et al.* (2011, 2012) used ERIC for bacterial typing to avoid redundancy during isolation of strains harvested from pig carcasses or smoked meat sausage. The same approach was used for describing a collection of *A. aquariorum* strains recovered from egg masses of midges

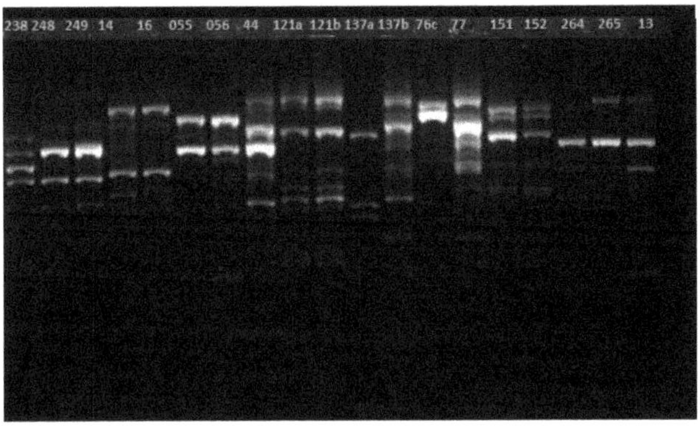

Figure 8.11 ERIC-PCR patterns of aeromonads belonging to various species (*A. veronii*: strains 77, 44, 137b, 151, 152; *A. hydrophila*: strains 76c; *A. media*: strain 137a; *A. caviae*: strains 14, 16, 121a, 121b, 238, 248, 249, 265; *A. aquariorum:* strains: 13, 055, 056, 264). Strains 121a and 121b, 055 and 056, 151 and 152, 248 and 249 show identical pattern.

(Figueras *et al.*, 2011), for characterizing class 1 integrons and their distribution in strains from diarrheic stools (Pérez-Valdespino *et al.*, 2009), or other human origins (Lee, M.F. *et al.*, 2008), and for characterizing *A. encheleia* strains from water (Nováková *et al.*, 2009). Other applications include strain typing in outbreak contexts in human or fish, for distinguishing sporadic cases from epidemic cases and specifying the extent of the epidemic (Davin-Regli *et al.*, 1998, Lee, M.F. *et al.*, 2008; Ye *et al.*, 2012). Finally, ERIC was used for comparing strains between water and clinical sources and the method showed that, to some extent, clinical and environmental strains may produce different clusters (Sechi *et al.*, 2002; Szczuka and Kaznowksi, 2004). An example of ERIC-PCR typing is shown in Fig. 8.11.

BOX element PCR

Described in the early 1990s, Box elements are interspersed throughout the genome of bacteria and have been used to genotype diverse bacterial species (Martin *et al.*, 1992; Versalovic *et al.*, 1994). In 2005, the BOX-PCR fingerprinting technique was evaluated for differentiating strains of some *Aeromonas* species (Tãcao *et al.*, 2005a). All of the 42 strains tested were typeable and the majority showed unique banding patterns. BOX-PCR fingerprinting was easy-to-perform, suitable for typing of *Aeromonas* strains, allowing differentiation at the strain level, and was considered as a useful complementary tool for epidemiologic studies of members of this genus. Other work confirmed its ability for typing at the *Aeromonas* strain level and as a potential tool for epidemiological purposes (Singh *et al.*, 2010). To our knowledge, however, no comparison of the discriminatory capacity against other typing methods (RAPD-, ERIC- or REP-PCR) is currently available.

Repetitive extragenic palindromic PCR (REP-PCR)

This kind of PCR-based fingerprinting has never been used alone for typing aeromonads. Szczuka and Kaznowski (2004) concluded that the method was less effective than RAPD or ERIC-PCR for isolates differentiating at the strain level. Further work is required to confirm this lower performance. Several works dealt with combinations including REP (see below).

Methods combining several types of repetitive elements (rep-PCR) including ERIC, REP or BOX

This typing method is generally designated as rep-PCR fingerprint (for repetitive element), and corresponds to the combination of the REP sequence, the ERIC sequence or the 154 bp BOX element. Therefore, rep-PCR should not be confused with REP-PCR limited to only one of these types of sequences. The method is based on the amplification of these highly variable regions in separate PCR reactions using any of the ERIC, REP and BOX primers to finally combine resulting data to form a composite fingerprint. The amount of information generated has been described to provide the best resolution for a fine discrimination of strains. To date, some studies on *Aeromonas* included ERIC and REP or BOX and REP. Beaz-Hidalgo *et al.* (2008) made promising attempts to identify *A. salmonicida* at the subspecies level using ERIC-PCR combined with REP-PCR, although some improvements are required. Soler *et al.* (2003) evaluated ERIC-PCR, REP-PCR and RFLP of the 16S-23S ISR for typing 26 isolates of *Aeromonas popoffii*. When the methods were independently studied, ERIC showed the highest discriminatory power, but the best combination was considered this of ERIC with REP, reinforced with the addition of ISR-RFLP (Soler *et al.*, 2003). Finally, some works combined REP and BOX elements for evaluating bacterial diversity either to assess plasmid transfer in liquid activated sludge studies or description of microbial community in bulking sludge (e.g. Goris *et al.*, 2003; Jin *et al.*, 2011).

Present trends and future prospects

Although during the past few years no single phenotypic or genotypic identification system assured a reliable diagnostic method for all *Aeromonas* strains, the present state of the art invites some optimism but, of course, moderate. No doubt a large part of current genetic techniques must evolve and develop within a period not too long, otherwise it will go to historical science overtaken by new ones that are at the door. The genus *Aeromonas* has been a role model to implement and validate the technological trends, although its pathogenicity has only been considered opportunistic. This situation is also changing for many. Nevertheless, when purposes are to defining taxonomic status, polyphasic data should undoubtedly be harmonized.

Aeromonas isolated in pure cultures may be now identified to the species level by sequencing a few housekeeping genes, however this method should be normalized and a curate database used as standard. Decisions to be taken include names and minimum number of target genes, minimum sequence lengths, or perhaps indicate an appropriated set of primers, sequence quality with zero tolerance, and the use of reference database as the frame work for a phylogenetic diagnostic. Taxa description is a different matter and standard recommendations should be followed.

As corresponds to expect from the 'genomic era', new alternatives obtaining the whole genome sequence must be attempted. The average nucleotide identity (ANI) seems currently the best option for a gold standard of classification (Richter and Rosselló-Móra, 2009) and this shortly will be applied to the genus *Aeromonas*. The high demand for low-cost sequencing has driven the development of high-throughput technologies, so-called next-generation sequencing that produce thousands or millions of sequences concurrently. Technologies include Pyrosequencing, Illumina, Ion Semiconductor, DNA nanoball, Nanopore, Tunnelling Currents and Microfluidic. So far, application of these methods has mainly been directed to metagenomic studies of microbial communities from complex natural water samples such as biofilms, pelagic or suspended bacterioplankton, and lotic systems from lakes, open sea, and rivers, for example the Amazon (Ghai *et al.*, 2011). The main difficulties to overcome include the relevant analysis of huge number of data.

Detection of *Aeromonas* strains is the most pragmatic aspect in systematics, relevant to diagnostics, not only clinical, but also in water, food, and environment. It is evident that future tools include microarrays and biosensors, including gene-expression (i.e. virulence, resistance), being the design the point to be questioned. However, real time PCR, even end-point PCR, is an approach to be supplemented in laboratories of control for the next few years. For this practice, considerable efforts to validate and harmonize methods, to follow guidelines of international norms for microbial control, and a large amount of 'know-how' transfer should be achieve. Between other aspects to cover, creation of new kit formats which facilitates access to reagents and the meaning of genomic counts must be included. It is still questioned (or unclear) how such kits may be used in clinical microbiology routine.

New genotyping methods should avoid the drawbacks experienced in the past time. The technology, a part of highly discriminative, should be simple, of course easy and cheap to run. However, the most relevant requirement by far is reproducibility and repeatability. The purpose should include strong efforts on harmonization and standardization of the method, make comparison of results generated in different laboratories as easy as possible, and therefore encourage its use in international surveillance. A novel molecular typing system based on rapidly evolving variable-number tandem repeat (VNTR) loci, method so-called multiple-locus variable-number of tandem-repeats analysis (MLVA), has emerged as a valuable method for subtyping bacterial pathogens and has been adopted in many laboratories as a critical component of their laboratory-based surveillance (Keim *et al.*, 2000). Development of a MLVA applied to species of the genus *Aeromonas* (to our knowledge never been published) is underway (Martínez-Murcia and co-workers)

After this review, the authors' intention is to provide a basis for rethinking and new technological and conceptual approaches in order to address the molecular diagnosis of these bacteria. It is now a matter of the reader, if desired, to fill in the gaps that may have been left behind or, what is much better, open new expectations on genetic techniques for the identification, detection, and typing, of strains of the genus *Aeromonas*.

Acknowledgements

The authors wish to acknowledge F. Roger, M.J. Figueras, G. Huys, and I. Sedlacek who contributed to some figures. MM wish to acknowledge the support obtained from L. Korenyuk, G. Bru, J.V. Escudero, Aaron Navarro, and particularly to B. Lamy who has starred in a real rescue with hard work and elegance. Part of the research leading to these results has received funding from the European Union Seventh Framework Programme ([FP7/2007-2013] [FP7/2007-2011]) under Grant agreement no: 311846.

References

Abdullah, A.I., Hart, C.A., and Winstanley, C. (2003). Molecular characterization and distribution of virulence-associated genes amongst *Aeromonas* isolates from Libya. J. Appl. Microbiol. 95, 1001–1007.

Aguilera-Arreola, M.G., Hernandez-Rodriguez, C., Zuniga, G., Figueras, M.J., and Castro-Escarpulli, G. (2005). *Aeromonas hydrophila* clinical and environmental ecotypes as revealed by genetic diversity and virulence genes. FEMS Microbiol. Lett. 242, 231–240.

Alavandi, S.V., Ananthan, S., and Pramod, N.P. (2001). Typing of *Aeromonas* isolates from children with diarrhoea and water samples by randomly amplified polymorphic DNA polymerase chain reaction and whole cell protein fingerprinting. Indian J. Med. Res. *113*, 85–97.

Alperi, A., Figueras, M.J., Inza, I., and Martínez-Murcia, A.J. (2008). Analysis of 16S rRNA gene mutations in a subset of *Aeromonas* strains and their impact in species delineation. Int. Microbiol. *11*, 185–194.

Altwegg, M. (1996). Subtyping methods for *Aeromonas* species. In The Genus *Aeromonas*, Austin, B., Altwegg, M., Gosling, P.J., and Joseph, S.W., eds (Wiley-Blackwell, New York, NY), pp. 109–125.

Altwegg, M. (1999). Subtyping methods for *Aeromonas* species. The Genus *Aeromonas*, Austin, B., Altwegg, M., Gosling, P.J., Joseph, S.W., eds (John Wiley & Sons, Chichester, UK), pp. 109–125.

Altwegg M., and Lüthy-Hottenstein J. (1991). Methods for the identification of DNA hybridization groups in the genus *Aeromonas*. Experientia *47*, 403–406.

Asfie, M., Yoshijima, T., and Sugita, H. (2003). Characterization of the goldfish fecal microflora by the fluorescent *in situ* hybridization method. Fish Sci. *69*, 21–26.

Ash, C., Martínez-Murcia, A.J., and Collins, M.D. (1993a). Molecular identification of *Aeromonas sobria* by polymerase chain reaction. Med. Microbiol. Lett. *2*, 80–86.

Ash, C., Martinez-Murcia, A.J., and Collins, M.D. (1993b). Identification of *A. schubertii* and *A. jandaei* by using a polymerase chain reaction-probe test. FEMS Microbiol. Lett. *108*, 151–155

Austin, B., Austin, D.A., Dalsgaard, I., Gudmundsdottir, B.K., Høie, S., Thornton, J.M., Larsen, J.L., O'Hici, B., and Powell, R. (1998). Characterization of atypical *Aeromonas salmonicida* by different methods. Syst. Appl. Microbiol. *21*, 50–64.

Azevedo, J.S., Ramos, I., Araújo, S., Oliveira, C.S., Correia, A., and Henriques, I.S. (2012). Spatial and temporal analysis of estuarine bacterioneuston and bacterioplankton using culture-dependent and culture-independent methodologies. Antonie Van Leeuwenhoek *101*, 819–835.

Bakker, E., and Qin, Y. (2006). Electrochemical sensors. Anal. Chem. *78*, 3965–3984.

Baloda, S.B., Krovacek, K., Eriksson, L., Linne, T., and Mansson, I. (1995). Detection of aerolysin gene in *Aeromonas* strains isolated from drinking water, fish and foods by the polymerase chain reaction. Compar. Immunol. Microbiol. Infect. Dis. *18*, 17–26.

Bauab, T.M., Levy, C.E., Rodrigues, J., and Falcão, D.P. (2003). Niche-specific association of *Aeromonas* ribotypes from human and environmental origin. Microbiol. Immunol. *47*, 7–16.

Beaz-Hidalgo, R., López-Romalde, S., Toranzo, A.E., and Romalde, J.L. (2008). Polymerase chain reaction amplification of repetitive intergenic consensus and repetitive extragenic palindromic sequences for molecular typing of *Pseudomonas anguilliseptica* and *Aeromonas salmonicida*. J. Aquat. Anim. Health. *20*, 75–85.

Bingen, E.H., Denamur, E., and Elion, J. (1994). Use of ribotyping in epidemiological surveillance of nosocomial outbreaks. Clin. Microbiol. Rev. *7*, 311–327.

Bomo, A.M., Storey, M.V., and Ashbolt, N.J. (2004). Detection, integration and persistence of aeromonads in water distribution pipe biofilms. J. Water Health. *2*, 83–96.

Bonadonna, L., Briancesco, R., Filetici, E., Manuppella, A., Pourshaban, M., and Semproni, M. (2002). Genomic heterogeneity of environmental and clinical aeromonads. New Microbiol. *2*, 21–29.

Borchardt, M.A., Stemper, M.E., and Standridge, J.H. (2003). *Aeromonas* isolates from human diarrheic stool and groundwater compared by pulsed-field gel electrophoresis. Emerg. Infect. Dis. *9*, 224–228.

Borrell, N., Acinas, S.G., Figueras, M.J., and Martinez-Murcia, A.J. (1997). Identification of *Aeromonas* clinical isolates by restriction fragment length polymorphism of PCR-amplified 16S rRNA genes. J. Clin. Microbiol. *35*, 1671–1674.

Calhau, V., Tacão, M., Morgado, F., Correia, A., and Henriques, I. (2010). PCR-DGGE-based methodologies to assess diversity and dynamics of *Aeromonas* communities. J. Appl. Microbiol. *108*, 611–623.

Cascon, A., Anguita, J., Hernanz, C., Sanchez, M., Fernandez, M., and Naharro, G. (1996). Identification of *Aeromonas hydrophila* hybridization group 1 by PCR assays. Appl. Environ. Microbiol. *62*, 1167–1170.

Chacon, M.R., Castro-Escarpulli, G., Soler, L., Guarro, J., and Figueras, M.J. (2002). A DNA probe specific for *Aeromonas* colonies. Diagn. Microbiol. Infect. Dis. *44*, 221–225.

Chacon, M.R., Figueras, M.J., Castro-Escarpulli, G., Soler, L., and Guarro, J. (2003). Distribution of virulence genes in clinical and environmental isolates of *Aeromonas* spp. Antonie van Leeuwenhoek *84*, 269–278.

Chomarat, M., Guérin-Faublée, V., Kodjo, A., Breysse, F., and Flandrois, J.P. (1998). Molecular analysis of the fish pathogen *Aeromonas salmonicida* subsp. *salmonicida* by pulsed-field gel electrophoresis. Rev. Med. Vet. *149*, 245–250.

Chu, W.H., and Lu, C.P. (2005). Multiplex PCR assay for the detection of pathogenic *Aeromonas hydrophila*. J. Fish Dis. *28*, 437–441.

Ciapini, A., Dei, R., Sacco, C., Ventura, S., Viti, C., and Giovannetti, L. (2002). Phenotypic and genotypic characterisation of *Aeromonas* isolates. Ann. of Microbiol. 52, 339–352.

Davin-Regli, A., Bollet, C., Chamorey, E., Colonna D'istria, V., and Cremieux, A. (1998). A cluster of cases of infections due to *Aeromonas hydrophila* revealed by combined RAPD and ERIC-PCR. J. Med. Microbiol. 47, 499–504.

Delamare, A.P.L., Artico, L.D., Grazziotin, F.G., Echeverrigaray, S., and da Costa, S.O.P. (2002). Total protein electrophoresis and RAPD fingerprinting analysis for the identification of *Aeromonas* at the species level. Braz. J. Microbiol. 33, 358–362.

Demarta, A., Tonolla, M., Caminada, A., Beretta, M., and Peduzzi, R. (2000). Epidemological relationship between *Aeromonas* strains isolated from symptomatic children and household environments as determined by ribotyping. Eur. J. Epidemiol. *16*, 447–453.

Demarta, A., Huys, G., Tonolla, M., Swings, J., and Peduzzi, R. (2004). Polyphasic taxonomic study of '*Aeromonas* eucrenophila-like' isolates from clinical and environmental sources. Syst. Appl. Microbiol. 27, 343–349.

Dimitroglou, A., Merrifield, D.L., Moate, R., Davies, S.J., Spring, P., Sweetman, J., and Bradley, G. (2009). Dietary mannan oligosaccharide supplementation modulates intestinal microbial ecology and improves gut morphology of rainbow trout, *Oncorhynchus mykiss* (Walbaum). J. Anim. Sci. 87, 3226–3234.

Dodd, H.N., and Pemberton, J.M. (1998). Construction of a physical and preliminary genetic map of *Aeromonas hydrophila* JMP636. Microbiology. *144*, 3087–3096.

Dorsch, M., Ashbolt, N.J., Cox, P.T., and Goodman, A.E. (1994). Rapid identification of *Aeromonas* species using 16S rRNA targeted oligonucleotide primers: a molecular approach based screening of environmental isolates. J. Appl. Bacteriol. 77, 722–726.

East, A.K., and Collins, M.D. (1993). Molecular characterization of DNA encoding 23S rRNA and 16S-23S rRNA intergenic spacer regions of *Aeromonas hydrophila*. FEMS Microbiol. Lett. *106*, 129–133.

Fadl, A.A., Galindo, C.L., Sha, J., Zhang, F., Garner, H.R., Wang, H.Q., and Chopra, A.K. (2007). Global transcriptional responses of wild-type *Aeromonas hydrophila* and its virulence-deficient mutant in a murine model of infection. Microb. Pathog. *42*, 193–203.

Figueras, M.J., Soler, L., Chacon, M.R., Guarro, J., and Martinez-Murcia, A.J. (2000). Extended method for discrimination of *Aeromonas* spp. by 16S rDNA RFLP analysis. Int. J. Syst. Evol. Microbiol. *50*, 2069–2073.

Figueras, M.J., Alperi, A., Guarro, J., and Martínez-Murcia, A.J. (2006). Genotyping of isolates included in the description of a novel species should be mandatory. Int. J. Syst. Evol. Microbiol. *56*, 1183–1184.

Figueras, M.J., Beaz-Hidalgo, R., Senderovich, Y., Laviad, S., and Halpern, M. (2011a). Re-identification of *Aeromonas* isolates from chironomid egg masses as the potential pathogenic bacteria *Aeromonas aquariorum*. Environ. Microbiol. Rep. 3, 239–244.

Figueras, M.J., Beaz-Hidalgo, R., Collado, L., and Martínez-Murcia, A.J. (2011b). Recommendations for a new bacterial species description based on analyses of the unrelated genera *Aeromonas* and *Arcobacter*. Bull. BISMiS *2*, 1–16.

Fontes, M.C., Saavedra, M.J., Martins, C., and Martínez-Murcia, A.J. (2011). Phylogenetic identification of *Aeromonas* from pigs slaughtered for consumption in slaughterhouses at the North of Portugal. Int. J. Food Microbiol. *146*, 118–122.

Fontes, M.C., Martins, C., Martínez-Murcia, A.J., and Saavedra, M.J. (2012). Phylogenetic diversity of *Aeromonas* from 'alheira,' a traditional Portuguese meat product. Foodborne Pathog. Dis. *9*, 713–718.

Fukushima, H., Tsunomori, Y., and Seki. R. (2003). Duplex real-time SYBR green PCR assays for detection of 17 species of food- or waterborne pathogens in stools. J. Clin. Microbiol. *41*, 5134–5146.

Galindo, C.L., Sha, J., Ribardo, D.A., Fadl, A.A., Pillai, L., and Chopra, A.K. (2003). Identification of *Aeromonas hydrophila* cytotoxic enterotoxin-induced genes in macrophages using microarrays. J. Biol. Chem. *278*, 40198–40212.

Galindo, C.L., Fadl, A.A., Sha, J., and Chopra, A.K. (2004). Microarray analysis of *Aeromonas hydrophila* cytotoxic enterotoxin-treated murine primary macrophages. Infect. Immun. 72, 5439–5445.

Galindo, C.L., Fadl, A.A., Sha, J., Pillai, L., Gutierrez, C. Jr., and Chopra, A.K. (2005). Microarray and proteomics analyses of human intestinal epithelial cells treated with the *Aeromonas hydrophila* cytotoxic enterotoxin. Infect. Immun. 73, 2628–2643.

Garcia, J.A., Larsen, J.L., Dalsgaard, I., and Pedersen, K. (2000). Pulsed-field gel electrophoresis analysis of *Aeromonas salmonicida* ssp. *salmonicida*. FEMS Microbiol. Lett. *190*, 163–166.

Ghai, R., Rodriguez-Valera, F., McMahon, K.D., Toyama, D., Rinke, R., Souza de Oliveira C.T., Wagner-Garcia, J., Pellon de Miranda, F., and Henrique-Silva, F. (2011). Metagenomics of the water column in the pristine upper course of the Amazon river. PLoS One *6*, e23785.

Giraud, E., Blanc, G., Bouju-Albert, A., Weill, F.X., and Donnay-Moreno, C. (2004). Mechanisms of quinolone resistance and clonal relationship among *Aeromonas salmonicida* strains isolated from reared fish with furunculosis. J. Med. Microbiol. *53*, 895–901.

Gonzalez-Rodriguez, M.N., Santos, J.A., Otero, A., and Garcia-Lopez, M.L. (2002). PCR detection of potentially pathogenic aeromonads in raw and cold-smoked freshwater fish. J. Appl. Microbiol. *93*, 675–680.

Goris, J., Boon, N., Lebbe, L., Verstraete, W., and Vos, P. (2003). Diversity of activated sludge bacteria receiving the 3-chloroaniline-degradative plasmid pC1gfp. FEMS Microbiol. Ecol. *46*, 221–230.

Graf J. (1999). Diverse restriction fragment length polymorphism patterns of the PCR-amplified 16S rRNA genes in *Aeromonas veronii* strains and possible misidentication of *Aeromonas* species. J. Clin. Microbiol. *37*, 3194–3197.

Grimont, F., and Grimont, P.A.D. (1986). Ribosomal ribonucleic acid gene restriction patterns as potential taxonomic tools. Ann. Inst. Pasteur Microbiol. *137*, 165–175.

Hänninen, M.L., and Siitonen, A. (1995). Distribution of *Aeromonas* phenospecies and genospecies among strains isolated from water, foods or from human clinical samples. Epidemiol. Infect. *115*, 39–50.

Hänninen, M.L., and Hirvelä-Koski, V. (1997). Molecular and phenotypic methods for the characterization of atypical *Aeromonas salmonicida*. Vet. Microbiol. *56*, 147–158.

Hänninen, M.L., and Hirvelä-Koski, V. (1999). Genetic diversity of atypical *Aeromonas salmonicida* studied by pulsed-field gel electrophoresis. Epidemiol. Infect. *123*, 299–307.

Hänninen, M.L., Ridell, J., and Hirvelä-Koski, V. (1995a). Phenotypic and molecular characteristics of *Aeromonas salmonicida* subsp. *salmonicida* isolated in southern and northern Finland. J. Appl. Bacteriol. *79*, 12–21.

Hänninen, M.L., Salmi, S., Mattila, L., Taipalinen, R., and Siitonen, A. (1995b). Association of *Aeromonas* spp. with travellers' diarrhoea in Finland. J. Med. Microbiol. *42*, 26–31.

Hong, S.W., Kim, I.S., Lee, J.S., and Chung, K.S. (2011a). Culture-based and denaturing gradient gel electrophoresis analysis of the bacterial community structure from the intestinal tracts of earthworms (*Eisenia fetida*). J. Microbiol. Biotechnol. *21*, 885–892.

Hong, S.W., Lee, J.S., and Chung, K.S. (2011b). Polymerase chain reaction-denaturing gradient gel electrophoresis analysis of bacterial community structure in the food, intestines, and feces of earthworms. J. Microbiol. *49*, 544–550.

Huang, W.M. (1996). Bacterial diversity based on type II DNA topoisomerase genes. Annu. Rev. Genet. *30*, 79–107.

Huber, I., Spanggaard, B., Appel, K.F., Rossen, L., Nielsen, T., and Gram, L. (2004). Phylogenetic analysis and *in situ* identification of the intestinal microbial community of rainbow trout (*Oncorhynchus mykiss*, Walbaum). J. Appl. Microbiol. *96*, 117–132.

Huys, G., Coopman, R., Janssen, P., and Kersters, K. (1996). High resolution genotypic analysis of the genus *Aeromonas* by AFLP fingerprinting. Int. J. Syst. Bacteriol. *46*, 572–580.

Huys, G., Kämpfer, P., Altwegg, M., Kersters, I., Lamb, A., Coopman, R., Lüthy-Hottenstein, J., Vancanneyt, M., Janssen, P., and Kersters, K. (1997). *Aeromonas popoffii* sp. nov., a mesophilic bacterium isolated from drinking water production plants and reservoirs. Int. J. Syst. Bacteriol. *47*, 1165–1171.

Huys, G., Kämpfer, P., Albert, M.J., Kühn, I., Denys, R., and Swings, J. (2002). *Aeromonas hydrophila* subsp. *dhakensis* subsp. nov., isolated from children with diarrhoea in Bangladesh, and extended description of *Aeromonas hydrophila* subsp. *hydrophila* (Chester 1901) Stanier 1943 (approved lists 1980). Int. J. Syst. Evol. Microbiol. *52*, 705–712.

Huys, G., Pearson, M., Kämpfer, P., Denys, R., Cnockaert, M., Inglis, V., and Swings, J. (2003). *Aeromonas hydrophila* subsp. *ranae* subsp. nov., isolated from septicaemic farmed frogs in Thailand. Int. J. Syst. Evol. Microbiol. *53*, 885–891.

Laganowska, M., and Kaznowski, A. (2004). Restriction fragment length polymorphism of 16S-23S rDNA intergenic spacer of *Aeromonas* spp. Syst. Appl. Microbiol. *27*, 549–557.

Janssen, P., Coopman, R., Huys, G., Swings, J., Bleeker, M., Vos, P., Zabeau, M., and Kersters, K. (1996). Evaluation of the DNA fingerprinting method AFLP as a new tool in bacterial taxonomy. Microbiology. *142*, 1881–1893.

Jezbera, J., Hornák, K., and Simek, K. (2005). Food selection by bacterivorous protists: insight from the analysis of the food vacuole content by means of fluorescence *in situ* hybridization. FEMS Microbiol. Ecol. *52*, 351–363.

Ji, N., Peng, B., Wang, G., Wang, S., and Peng, X. (2004). Universal primer PCR with DGGE for rapid detection of bacterial pathogens. J. Microbiol. Methods *57*, 409–413.

Jin, D., Wang, P., Bai, Z., Wang, X., Peng, H., Qi, R., Yu, Z., and Zhuang, G. (2011). Analysis of bacterial community in bulking sludge using culture-dependent and -independent approaches. J. Environ. Sci. (China). *23*, 1880–1887.

Kannan, S., Suresh Kanna, P., Karkuzhali, K., Chattopadhyay, U.K., and Pal, D. (2001). Direct detection of diarrheagenic *Aeromonas* from faeces by polymerase chain reaction (PCR) targeting aerolysin toxin gene. Eur. Rev. Med. Pharmacol. Sci. *5*, 91–94.

Keim, P., Price, L.B., Klevytska, A.M., Smith, K.L., Schupp, J.M., Okinaka, R., Jackson, P.J., and Hugh-Jones, M.E. (2000). Multiple-locus variable-number tandem repeat analysis reveals genetic relationships within *Bacillus anthracis*. J. Bacteriol. *182*, 2928–2936.

Kenzaka, T., Yamaguchi, N., Tani, K., and Nasu, M. (1998). rRNA-targeted fluorescent in situ hybridization analysis of bacterial community structure in river water. Microbiology. *144*, 2085–2093.

Kenzaka, T., Tamaki, S., Yamaguchi, N., Tani, K., and Nasu, M. (2005). Recognition of individual genes in diverse microorganisms by cycling primed in situ amplification. Appl. Environ. Microbiol. *71*, 7236–7244.

Khajanchi, B.K., Fadl, A.A., Borchardt, M.A., Berg, R.L., Horneman, A.J., Stemper, M.E., Joseph, S.W., Moyer, N.P., Sha, J., and Chopra, A.K. (2010). Distribution of virulence factors and molecular fingerprinting of *Aeromonas* species isolates from water and clinical samples: suggestive evidence of water-to-human transmission. Appl. Environ. Microbiol. *76*, 2313–2325.

Khan, A.A., and Cerniglia, C.E. (1997). Rapid and sensitive method for the detection of *Aeromonas caviae* and *Aeromonas trota* by polymerase chain reaction. Lett. Appl. Microbiol. *24*, 233–239.

Khan, A.A., Nawaz, M.S., Khan, S.A., and Cerniglia, C.E. (1999). Identification of *Aeromonas trota* (hybridization group 13) by amplification of the aerolysin gene using polymerase chain reaction. MCP. *13*, 93–98.

Khushiramani, R., Girisha, S.K., Karunasagar, I., and Karunasagar, I. (2009). Evaluation of a digoxigenin-labelled probe for detection of *Aeromonas* spp. Lett. Appl. Microbiol. *48*, 383–385.

Kikuchi, Y., and Graf, J. (2007). Spatial and temporal population dynamics of a naturally occurring two-species microbial community inside the digestive tract of the medicinal leech. Appl. Environ. Microbiol. *73*, 1984–1991.

Kingombe, C.I., Huys, G., Tonolla, M., Albert, M.J., Swings, J., Peduzzi, R., and Jemmi, T. (1999). PCR detection, characterization, and distribution of virulence genes in *Aeromonas* spp. AEM *65*, 5293–5302.

Kingombe, C.I., Huys, G., Howald, D., Luthi, E., Swings, J., and Jemmi, T. (2004). The usefulness of molecular techniques to assess the presence of *Aeromonas* spp. harboring virulence markers in foods. Int. J. Food Microbiol. *94*, 113–121.

Kong, R.Y., Pelling. A., So, C.L., and Wu, R.S. (1999). Identification of oligonucleotide primers targeted at the 16S-23S rDNA intergenic spacers for genus- and species-specific detection of aeromonads. Mar. Pollut. Bull. *38*, 802–808.

Kong, R.Y., Lee, S.K., Law, T.W., Law, S.H., and Wu, R.S. (2002). Rapid detection of six types of bacterial pathogens in marine waters by multiplex PCR. Water Res. *36*, 2802–2812.

Krejcí, E., Sedlácek, I., and Baudisová, D. (2009). Classification of brown pigmented aeromonads isolated from river water. Folia Microbiol. (Praha). *54*, 123–129.

Küpfer, M., Kuhnert, P., Korczak, B.M., Peduzzi, R., and Demarta, A. (2006). Genetic relationships of *Aeromonas* strains inferred from 16S rRNA, *gyrB* and *rpoB* gene sequences. Int. J. Syst. Evol. Microbiol. *56*, 2743–2751.

Lafarge, V., Ogier, J.C., Girard, V., Maladen, V., Leveau, J.Y., Gruss, A., and Delacroix-Buchet, A. (2004). Raw cow milk bacterial population shifts attributable to refrigeration. Appl. Environ. Microbiol. *70*, 5644–5650.

Lamy, B., Laurent, F., and Kodjo, A. (2010). Validation of partial *rpoB* gene sequence as a tool for phylogenetic identification of aeromonads isolated from environmental sources. Can. J. Microbiol. *56*, 217–228.

Lee, D.Y., Lauder, H., Cruwys, H., Falletta, P., and Beaudette, L.A. (2008). Development and application of an oligonucleotide microarray and real-time quantitative PCR for detection of wastewater bacterial pathogens. Sci. Total Environ. *398*, 203–211.

Lee, M.F., Peng, C.F., Lin, Y.H., Lin, S.R., and Chen, Y.H. (2008). Molecular diversity of class 1 integrons in human isolates of *Aeromonas* spp. from southern Taiwan. Jpn. J. Infect. Dis. *61*, 343–349.

Li, M.Y., Zhou, G.H., Xu, X.L., Li, C.B., and Zhu, W.Y. (2006). Changes of bacterial diversity and main flora in chilled pork during storage using PCR-DGGE. Food Microbiol. *23*, 607–611.

Lucarelli, F., Tombelli, S., Minunni, M., Marrazza, G., and Mascini, M. (2008). Electrochemical and piezoelectric DNA biosensors for hybridisation detection. Anal. Chim. Acta *609*, 139–159.

Liu, J., Janaki, L., and Houpt, E. (2012). Precaution for using nucleic acid-based methods to detect *Aeromonas*. J. Clin. Microbiol. *50*, 1829.

Lund, V., Espelid, S., and Mikkelsen, H. (2003). Vaccine efficacy in spotted wolffish *Anarhichas* minor: relationship to molecular variation in A-layer protein of atypical *Aeromonas salmonicida*. Dis. Aquat. Organ. *56*, 31–42.

Maiden, M.C., Bygraves, J.A., Feil, E., Morelli, G., Russell, J.E., Urwin, R., Zhang, Q., Zhou, J., Zurth, K., Caugant, D.A., *et al.* (1998). Multilocus sequence typing: a portable approach to the identification of clones within populations of pathogenic microorganisms. Proc. Natl. Acad. Sci. U.S.A. *95*, 3140–3145.

Maiti, B., Raghunath, P., Karunasagar, I., and Karunasagar, I. (2009). Typing of clinical and environmental strains of *Aeromonas* spp. using two PCR based methods and whole cell protein analysis. J. Microbiol. Methods. *78*, 312–318.

Martin, B., Humbert, O., Camara, M., Guuenzi, E., Walker, J., Mitchell, T., Andrew, P., Prudhomme, M., Alloing, G., Hakenbeck, R., *et al.* (1992). A highly conserved repeated DNA element located in the chromosome of *Streptococcus pneumoniae*. Nucleic Acids Res. *20*, 3479–3483.

Martinetti, L.G., and Altwegg, M. (1992). rRNA gene restriction patterns as taxonomic tools for the genus *Aeromonas*. Int. J. Syst. Bacteriol. *42*, 384–389.

Martinez-Murcia, A.J. (1993). Phylogenetic analysis of the genera *Leuconostoc* and *Aeromonas*. PhD thesis, University of Reading. London British Library.

Martinez-Murcia, A.J., Benlloch, S., and Collins, M.D. (1992). Phylogenetic interrelationships of members of the genera *Aeromonas* and *Plesiomonas* as determined by 16S ribosomal DNA sequencing lack of congruence with results of DNA-DNA hybridizations. Int. J. Syst. Bacteriol. *42*, 412–421.

Martínez-Murcia, A.J., Boan, I.F., and Rodrıguez-Valera, F. (1995). Evaluation of the authenticity of haloarchaeal strains by randomamplified polymorphic DNA. Lett. Appl. Microbiol. *21*, 106–108.

Martínez-Murcia, A.J., Antón, A.I., and Rodriguez-Valera, F. (1999). Patterns of sequence variation in two regions of the 16S rRNA multigene family of *Escherichia coli*. Int. J. Syst. Bacteriol. *49*, 601–610.

Martínez-Murcia, A.J., Borrell, N., and Figueras, M.J. (2000). Typing of clinical and environmental *Aeromonas veronii* strains based on the 16S-23S rDNA spacers. FEMS Immunol. Med. Microbiol. *28*, 225–232.

Martínez-Murcia, A.J., Soler, L., Saavedra, M.J., Chacón, M.R., Guarro, J., Stackebrandt, E., and Figueras, M.J. (2005). Phenotypic, genotypic, and phylogenetic discrepancies to differentiate *Aeromonas salmonicida* from *Aeromonas bestiarum*. Int. Microbiol. *8*, 259–269.

Martínez-Murcia, A.J., Figueras, M.J., Saavedra, M.J., and Stackebrandt, E. (2007). The recently proposed *Aeromonas sharmana* sp. nov., isolate GPTSA-6T, is not a member of the genus *Aeromonas*. Int. Microbiol. *10*, 61–64.

Martínez-Murcia, A.J., Saavedra, M.J., Mota, V.R., Maier, T., Stackebrandt, E., and Cousin, S. (2008). *Aeromonas aquariorum* sp. nov., isolated from aquaria of ornamental fish. Int. J. Syst. Evol. Microbiol. *58*, 1169–1175.

Martínez-Murcia, A.J., Monera, A., Alperi, A., Figueras, M.J., and Saavedra, M.J. (2009). Phylogenetic evidence suggests that strains of *Aeromonas hydrophila* subsp. *dhakensis* belong to the species *Aeromonas aquariorum* sp. nov. Curr. Microbiol. *58*, 76–80.

Martínez-Murcia, A.J., Monera, A., Saavedra, M.J., Oncina, R., Lopez-Alvarez, M., Lara, E., and Figueras, M.J. (2011). Multilocus phylogenetic analysis of the genus *Aeromonas*. Syst. Appl. Microbiol. *34*, 189–199.

Martínez-Murcia, A., Beaz-Hidalgo, R., Saavedra, M.J., Figueras, M.J., and Sedlacek, I. (2013). *Aeromonas cavernicola* sp. nov., isolated from fresh water of a brook in a cavern. Curr. Microbiol. *66*, 197–204

Martino, M.E., Fasolato, L., Montemurro, F., Rosteghin, M., Manfrin, A., Patarnello, T., Novelli, E., and Cardazzo, B. (2011). Determination of microbial diversity of *Aeromonas* strains on the basis of multilocus sequence typing, phenotype, and presence of putative virulence genes. Appl. Environ. Microbiol. *77*, 4986–5000.

Mendes-Marques, C.L., Nascimento, L.M., Theophilo, G.N., Hofer, E., Melo Neto, O.P., and Leal, N.C. (2012). Molecular characterization of *Aeromonas* spp. and *Vibrio cholerae* O1 isolated during a diarrhea outbreak. Rev. Inst. Med. Trop. Sao Paulo. *54*, 299–304.

Millán, A., Gómez-Tato, A., Pardo, B.G., Fernández, C., Bouza, C., Vera, M., Alvarez-Dios, J.A., Cabaleiro, S., Lamas, J., Lemos, M.L., and Martínez, P. (2011). Gene expression profiles of the spleen, liver, and head kidney in turbot (*Scophthalmus maximus*) along the infection process with *Aeromonas salmonicida* using an immune-enriched oligo-microarray. Mar. Biotechnol. *13*, 1099–1114.

Miñana-Galbis, D., Urbizu-Serrano, A., Farfán, M., Fusté, M.C., and Lorén, J.G. (2009). Phylogenetic analysis and identification of *Aeromonas* species based on sequencing of the *cpn60* universal target. Int. J. Syst. Evol. Microbiol. *59*, 1976–1983.

Miyata, M., Aoki, T., Inglis, V., Yoshida, T., and Endo, M. (1995). RAPD analysis of *Aeromonas salmonicida* and *Aeromonas hydrophila*. J. Appl. Bacteriol. *79*, 181–185.

Morandi, A., Zhaxybayeva, O., Gogarten, J.P., and Graf, J. (2005). Evolutionary and diagnostic implications of intragenomic heterogeneity in the 16S rRNA gene in *Aeromonas* strains. J. Bacteriol. *187*, 6561–6564.

Moyer, N.P., Martinetti, G., Luthy-Hottenstein, J., and Altwegg, M. (1992). Value of rRNA gene restriction patterns of *Aeromonas* spp. for epidemiological investigations. Curr. Microbiol. *24*, 15–22.

Nam, I.Y., and Joh, K. (2007). Rapid detection of virulence factors of *Aeromonas* isolated from a trout farm by hexaplex-PCR. J. Microbiol. *45*, 297–304.

Nash, J.H., Findlay, W.A., Luebbert, C.C., Mykytczuk, O.L., Foote, S.J., Taboada, E.N., Carrillo, C.D., Boyd, J.M., Colquhoun, D.J., Reith, M.E., and Brown, L.L. (2006). Comparative genomics profiling of clinical isolates of *Aeromonas salmonicida* using DNA microarrays. BMC Genomics 7, 7–43.

Nayduch, D., Honko, A., Noblet, G.P., and Stutzenberger, F. (2001). Detection of *Aeromonas caviae* in the common housefly *Musca domestica* by culture and polymerase chain reaction. Epidemiol. Infect. *127*, 561–566.

Ndi, O.L., and Barton, M.D. (2011). Incidence of class 1 integron and other antibiotic resistance determinants in *Aeromonas* spp. from rainbow trout farms in Australia. J. Fish Dis. *34*, 589–599.

Neilan, B.A. (1995). Identification and phylogenetic analysis of toxigenic cyanobacteria by multiplex randomly amplified polymorphic DNA PCR. Appl. Environ. Microbiol. *61*, 2286–2291.

Nhung, P.H., Hata, H., Ohkusu, K., Noda, M., Shah, M.M., Goto, K., and Ezaki, T. (2007). Use of the novel phylogenetic marker dnaJ and DNA-DNA hybridization to clarify interrelationships within the genus *Aeromonas*. Int. J. Syst. Evol. Microbiol. *57*, 1232–1237.

Nováková, D., Svec, P., and Sedlácek, I. (2009). Characterization of *Aeromonas encheleia* strains isolated from aquatic environments in the Czech Republic. Lett. Appl. Microbiol. *48*, 289–294.

Oakey, H.J., Gibson, L.F., and George, A.M. (1996). Differentiation of *Aeromonas* genomo-species using random amplified polymorphic DNA polymerase chain reaction (RAPD-PCR). J. Appl. Bacteriol. *80*, 402–410.

Oakey, H.J., Ellis, J.T., and Gibson, L.F. (1998). The development of random DNA probes specific for *Aeromonas salmonicida*. J. Appl. Microbiol. *84*, 37–46.

Oakey, H.J., Gibson, L.F., and George, A.M. (1999). DNA probes specific for *Aeromonas hydrophila* (HG1). J. Appl. Microbiol. *86*, 187–193.

O'hIci, B., Olivier, G., and Powell, R. (2000). Genetic diversity of the fish pathogen *Aeromonas salmonicida* demonstrated by random amplified polymorphic DNA and pulsed-field gel electrophoresis analyses. Dis. Aquat. Organ. *39*, 109–119.

Ozbas, Z.Y., Lehner, A., and Wagner, M. (2000). Development of a multiplex and semi-nested PCR assay for detection of *Yersinia enterocolitica* and *Aeromonas hydrophila* in raw milk. Food Microbiol. *17*, 197–203.

Pablos, M., Remacha, M.A., Rodríguez-Calleja, J.M., Santos, J.A., Otero, A., and García-López, M.L. (2010). Identity, virulence genes, and clonal relatedness of *Aeromonas* isolates from patients with diarrhea and drinking water. Eur. J. Clin. Microbiol. Infect. Dis. *29*, 1163–1172.

Pedersen, K., Dalsgaard, I., and Larsen, J.L. (1996). Characterization of atypical *Aeromonas salmonicida* isolates by ribotyping and plasmid profiling. J. Appl. Bacteriol. *80*, 37–44.

Peng, X.X., Zhang, J.Y., Wang, S.Y., Lin, Z.L., and Zhang, W.Y. (2002). Immuno-capture PCR for detection of *Aeromonas hydrophila*. J. Microbiol. Methods *49*, 335–338.

Pérez-Valdespino, A., Fernández-Rendón, E., and Curiel-Quesada, E. (2009). Detection and characterization of class 1 integrons in *Aeromonas* spp. isolated from human diarrheic stool in Mexico. J. Basic Microbiol. *49*, 572–578.

Perkins, S.L., Budinoff, R.B., and Siddall, M.E. (2005). New gammaproteobacteria associated with blood-feeding leeches and a broad phylogenetic analysis of leech endosymbionts. Appl. Environ. Microbiol. *71*, 5219–5224.

Pidiyar, V.J., Jangid, K., Patole, M.S., and Shouche, Y.S. (2003). Analysis of 16S-23S intergenic spacer regions and rrn operon copy number of *Aeromonas culicicola* MTCC 3249^T. DNA Seq. *14*, 183–194.

Popoff, M. (1984). Genus III. *Aeromonas*. In Bergey's Manual of Systematic Bacteriology, vol.1, Krieg, N.R. ed. (Williams & Wilkins Co., Baltimore, MA), pp. 545–548.

Pridgeon, J.W., and Klesius, P.H. (2011). Molecular identification and virulence of three *Aeromonas hydrophila* isolates cultured from infected channel catfish during a disease outbreak in west Alabama (USA) in 2009. Dis. Aquat. Organ. *94*, 249–253

Pridgeon, J.W., Klesius, P.H., Song, L., Zhang, D., Kojima, K., and Mobley, J.A. (2013). Identification, virulence, and mass spectrometry of toxic ECP fractions of West Alabama isolates of *Aeromonas hydrophila* obtained from a 2010 disease outbreak. Vet. Microbiol. *164*, 336–343.

Quinn, R.A., and Stevenson, R.M. (2012). Denaturing gradient gel electrophoresis for nonlethal detection of *Aeromonas salmonicida* in salmonid mucus and its potential for other bacterial fish pathogens. Can. J. Microbiol. *58*, 563–571.

Rahman, M., Colque-Navarro, P., Kühn, I., Huys, G., Swings, J., and Möllby, R. (2002). Identification and characterization of pathogenic *Aeromonas veronii* biovar *sobria* associated with epizootic ulcerative syndrome in fish in Bangladesh. Appl. Environ. Microbiol. *68*, 650–655.

Reith, M.E., Singh, R.K., Curtis, B., Boyd, J.M., Bouevitch, A., Kimball, J., Munholland, J., Murphy, C., Sarty, D., Williams, J., *et al.* (2008). The genome of *Aeromonas salmonicida* subsp. *salmonicida* A449: insights into the evolution of a fish pathogen. BMC Genomics 9, 427.

Richter, M., and Rosselló-Móra, R. (2009). Shifting the genomic gold standard for the prokaryotic species definition. Proc. Natl. Acad. Sci. U.S.A. *106*, 19126–19131.

Roger, F., Marchandin, H., Jumas-Bilak, E., Kodjo, A., ColBVH study group, and Lamy, B. (2012a). Multilocus genetics to reconstruct aeromonad evolution. BMC Microbiol. *12*, 62.

Roger, F., Lamy, B., Jumas-Bilak, E., Kodjo, A., Carmagnol, F., and Marchandin, H. (2012b). Ribosomal multi-operon diversity: an original perspective on the genus *Aeromonas*. PLoS One 7, e46268.

Schleifer, K.H. (2009). Classification of bacteria and archaea: past, present and future. Syst. Appl. Microbiol. *32*, 533–542.

Sechi, L.A., Deriu, A., Falchi, M.P., Fadda, G., and Zanetti, S. (2002). Distribution of virulence genes in *Aeromonas* spp. isolated from Sardinian waters and from patients with diarrhoea. J. Appl. Microbiol. *92*, 221–227.

Sedláček, I., Krejčí, E., Andělová, A., Sedláčková, M., Porazilová, I., and Holochová, P. (2012). *Aeromonas hydrophila* subsp. *dhakensis*--a causative agent of gastroenteritis imported into the Czech Republic. Ann. Agric. Environ. Med. *19*, 409–413.

Sen, K. (2005). Development of a rapid identification method for *Aeromonas* species by multiplex PCR. Can. J. Microbiol. *51*, 957–966.

Sen, K., and Rodgers, M. (2004). Distribution of six virulence factors in *Aeromonas* species isolated from U.S. drinking water utilities: a PCR identification. J. Appl. Microbiol. *97*, 1077–1086.

Sepe, A., Barbieri, P., Peduzzi, R., and Demarta, A. (2008). Evaluation of *recA* sequencing for the classification of *Aeromonas* strains at the genotype level. Lett. Appl. Microbiol. *46*, 439–444.

Seshadri, R., Joseph, S.W., Chopra, A.K., Sha, J., Shaw, J., Graf, J., Haft, D., Wu, M., Ren, Q., Rosovitz, M.J., *et al.* (2006). Genome sequence of *Aeromonas hydrophila* ATCC 7966T: jack of all trades. J. Bacteriol. *188*, 8272–8282.

Shannon, K.E., Lee, D.Y., Trevors, J.T., and Beaudette, L.A. (2007). Application of real-time quantitative PCR for the detection of selected bacterial pathogens during municipal wastewater treatment. Sci. Total Environ. *382*, 121–129.

Silver, A.C., Kikuchi, Y., Fadl, A.A., Sha, J., Chopra, A.K., and Graf, J. (2007). Interaction between innate immune cells and a bacterial type III secretion system in mutualistic and pathogenic associations. Proc. Natl. Acad. Sci. U.S.A. *104*, 9481–9486.

Singh, V., Chaudhary, D.K., Mani, I., Somvanshi, P., Rathore, G., and Sood, N. (2010). Genotyping of *Aeromonas hydrophila* by box elements. Mikrobiologiia. *79*, 390–393.

Sneath, P.H.A. (1993). Evidence from *Aeromonas* for genetic crossing-over in ribosomal sequences. Int. J. Syst. Bacteriol. *43*, 626–629.

Soler, L., Figueras, M.J., Chacon, M.R., Guarro, J., and Martinez-Murcia, A.J. (2003). Comparison of three molecular methods for typing *Aeromonas popoffii* isolates. Antonie van Leeuwenhoek *83*, 341–349.

Soler, L., Yáñez, M.A., Chacon, M.R., Aguilera-Arreola, M.G., Catalán, V., Figueras, M.J., and Martínez-Murcia, A.J. (2004). Phylogenetic analysis of the genus *Aeromonas* based on two housekeeping genes. Int. J. Syst. Evol. Microbiol. *54*, 1511–1519.

Stackebrandt, E., and Ebers, J. (2006). Taxonomic parameters revisited: tarnished gold standards. Microbiol. Today *33*, 152–155.

Stackebrandt, E., Frederiksen, W., Garrity, G.M., Grimont, P.A.D., Kämpfer, P., Maiden, M.C.J., Nesme, X., Rosselló-Móra, R., Swings, J., Trüper, H.G., Vauterin, L., Ward, A.C., and Whitman, W.B. (2002). Report of the ad hoc committee for the reevaluation of the species definition in bacteriology. Int. J. Syst. Evol. Microbiol. *52*, 1043–1047.

Stine, O.C., Carnahan, A., Singh, R., Powell, J., Furuno, J.P., Dorsey, A., Silbergeld, E., Williams, H.N., and Morris, J.G. (2003). Characterization of microbial communities from coastal waters using microarrays. Environ. Monit. Assess. *81*, 327–336.

Szczuka, E., and Kaznowski, A. (2004). Typing of clinical and environmental *Aeromonas* spp. strains by random amplified polymorphic DNA PCR, repetitive extragenic palindromic PCR, and enterobacterial repetitive intergenic consensus sequence PCR. J. Clin. Microbiol. *42*, 220–228.

Tacão, M., Alves, A., Saavedra, M.J., and Correia, A. (2005a). BOX-PCR is an adequate tool for typing *Aeromonas* spp. Antonie Van Leeuwenhoek *88*, 173–179.

Tacão, M., Moura, A., Alves, A., Henriques, I., Saavedra, M.J., and Correia, A. (2005b). Evaluation of 16S rDNA- and *gyrB*-DGGE for typing members of the genus *Aeromonas*. FEMS Microbiol. Lett. *246*, 11–18.

Talon, D., Dupont, M.J., Lesne, J., Thouverez, M., and Michel-Briand, Y. (1996). Pulsed-field gel electrophoresis as an epidemiological tool for clonal identification of *Aeromonas hydrophila*. J. Appl. Bacteriol. *80*, 277–282.

Talon, D., Mulin, B., and Thouverez, M. (1998). Clonal identification of *Aeromonas hydrophila* strains using randomly amplified polymorphic DNA analysis. Eur. J. Epidemiol. *14*, 305–310.

Tichoniuk, M., Gwiasdowska, D., Ligaj, M., and Filipiak, M. (2010). Electrochemical detection of foodborne pathogen *Aeromonas hydrophila* by DNA hybridization biosensor. Biosens. Bioelectron. *26*, 1618–1623.

Tindall, B., Rosselló-Móra, J.R., Busse, H.J., Ludwig, W., and Kämpfer, P. (2010). Notes on the characterization of prokaryote strains for taxonomic purposes. Int. J. Syst. Evol. Microbiol. *60*, 249–266.

Tombelli, S., Mascini, M., Sacco, C., and Turner, A.P.F. (2000). A DNA piezoelectric biosensor assay coupled with a polymerase chain reaction for bacterial toxicity determination in environmental samples. Anal. Chim. Acta *418*, 1–9.

Trakhna, F., Harf-Monteil, C., Abdel-Nour, A., Maaroufi, A., and Gadonna-Widehem, P. (2009). Rapid *Aeromonas hydrophila* identification by TaqMan PCR assay: comparison with a phenotypic method. Lett. Appl. Microbiol. *49*, 186–190.

Umelo, E., and Trust, T.J. (1998). Physical map of the chromosome of *Aeromonas salmonicida* and genomic comparisons between *Aeromonas* strains. Microbiology *144*, 2141–2149.

Vauterin, L., and Vauterin, P. (2006). Integrated databasing and analysis. In Molecular Identification, Systematics, and Population Structure of Prokaryotes, Stackebrandt, E. ed. (Springer-Verlag, Berlin Heidelberg, Germany), pp. 141–217.

Verner-Jeffreys, D.W., Welch, T.J., Schwarz, T., Pond, M.J., Woodward, M.J., Haig, S.J., Rimmer, G.S., Roberts, E., Morrison, V., and Baker-Austin, C. (2009). High prevalence of multidrug-tolerant bacteria and associated antimicrobial resistance genes isolated from ornamental fish and their carriage water. PLoS One. *21*, e8388.

Versalovic, J., Koeuth, T., and Lupski, J.R. (1991). Distribution of repetitive DNA sequences in eubacteria and application to fingerprinting. Nucleic Acids Res. *19*, 6823–6831.

Versalovic, J., Schneider, M., de Bruijn, F.J., and Lupski, J.R. (1994). Genomic fingerprinting of bacteria using the repetitive sequence-based polymerase chain reaction. Methods Mol. Cell. Biol. *5*, 25–40.

Villari, P., Crispino, M., Montuori, P., and Stanzione, S. (2000). Prevalence and molecular characterization of *Aeromonas* spp. in ready-to-eat foods in Italy. J. Food Prot. *63*, 1754–1757.

Villari, P., Crispino, M., Montuori, P., and Boccia, S. (2003). Molecular typing of *Aeromonas* isolates in natural mineral waters. Appl. Environ. Microbiol. *69*, 697–701.

Wang, G., Clark, C.G., Liu, C., Pucknell, C., Munro, C.K., Kruk, T.M., Caldeira, R., Woodward, D.L., and Rodgers, F.G. (2003). Detection and characterization of the hemolysin genes in *Aeromonas hydrophila* and *Aeromonas sobria* by multiplex PCR. J. Clin. Microbiol. *41*, 1048–1054.

Wei, X.N., Zheng, Z.J., Zhang, L.H., Qu, F., and Huang, X. (2008). Sensitive and rapid detection of *Aeromonas caviae* in stool samples by loop-mediated isothermal amplification. Diagn. Microbiol. Infect. Dis. *60*, 113–116.

Yamamoto, S., and Harayama, S. (1995). PCR amplification and direct sequencing of *gyrB* genes with universal primers and their application to the detection and taxonomic analysis of *Pseudomonas putida* strains. Appl. Environ. Microbiol. *61*, 1104–1109.

Yamamoto, S., and Harayama, S. (1996). Phylogenetic analysis of *Acinetobacter* strains based on the nucleotide sequences of *gyrB* genes and on the amino acid sequences of their products. Int. J. Syst. Bacteriol. *46*, 506–511.

Yamamoto, S., and Harayama, S. (1998). Phylogenetic relationships of *Pseudomonas putida* strains deduced from the nucleotide sequences of *gyrB*, *rpoD* and 16S rRNA genes. Int. J. Syst. Bacteriol. *48*, 813–819.

Yáñez, M.A., Catalán, V., Apráiz, D., Figueras, M.J., and Martínez-Murcia, A.J. (2003). Phylogenetic analysis of members of the genus *Aeromonas* based on *gyrB* gene sequences. Int. J. Syst. Evol. Microbiol. *53*, 875–883.

Ye, Y., Jiang, Y., Fan, T., Jiang, Q., Cheng, Y., Lu, J., and Lin, L. (2012). Resistance characterization, virulence factors, and ERIC-PCR fingerprinting of *Aeromonas veronii* strains isolated from diseased *Trionyx sinensis*. Foodborne Pathog. Dis. *9*, 1053–1055.

Zhou, G., Wen, S., Liu, Y., Li, R., Zhong, X., Feng, L., Wang, L., and Cao, B. (2011). Development of a DNA microarray for detection and identification of *Legionella pneumophila* and ten other pathogens in drinking water. Int. J. Food Microbiol. *145*, 293–300.

Zuckerkandl, E., and Pauling, L. (1965). Evolutionary divergence and convergence in proteins. In Evolving Genes and Proteins, Bryson, V., and Vogel, H.J., eds (Academic Press, New York, NY), pp. 97–166.

Occurrence and Virulence Potential of *Aeromonas* in Food and Water

9

Christopher J. Grim

Abstract

Aeromonas is ubiquitous in surface waters. In terms of water distribution systems, elevated temperatures in surface waters lead to proliferation and regrowth of the organisms, especially in systems that are prone to biofilm development and in the extreme periphery, where free chlorine levels are low. Maintenance of chlorine levels at 0.3 mg/l or above, use of ground water sources, and biofilm and biofouling control are essential to prevent regrowth of *Aeromonas* in these systems. Food commodities are easily contaminated with *Aeromonas*, likely from exposure to contaminated water. In contrast to the seasonal distribution in surface waters, the psychrotrophic tolerance of aeromonads allows for their survival and proliferation in refrigerated foods.

Introduction

Aeromonas is generally considered to be ubiquitous within the microbial biosphere. In particular, aeromonads are autochthonous members of most freshwater aquatic environments, including surface waters such as ponds, lakes and rivers; ground water; chlorinated and non-chlorinated drinking water; bottled water; sewage and sewage effluents (Chauret *et al.*, 2001; Clark *et al.*, 1982; Fliermans *et al.*, 1977; Villari *et al.*, 2003). They can also tolerate sodium chloride concentrations as high as 2.5 to 3%; therefore, they are also ideally suited for survival in bays and estuaries, where freshwater from rivers merge and mix with seawater. Due to their distribution in aquatic systems, aeromonads can colonize most aquatic animals and plants, and, as a consequence, may cause a spectrum of diseases in these organisms. They are not generally considered as part of the normal flora of human beings, however, small percentages (< 1%) of healthy adults are usually found to be carriers (Araujo *et al.*, 1991). Humans typically become infected or colonized by *Aeromonas* through one of two routes, consumption or exposure to 'contaminated' drinking or natural waters, and consumption of food contaminated with the organism due to environmental exposure. Several topics will be presented in this chapter as they relate to the distribution of *Aeromonas* in food and water. For additional information, please consult the following reviews (Ghenghesh *et al.*, 2008; Kirov, 1993; Sartory, 2002; Stratev *et al.*, 2012; USEPA, 2006).

Occurrence of *Aeromonas* in water

Aeromonas spp. are ubiquitous members of the aquatic environment and have been isolated from both eutrophic and oligotrophic waters. Their abundance has been shown to increase during the summer and late summer season in surface waters (Burke *et al.*, 1984b; Rippey and Cabelli, 1989). Exposure or consumption of contaminated water has been determined to be a major risk factor for *Aeromonas* infection (Kirov, 1993). For example, Krovacek *et al.* (Krovacek *et al.*, 1989) reported that the use of private, untreated well water was the causative agent of diarrhoea of an infant in Sweden. Levels of *A. hydrophila* were found to be as high as 6×10^4 cfu/100 ml, and the population of *Aeromonas* persisted in the well over a long period of time (Krovacek *et al.*, 1989). Table 9.1. provides some reported densities of *Aeromonas* from various water sources.

One of the first ecological studies on the distribution of *Aeromonas* from aquatic sources was reported by Hazen *et al.* (Hazen *et al.*, 1978). Sampling sites for this study included rivers, springs, pond, harbours, and bays across the United States, including Puerto Rico. Overall, the authors found that 92% (135/147) of environmental water samples harboured *Aeromonas*. Of the 12 sites for which *Aeromonas* was not isolated, seven were considered 'extreme environments', i.e. hypersaline lakes (300‰), geothermal springs (66°C and 47°C), and extremely polluted rivers (106 and 400 JTU, Jackson turbidity units). A general trend of lower variation in abundance for freshwater sites, and, conversely, a higher but more variable density from saline sites was observed. Further, lotic (flowing) sites had higher densities of *Aeromonas* than lentic (still) sites. *Aeromonas* was isolated from sites with temperature ranges between 4° and 45°C, 0 and 100 JTU (turbidity), and pH 5.2 to 9.8; the only physical parameter that correlated with abundance was conductivity.

Several microbiological surveys of *Aeromonas* from aquatic sites would quickly follow, from the US and abroad. For example, Kaper *et al.* (1981) conducted an intensive ecological surveillance of *Aeromonas* from the Chesapeake Bay, MD. They concluded the organisms to be ubiquitously distributed, both in the water column and sediment, along the course of

Table 9.1 Abundance of *Aeromonas* spp. from natural water bodies

Study	Geographic region	Water source	Abundance, range (cfu/100 ml)
Maleeja *et al.* (2003)	Tunisia	Municipal water effluent	1.5×10^5–2.2×10^8
Maleeja *et al.* (2003)	Sfax, Mediterranean	Seawater	$<8.0 \times 10^3$
Maasa *et al.* (2001)	Lecce, Italy	Private well water	30×10^2–1.6×10^8
Havelaar *et al.* (1990)	Netherlands	River water	$>1.0 \times 10^4$
Havelaar *et al.* (1990)	Netherlands	Reservoir	1.0×10^3 –1.0×10^4
Kaper *et al.* (1981)	Chesapeake Bay, MD[a]	Water	$\leq 5 \times 10^5$
Kaper *et al.* (1981)	Chesapeake Bay, MD	Sediment	4.6×10^2 (per gram)
Borrell *et al.* (1998)	Catalonia, Spain	Lake and reservoirs	Up to 6.9×10^3
Borrell *et al.* (1998)	Catalonia, Spain	River water	Up to 3.4×10^4
Monfort and Baleux (1991)	Southern France	Thau Lagoon (sewage input)	1×10^6–1×10^2
Weinbauer and Hofle (1998)	Northern Germany	Lake Plußsee (eutrophic)	Up to 2×10^2

[a]For estimates of abundance of *Aeromonas* spp. from US waters, see the comprehensive survey reported by Hazen *et al.* (1978).

the vast estuary. In this study, abundance of *Aeromonas* was lower in more saline sites, and during winter months.

Following early studies that reported the high abundance of *Aeromonas* in natural water bodies, several researchers began to investigate drinking water systems, especially those that were fed by surface water. In Australia, Burke *et al.* (1984b) conducted a weekly ecological surveillance of *Aeromonas* over a period of one year, targeting the municipal water supply of Perth. This report represented one of the first microbiological surveys of drinking water. *Aeromonas* was isolated from untreated surface water for all but 2 weeks (96%) out of the year-long study. In contrast, untreated ground water yielded *Aeromonas* for only 10 weeks (19%), confined to the period of late summer to mid-Autumn. Following primary chlorine treatment, the frequency of isolation of *Aeromonas* dropped to 69% for surface water and *Aeromonas* was not detected in conventionally treated (alum coagulation, sedimentation, and sand filtration) ground water. However, when these treated waters were held in service reservoirs, abundance of *Aeromonas* increased to levels near that of untreated water and the organism was detected at a frequency similar to that of untreated water. In some instances, the density of *Aeromonas* was higher in the service reservoirs than levels in the untreated surface and ground water. Chlorination of the service reservoir effluents again decreased the frequency and abundance of *Aeromonas*; however, residual chorine levels decreased as a function of distance from the treatment facility and sampling at points along the distribution system mimicked conditions in the untreated waters. By comparison, *E. coli* and coliform counts were essentially negative at points along the distribution system following the two chemical treatments. The authors reported a correlation of abundance of *Aeromonas* spp. with water temperature, for those samples that were free of chlorine or at levels below 0.3 mg/l. Essentially, *Aeromonas* was isolated from unchlorinated samples when the water temperature was above 14.5°C.

Additional reports of *Aeromonas* abundance and distribution from drinking water sources have followed. Di Bari *et al.* (2007) reported that 39% of drinking water samples from wells, springs, and fountains in São Paulo, Brazil, were positive for *Aeromonas*. Interestingly, springs (64%) and, especially, fountains (100%) had a higher frequency of *Aeromonas*, as compared to that of the more-protected below-ground wells (30%). In agreement with Burke *et al.* (1984b), there was no correlation between *Aeromonas* and indicators of faecal contamination; i.e. coliforms and thermotolerant coliforms. A small number of mineral water bottles were tested and all were negative for *Aeromonas*. Another report from Brazil (de Oliveira Scoaris *et al.*, 2008a) examined bottled mineral water, well (Maringá) and municipal tap water (Londrina). The authors reported much lower frequencies than Di Bari *et al.* (2007), 8.3% for wells and 6.5% for tap water, but 12.7% for bottled mineral water. There was no correlation between *Aeromonas* and the faecal indicator organism, in this case, total coliform, for samples from bottled mineral water; however, there was a significant correlation for well and tap water. From Argentina, Fernández *et al.* (2000) reported the isolation of *A. hydrophila* from chlorinated drinking water in Buenos Aires that otherwise conformed to nationwide water quality standards. Borrell *et al.* (1998) found that 96% of lakes and reservoirs samples and 88% of river water samples from Catalonia, Spain, harboured *Aeromonas*. These water types also had the highest diversity of aeromonads, as compared to seawater, and untreated drinking water, which also exhibited a high frequency of *Aeromonas*, 83.1 and 64.4%, respectively. In contrast, only 10% of treated drinking water samples were positive for *Aeromonas*.

There have also been a number of reports from European countries, establishing the widespread distribution of *Aeromonas* in drinking water systems and sources, irrespective of hygiene levels. Havelaar *et al.* (1990) conducted a 1½ year surveillance of 20 different drinking water systems, untreated and treated waters, in the Netherlands. The results are strikingly similar to that of Burke and co-authors conducted in Australia (Burke *et al.*, 1984b). In this study, river water and reservoirs were the main sources and both yielded high counts of *Aeromonas*, with river sources being slightly higher. Groundwater source water, as well as river water that had standard sand filtration treatment was essentially free of aeromonads. However, water samples from distribution points demonstrated regrowth of *Aeromonas*, to levels lower than that of the source water, but which showed a correlation to temperature and residence time.

Stojek and Dutkiewicz (2011) conducted a 3-year summer month sampling of private wells, untreated private well water supply systems, and the municipal water of Lublin, Poland and surrounding rural communities. Although this study was primarily aimed at assessing the population dynamics of *Legionella* spp., water samples were assayed for other gram-negative pathogens, such as *Aeromonas*. Overall, 15.6% of 320 water samples tested positive for *Aeromonas*. Among all other water borne pathogens, the frequency of *Aeromonas* isolation was second to only *Pseudomonas* spp. In this study, *Aeromonas* was only isolated one time from private household wells ($n=45$). In contrast, the organisms were isolated from 28% (31/109) of well water supply systems and 11% (18/166) of municipal treated water system samples. Knochel and Jeppesen (1990) also reported a low frequency, 28% ($n=130$) of *Aeromonas* from ground water sources from Copenhagen, Denmark. Abundance of *Aeromonas* spp. was low, as was that for heterotrophic plate counts, but, again, indicative that *Aeromonas* can be isolated from 'good quality' drinking water.

Kuhn *et al.* (1997) reported 46% of water samples from 13 water distribution systems in Sweden harboured *Aeromonas*. Abundance and frequency was higher for surface water sources as opposed to ground water, in agreement with other reports. The highest abundance of *Aeromonas* for untreated water was in June, while August yielded the highest numbers from the end-of-distribution points (taps). In this study, lack of agreement or, the absence of a discernible trend, between densities of source water and within distribution system points to the contribution of resident *Aeromonas* populations, most likely as biofilms (discussed later). Krovacek *et al.* (1992) had previously conducted a surveillance study on three municipal water systems in Sweden, and found that 85% (23/27) of water samples were positive for presumptive *Aeromonas*.

Legnani *et al.* (1998) reported that 22% of water samples of a drinking water system from the northeast region of Italy in the Dolomite Mountains harboured *Aeromonas*. As with other reports, surface water sources yielded a higher frequency of *Aeromonas* isolation (40%), as compared to ground water sources (24.9% for springs, 28.6% for wells). The reservoir water that feeds this system is not treated by filtration, and only chemically treated with chlorine when water quality indicators are below standards. Massa *et al.* (2001) examined natural mineral water and private wells of Lecce, Italy for the presence of *Aeromonas* spp. The authors did not detect the organisms in any of the 60 mineral water samples from 15 brands, using direct and enrichment culture methods. Conversely, 20% of well sites (5/20) harboured *Aeromonas*.

Very few studies have been reported on *Aeromonas* spp. in water distribution systems in the United States. Chauret *et al.* (2001) conducted a 16 month study of a municipal water system, which services Muncie, Indiana. *Aeromonas* spp. was detected in all source water (White River) samples, regardless of whether the samples were collected distantly or proximally to the plant intake. According to biochemical characterization, 75% of *Aeromonas* spp. isolates were *A. hydrophila*. Following granular activated carbon filtration, 86% of water samples were still positive for the presence of *Aeromonas* spp., 60% belonging to *A. hydrophila*. The effluent, which was treated with chloramine, was found to be free of *Aeromonas* spp., or rather at levels below the detection limit of the membrane filtration method. Samples along the distribution system yielded presumptive *Aeromonas* spp., but none were *A. hydrophila*. It is important to note that the residual chemical levels were in the range of 0.4–2.9 mg/l (Chauret *et al.*, 2001). These results agree with several previous reports on the efficacy of chlorination and residual chlorine levels (Gavriel *et al.*, 1998; Kuhn *et al.*, 1997; Sisti *et al.*, 1998). This study also points out the ineffectiveness of activated carbon filter systems. These types of filters effectively reduce free chlorine levels to as low as 0.1 mg/l. In contrast, high free chlorine levels are maintained in sand-based gravity filter beds.

Many authors have reported essentially no aeromonads in mineral water source waters and/or bottled products, although enterics and *Pseudomonas aeruginosa* appear to be common components of the microbial flora (Havelaar *et al.*, 1990; Loy *et al.*, 2005; Massa *et al.*, 2001). For example, Hunter and Burge (1987) examined 58 bottles of mineral water from retailers in Wales and found no aeromonads, or coliforms, in any of the bottles. Havelaar *et al.* (1990) hypothesized that the natural microbial flora of mineral source water inhibits the growth and survival of aeromonads in some way; as evidenced by experimental studies using untreated and pasteurized mineral water. This observation was later confirmed by Messi *et al.* (2002).

Conversely, two studies from Spain reported the detection of *Aeromonas* spp. from mineral water sources. Gonzalez *et al.* (1987) isolated *Aeromonas* from one of three sources tested, and Quevedo-Sarmiento *et al.* (1986) reported that 10% of culturable isolates from mineral water source samples were aeromonads. Recently, Daood (2008) reported that 9% of local and imported bottled mineral water products of Syria harboured *Aeromonas* spp., using a non-selective minimal media.

The consensus from several reports, such as those above, is that distribution of *Aeromonas* is not correlated with faecal contamination indicators; however, the presence of both types of organisms can occur in polluted waters. Conversely, many researchers have reported a seasonal pattern of abundance of *Aeromonas* in natural and municipal water (Burke *et al.*, 1984b; Havelaar *et al.*, 1990; Rippey and Cabelli, 1989).

There have been a limited number of reports on *Aeromonas* in sewage effluents. For example, Maalej *et al.* (2003) conducted a one year ecological surveillance program of detection and enumeration of *Aeromonas* in Tunisia, examining both treated municipal effluent water and seawater (Mediterranean Sea). The authors reported high densities of *Aeromonas* in urban effluents from Tunisia, while natural seawater samples from the same region were much lower. Interestingly, in this study, *Aeromonas* displayed a contrasting seasonal pattern between the two types of water sources. Aeromonads displayed a typical late summer to autumn increase when isolated from the marine waters, in agreement with trends reported for surface water sources. However, abundance of aeromonads in urban effluent samples peaked during the cold season, November to April. This reverse seasonal trend for *Aeromonas* in sewage samples was also reported by Stampi *et al.* (2000).

A recent report from Brazil (Lins, São Paulo) found that 72% of influent samples of a 'sanitary sewage stabilization pond treatment system' harboured *Aeromonas* (Matrone-Rocha *et al.*, 2010). The treatment system was found to be marginally effective at reducing the load of *Aeromonas*. In general, the abundance of *Aeromonas* decreased; however, 55% of samples from anaerobic lagoon effluents and 48% of samples from facultative lagoons still tested positive for *Aeromonas* by MPN. Further, two recent reports found that *Aeromonas* was one of the dominant bacterial signatures in sewage samples (Vandewalle *et al.*, 2012; Ye and Zhang, 2011).

Metagenomics perspective

While there have been numerous ecological surveillance studies specifically targeting *Aeromonas* spp. from natural and drinking water sources, *Aeromonas* spp. have also been detected in several microbial community studies. Prior to the use of high-throughput pyrosequencing, or when this methodology is unavailable, the method of choice has been PCR amplification and sequencing of clone libraries isolated from study samples. This approach is more labour intensive and often results in very low sequencing depth (typically 100–1000 clones); however, we can still capture a glimpse of the microbial community. Someya *et al.* (2011) conducted a microbial community investigation of chitinolytic bacteria from the rhizospheres of agronomic plants of the Ibaraki Prefecture, Japan. Although the authors analysed 36 samples from 16 different species of plants from 6 geographically separated fields, only 100 16S rRNA sequences were obtained. One *Aeromonas* spp. sequence was reported, which was 99% identical to a strain of *A. hydrophila* isolated from blood of a septicaemia patient. The sequence came from a colony isolated from the rhizosphere from a taro plant, which is an indigenous perennial, grown from corm, from Southeast Asia. Interestingly, this type of plant is often cultivated in similar growth conditions to rice, i.e. it can be grown in very moist soil or even paddy fields, which are typically flooded by heavy rainfall or natural irrigation. Cultivation conditions are most likely responsible for the isolation of *A. hydrophila* from this plant rhizosphere, and this raises concern for contamination from natural or diverted irrigation for similarly grown food crops.

This method has also been employed when very little community DNA was isolated or intrinsic inhibition prevented direct amplification of 16S rRNA for next-generation sequencing. Dingsdag and Coleman (2012) analysed bacterial community composition from food court tables from a shopping mall in Sydney, Australia. Extracted DNA from environmental swabs failed to produce a 16S rRNA amplicon of sufficient yield for sequencing, so samples were first cultivated on a non-selective minimal media and then pooled cell suspensions were subjected to pyrosequencing. Additionally, a cleaning cloth that had been used over several weeks to clean the tables was also analysed by direct pyrosequencing. *Aeromonas* spp. has the third most abundant (14.1%) genus level taxonomic unit found on the clean cloth, when the community was directly interrogated, without a cultivation step. The same cloth was analysed by first plating on R2A media, in parallel with environmental swabs from the food court tables. The abundance of *Aeromonas* spp. 16S rRNA signatures dropped to 0.09%, apparently from inter-specific competition on the culture medium. With regard to the table swabs, Gram-positive organisms were most abundant, most likely due to their improved desiccation resistance. *Aeromonas* spp. 16S rRNA signatures constituted 0.6% of this microbial community. This is not surprising when we consider that the cleaning cloth is consistently moist and the food court tables are usually dry. It is unclear from the few environmental samples analysed as to the origin of the *Aeromonas* spp. signatures on the food court tables. The authors concluded that *Aeromonas* spp. were present on the food court tables from which patrons were consuming food and that improper hygiene with regard to repeated use of the cleaning cloth likely re-introduced *Aeromonas* onto the tables with regular frequency.

Kwon *et al.* (2011) conducted a metagenomic study on the microbial communities associated with a drinking water treatment pilot submerged membrane-filtration system from Seoul, South Korea. Three different types of samples were analysed, untreated water from a raw water equalization basin, particulate matter from the membrane tank, and from the surface of the filtration membrane. The authors found that the untreated source water was dominated by Betaproteobacteria, while both communities from the membrane tank were dominated by the Alphaproteobacteria. In this study, the large number of sequences analysed (44,000 reads across all three communities), using pyrosequencing, allowed the detection of rare members of the microbial communities. One *Aeromonas* spp. signature was detected from the source water and one *A. sobria* signature was detected from the particulate matter in the membrane tank. Other pathogens were present, such as *Legionella* spp., *Pseudomonas* and *Chromobacterium*, at slightly higher levels. Although *Aeromonas* spp. were found as rare members of the microbial community from this pilot study system, previous work reported that longer residence time, or repeated intrusion-event inoculations of *Aeromonas* may be needed for establishment of *Aeromonas* in biofilms in drinking water distribution systems (Chauret *et al.*, 2001). Additionally, chlorine was routinely used, at high levels, to periodically clean the filter membrane.

A recent survey of the microbial population associated with freshly harvested cilantro revealed that almost 60% (35/59) of samples contained *Aeromonas* (K. Jarvis, unpublished data). However the abundance of aeromonads was low, typically below 0.1%, as measured as a percentage of 16S rRNA signatures. In fact, the majority of positive samples were from enriched sub-samples, using buffered peptone water incubated overnight at 37°C, as compared to paired un-enriched aliquots. By culture method, representative strains were found to be primarily *A. veronii* biovar *sobria*, determined by *rpoD* sequence analysis, although *A. veronii* biovar *veronii* was also isolated (C. Grim, unpublished data).

Aeromonas and regrowth in reclaimed water

The use of reclaimed (formerly 'black', or containing human waste) and greywater is becoming common, as the 'green revolution' has sought ways to preserve and replenish freshwater reserves (USEPA, 2012; WHO, 2006). Both types of water are typically treated in some way prior to usage or storage, and these include mechanical systems, such as sand/gravel bed filtration, UV irradiation, and membrane filtration; and biological treatment, such as activated sludge systems and biofilters. Often, the quality of reclaimed and repurposed greywater is assessed through the use of indicator organisms, most notably coliforms, faecal coliforms, *E. coli* and enterococci. However, as has been documented for *Aeromonas*, abundance of opportunistic pathogens, such as *Legionella* and *Mycobacteria*, does not always correlate with that of faecal indicators.

Although the practice is in its infancy, there have been several studies on the quality of water from reclaimed systems (Narasimhan, 2005; Ryu *et al.*, 2005). Jjemba *et al.* (2010) analysed water from four different reclaimed water distribution systems, assaying specifically for *Aeromonas, Legionella, Mycobacterium, Pseudomonas*, and *E. coli* O157:H7. The authors analysed water from each system's effluent, reservoir, and at three different points within the distribution system, downstream of the reservoirs, sampling once for each season for one year. Two systems used conventional water treatment, namely trickling filters with tertiary sand filtration, and activated sludge with secondary filtration. The other two systems used membrane bioreactors (MBR), which combine activated sludge treatment followed by membrane filtration in a small footprint single reactor. Three of the systems used chlorine in the storage reservoir and the fourth used a combination of UV irradiation and ozone. All four types of systems were effective at reducing the total and indicator bacteria, and opportunistic pathogens, as measured at the effluent. However, bacterial densities were found to increase within each distribution centre, by a log-10 fold, or higher. Some systems also showed regrowth of the pathogenic microbial population in the storage reservoirs. Specifically for *Aeromonas*, the abundance at the effluent of each system was around or below the limit of detection, which was 1×10^2 cfu/100 ml. Regrowth was observed within the distribution system for all four treatment plants. Regrowth within the storage reservoirs was not observed for the two MBR systems, indicating their effectiveness for removal of *Aeromonas* spp. from black- and greywater.

Similar results have been reported by other researchers. Schets *et al.* (2010) conducted a 3-year study of roof-collected rainwater systems in the Netherlands and found that storage reservoirs were often contaminated with opportunistic pathogenic microbes, including *Campylobacter, Giardia, Legionella* and *Aeromonas*. Al-Bahry *et al.* (2009) found that *Aeromonas* spp. was one of the dominant heterotrophic bacteria, along with *Enterobacter* and *Pseudomonas* spp., isolated from tertiary treated effluent water which was intended for irrigation in Oman.

While reclaimed water is not usually intended to replace or replenish potable water, and thus not consumed directly by humans, in many instances it is used for irrigation. While irrigation usually implies that for landscape use, special attention needs to be directed towards reclaimed and recycled water used in irrigation for agronomic purposes. While surface waters and redirected rainfall may contain *Aeromonas*, regrowth of the organisms in reclaimed water may deliver a higher abundance of pathogenic species onto crops, such as leafy greens, which may be consumed with very little treatment after harvesting.

Mechanisms of survival in water

Microbes in the aquatic environment are rarely planktonic, but rather are associated with diverse surfaces, including detritus, abiotic aggregates, living tissues, and industrial or potable water system piping. This attached lifestyle, termed biofilm, aids bacteria in acquiring nutrients, escaping predation and from being physically displaced (Johnson, 2008). Biofilms begin as a monolayer of cells, which attach by various means and appendages. As biofilms 'mature', microbial cells become surrounded by a matrix of exopolysaccharide (EPS). EPS, secreted by bacterial cells, forms into complex three dimensional structures, with mushroom-shaped pillars and internal canals, which allow intrusion of nutrient rich waters, but afford some protection from harsh chemicals, disinfectants, and antimicrobials. Individual cells or small aggregates are constantly shed from mature biofilms, acting as a seeding mechanism for further biofilm or biofouling formation. Biofilms can be composed of a single species, but more commonly are composed of a community of mixed microbial species. The diversity of microorganisms within a biofilm, including heterotrophic bacteria, may promote the growth of singular species, by secretion of essential extracellular compounds, particularly those that may be growth-limiting (Cotuk *et al.*, 2005; Moritz *et al.*, 2010). In contrast to synergistic effects, antagonistic effects can also occur (Cotuk *et al.*, 2005).

Biofilm formation can be viewed as a survival mechanism for bacteria; they can provide protection from a variety of physiochemical stresses, including UV light, oxidative stress, thermal stress, dehydration, biocidal agents, and predation (Johnson, 2007) (Temmerman *et al.*, 2006). This is particularly true for bacteria that live in diverse aquatic environments, such as lakes, rivers, seas, and oceans.

Biofilm formation by can also play a role in the transmission of opportunistic waterborne pathogens (Lehtola *et al.*, 2007). Biofilms in drinking water containing *E. coli*, *Aeromonas*, and *Pseudomonas*, harbour potential risk to human health if pathogenic strains are involved (September *et al.*, 2007). The number of cells in a biofilm may reach as high as 10^9 cells, which in most cases translates to an infectious dose of a pathogen. For example, biofilms have been implicated as an important mechanism in the epidemiology of *V. cholerae* from surface waters in a cholera-endemic area in Bangladesh. Faruque *et al.* (2006) observed high concentrations of *V. cholerae* in environmental waters, as *in vivo*-formed biofilms, and suggested that the shedding of these clumps of cells appears to be an efficient way of delivering an infectious dose of toxigenic strains back into the environment (Faruque *et al.*, 2006). A concurrent epidemiological and ecological surveillance conducted in the coastal villages of Bay of Bengal, where seasonal outbreaks of cholera occur annually, also revealed the presence of biofilm-bound non-culturable toxigenic *V. cholerae* O1 in surface waters used for drinking water and other domestic purposes during inter-epidemic periods (Alam *et al.*, 2006).

Biofilm formation by *Aeromonas* has been well documented. Aeromonads, like many aquatic bacteria, such as *V. cholerae*, are efficient colonizers of biotic and abiotic surfaces (Fig. 9.1). *Aeromonas* are often found as bacterial components of biofilms from water distribution systems and food processing equipment (de Oliveira Scoaris *et al.*, 2008b). Primarily, this is attributed to their ubiquitous distribution in fresh water environments. Aeromonads are also efficient degraders of chitin, accomplished via the action of secreted chitinase(s) (Janda, 1985; Lan *et al.*, 2008). Chitin, a polymer of β-1,4-*N*-acetyl-D-glucosamine (GlcNAc), is

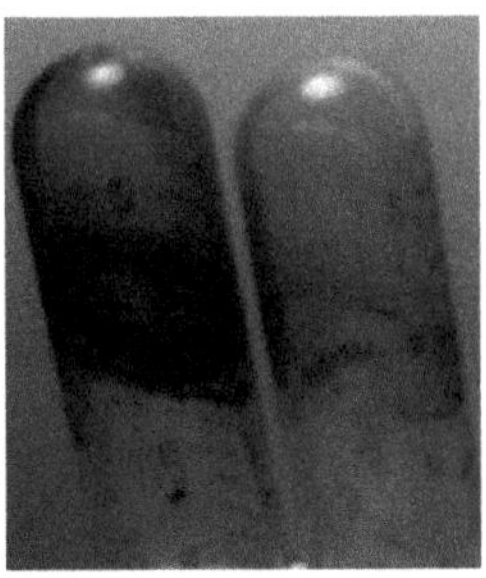

Figure 9.1 A crystal violet tube-based assay detecting biofilm production by *A. hydrophila* strain SSU and ATCC 7966. Photo courtesy of Dr Ashok Chopra, UTMB.

the most abundant polymer in aquatic environments. Its initial degradation is an important step in carbon recycling in aquatic environments. Jagmann *et al.* (2012) reported that chitinolytic enzymes could only be detected from biofilm fraction, and not the free-swimming fraction, of broth cultures of *A. hydrophila*, indicating that biofilm formation may prevent sequestration of chitin degradation products by competing bacteria in the water column.

Physiologically, several factors are responsible for biofilm formation of *Aeromonas*. Aeromonads possess a number of surface appendages that includes polar (Kirov *et al.*, 2004) and lateral flagella (Gavin *et al.*, 2003; Merino *et al.*, 2003, 2006), type IV pili (TFP) as well as chaperone-usher (CU) fimbriae (Seshadri *et al.*, 2006).

Flagella are important elements in the initial stage of biofilm formation, i.e. taxis and initial contact and adherence (Enos-Berlage *et al.*, 2005; Kirov *et al.*, 2004; O'Toole and Kolter, 1998a,b). Most mesophilic aeromonads are motile by polar flagella and its role in adherence to biotic surfaces has been confirmed (Merino *et al.*, 1997; Rabaan *et al.*, 2001). Interestingly, for *V. cholerae*, null mutants of *flaA*, the polar flagella structural subunit gene, were still able to form mature biofilms, by sedimentation of auto-aggregates and then typical biofilm maturation processes (Watnick *et al.*, 2001).

Lateral flagella are important in the migration, or swarming, along surfaces, permitting faster colonization (Fraser and Hughes, 1999) (Fig. 9.2). The distribution of lateral flagella genes among *Aeromonas* spp. is less clear than for polar flagella. Shimada *et al.* (1985) found that less than 20% of *Aeromonas* isolates exhibit lateral flagella. Kirov *et al.* (2002) reported that 49% of clinical and 70% of *Aeromonas* spp. isolates from fish and amphibians possess lateral flagella, while Gavin *et al* (2003) reported 70% of fish, 55% of food, and 62% of clinical isolates possess lateral flagella. Meanwhile, Merino *et al.* (2003) found that 100% of *A. salmonicida* isolates possessed lateral flagella genes, and Aguilera-Arreola (2007) reported 84% of clinical and environmental *A. hydrophila* isolates encode the *lafA* (lateral flagella subunit) gene. Santos *et al.* (2011) reported that both polar and lateral flagella must be present for strong biofilm adherence. This report is confounded by the fact that no *flaA*$^-$ (polar) *lafA*$^+$ (lateral) genotype strains were isolated or used in the study. A close examination of the data reveals a correlation of strong biofilm adherence with the presence of lateral flagella genes.

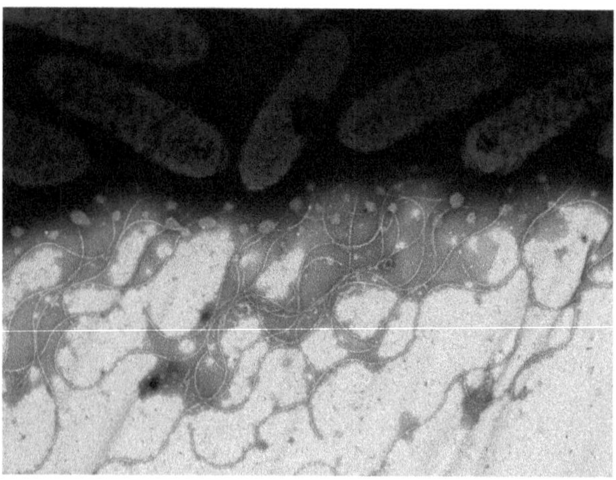

Figure 9.2 Transmission electron photomicrograph of *A. hydrophila* displaying lateral flagella.

TFP of *Aeromonas* spp. include the bundle-forming TFP (Bfp) (Kirov and Sanderson, 1996), which is homologous to the mannose sensitive haemagglutinin (MSHA) of the vibrios and appears to be conserved within the genus (C. Grim, unpublished), the Tap (type IV *Aeromonas* pilus) TFP (Barnett and Kirov, 1999; Barnett *et al.*, 1997), which is homologous to TFP of *P. aeruginosa* and *Neisseria* spp., and the tight adherence (*tad*) Flp pilus (Seshadri *et al.*, 2006). The Bfp TFP of *Aeromonas* spp. has been implicated as an important intestinal colonization factor, as most *Aeromonas* diarrheal isolates are hyper-piliated with this type of pilus (Kirov *et al.*, 1999). While the homologous MSHA of *V. cholerae* is not involved in pathogenicity, it has been shown to be important in biofilm formation on abiotic surfaces (Watnick *et al.*, 1999). Additionally, stable hyper-piliated *A. caviae* mutants (of TFP) have been reported to exhibit enhanced biofilm formation (Bechet and Blondeau, 2003).

The genomes of several *Aeromonas* spp. contain multiple C/U fimbriae, including α and π (P-fimbriae) type (C. Grim, unpublished data); however, no functional studies have been conducted to assess their role in biofilm formation. Honma and Nakasone (1990) reported that C/U fimbriae do not aid in intestinal colonization, but play a role in cell–cell aggregation. Thus, it is highly possible that these short, rigid fimbriae play a role in biofilm formation.

Other factors that have been implicated in biofilm formation include exopolysaccharide (EPS), which is the primary component of most biofilms, capsular polysaccharide, cellulose biosynthesis gene cluster, quorum sensing regulatory networks, C-di-GMP levels and other global regulators. There have not been a lot of studies to date that have addressed these factors in the context of biofilm formation of *Aeromonas* spp. The notable exception is studies conducted on the role of quorum sensing regulation on biofilm formation. Biofilm formation is a developmental process in which a quorum sensing signal molecule, autoinducer, functions to induce the secretion of EPS and leads to the formation of the characteristic three-dimensional biofilm architecture (Nadell *et al.*, 2008). Lynch *et al.* (2002) found that an isogenic Δ*ahyI* (autoinducer synthase) mutant of *A. hydrophila* AH-1 failed to produce mature biofilms. In agreement, Khajanchi *et al.* (2009) reported that a quorum-sensing double knock-out isogenic mutant, Δ*ahyRI* (transcriptional regulator and autoinducer synthase), exhibited a much reduced biofilm formation, as compared to wild type *A. hydrophila* SSU.

It is expected that all *Aeromonas* spp. share conserved regulatory proteins and signalling systems; however, it would not be unexpected if the biofilm regulatory network has unique elements in some or even each individual species. These differences are probably a reflection of the importance of the biofilm lifestyle to each *Aeromonas* spp., especially in the context of demonstrated *in vivo* and *ex vivo* alternating life cycles and differences in environmental niche occupation between species. The same can be said of the variability of bacterial appendages among *Aeromonas* spp. isolates. Further complicating the picture is the concept that only certain species of *Aeromonas* are responsible for human disease and that each of these species harbour virulent pathotypes, as well as benign environmental strains.

In addition to escaping predation through biofilm formation, some bacteria are able to survive intracellularly and even grow inside amoebic cells as endosymbionts. Bacteria are a main food source for free-living amoebae (FLA) and other bacteriovirus protists inhabiting aquatic systems. However, for several bacteria, intracellular growth or survival in higher-level microorganisms has also been reported. This association has been proposed to favour the survival and spread of bacteria in the aquatic environment and foods, and is even implicated as an important factor in the epidemiology of their respective diseases. Free-living amoebae (FLA) are extremely fit members of the aquatic environment and are resistant to chlorine and other chemical disinfection processes, desiccation, and extreme temperature fluctuations. In that regard, FLA can be viewed as the ideal vectors, or 'Trojan horses', for pathogenic bacteria, providing both protection against adverse environmental conditions and transportation. For example, *Legionella* has been isolated from the free-living amoebae, *Acanthamoeba*, *Naegleria*, and *Hartmanella* (Abu Kwaik, 1996; Barker *et al.*, 1992; Newsome *et al.*, 1985; Wadowsky *et al.*, 1991). Further, Barbaree *et al.* (1986) observed that FLAs and *Legionella* spp. were isolated from the same infection source during one epidemic of Legionnaires' disease. The authors found that the isolated amoebas could support the growth of *Legionella* intracellularly.

In a recent study, Anacarso *et al.* (2012) assessed the ability of a number of foodborne and opportunistic pathogens to survive and proliferate inside the FLA, *Acanthamoeba polyphaga*. Of the bacteria tested, *Yersinia enterocolitica* and *A. hydrophila* exhibited an initial proliferation, once internalized in the amoeba, followed by killing and ingestion of the bacteria. Conversely, *Enterococcus faecalis*, *Staphylococcus aureus*, *Listeria monocytogenes*, and *Salmonella* Enteritidis exhibited persistence and growth inside the amoeba. Although this is only a single report, *Acanthamoeba* is one of the most common protist of soil and fresh water.

Regulations of drinking water for *Aeromonas*

Most countries do not have safe drinking water guidelines for *Aeromonas*, despite the ubiquitous nature of the organism in aquatic environments, and an overwhelming number of reports indicating that environmental strains harbour the same virulence determinants as clinical isolates (discussed in detail later). Further, several documented outbreaks of *Aeromonas*-associated food poisoning have been reported, mostly from seafood (Kirov, 1993). Still, *Aeromonas* is largely ignored as a potential opportunistic pathogen in drinking water systems, primarily due to a lack of convincing evidence as to the epidemic potential of *Aeromonas* spp. from water.

Notable exceptions include Italy and the Netherlands. One of the first countries to regulate levels of *Aeromonas* in drinking water was the Netherlands. Health officials instituted *Aeromonas* spp. maximum levels for drinking water in 1985, which are a one-year mean of 20 cfu/100 ml in water leaving treatment facilities, and 1-year 90th percentile value of 200 cfu/100 ml for drinking-water collected from the distribution system (Trouwborst, 1992). By comparison, Italian law states that levels of *A. hydrophila* may not exceed 10 cfu/ml at the point source for natural drinking water and 100 cfu/mL during the marketing of such drinking water (Italiana, 1997). In addition, Canada and other countries have established microbiological quality standards for bottled water that include presence of *Aeromonas* (Warburton, 1993; Warburton *et al.*, 1992, 1994).

A. hydrophila was placed on the US Environmental Protection Agency (EPA) Contaminant Candidate List (CCL1) in 1998 (USEPA, 2005). The CCL is a list of chemical and microbiological contaminants of public health concern that are not currently regulated, but may need to be in the future, under the Safe Drinking Water Act amendment of 1996. When the second version of the Drinking Water CCL was released on 23 February 2005, *A. hydrophila* remained on the list (USEPA, 2005). In light of results from occurrence monitoring which the agency conducted in 2003 (Egorov *et al.*, 2011), *Aeromonas* has been removed from the latest version, CCL3 (USEPA, 2009). However, *Aeromonas* remains a component of the Unregulated Contaminant Monitoring Program, and utilities are required to report positive results, as a result of the Consumer Confidence Report (CCR) Rule. Regardless of results, all surveillance data conducted by a utility must be made available upon request to fulfil the Public Notification (PN) Rule.

Aeromonas extraintestinal infections due to exposure to water

In addition to exposure to *Aeromonas* spp. through consumption of contaminated drinking water, *Aeromonas* infections as a result of water-related recreational activities and wound infections from traumas (Haburchak, 1996) and bitings (Angel *et al.*, 2002) that have occurred during those activities have been frequently reported. These activities, which include swimming, fishing, boating and kayaking, water and jet skiing, occur most commonly on lakes and rivers, during the warm summer months – the same conditions that favour proliferation of *Aeromonas* spp. in the environment.

As these types of activities and access to them increase, it is apparent that humans will be at an increasing risk to post-traumatic wound infections by *Aeromonas*. Typically, cellulitis is the most common disease presentation from these type of infections (Behera *et al.*, 2011); however, necrotizing fasciitis is common in patients who have underlying health conditions. For example, Markov *et al.* (2007) reported a case of necrotizing fasciitis and myonecrosis due to *A. hydrophila* from a 64-year-old Bulgarian man who experienced trauma to his lower right leg while taking a mud bath in brackish water. Other bacteria were cultured from the wound and included *Enterococcus* sp., *S. aureus*, *Klebsiella oxytoca*, *Morganella morganii* and *E. coli*; however, the haemoculture revealed a monomicrobic *A. hydrophila* infection. The patient suffered from type II diabetes and the condition worsened such that amputation was required.

Shak *et al.* (2011) reported an unusual case of cellulitis involving a 35-year-old man who suffered a posterior right leg laceration as the patient fell through a wooden dock over the Chattahoochee River in Georgia, USA. The medical history of the patient was 'unremarkable'. Again, the wound culture showed a polymicrobic bacterial community, which included *E. coli*, *Enterococcus* spp., *Streptococcus* spp., *Clostridium perfringens* and two strains of *A. hydrophila*. In this case repeated surgical debridement and intensive antibiotic treatment was required due to the severity of the infection and one of the two strains being resistant to aminoglycosides and several cephalosporins.

These reports highlight the challenges associated with *Aeromonas* wound infections from water exposure. First, it is expected that most trauma wounds exposed to natural fresh and brackish water will contain polymicrobic infections. Further, several authors have speculated that the presence of certain mixtures of bacterial species may result in synergistic infections. Second, *Aeromonas* spp. are often resistant to commonly used antibiotics for wound infections due to the presence of multiple β-lactamases, some which impart extended spectrum resistance. Lastly, *Aeromonas* is rarely considered as the primary pathogen, or even as an opportunistic pathogen in these cases. For example, a review of literature specific to burn victims revealed *A. hydrophila* as an emerging pathogen in 76% of publications (Azzopardi *et al.*, 2011). Application of untreated water onto the burn wound was found to be the predominant route of exposure. Taken together, diagnosis and effective antibiotic treatment in clinical cases involving *Aeromonas* wound infections can often be delayed, to the point that a positive outcome is not achievable.

Occurrence of *Aeromonas* in food

In addition to drinking water being implicated as a vehicle of transmission of *Aeromonas* related gastroenteritis, food has also been implicated. While the association of aeromonads with fish and seafood spoilage and contamination, due to environmental exposure, has a long history, the link between *Aeromonas* and other food commodities were first reported in several early papers, primarily from Russian and Eastern European researchers, which implicated contamination of foods such as 'mother's milk', poultry, minced meat, and hard-boiled eggs (Bomar, 1979; Grafova, 1981, 1982; Jablonska-Strynkowska and Potapa, 1983; Kalina, 1977; Kleeberger, 1977; Maciejska-Roczan, 1979; Szollosy *et al.*, 1974; Vitkov, 1974).

Amidst these reports and an increasing number of reports of travellers' diarrhoea, food poisoning, and gastroenteritis in which *Aeromonas* was suspected or associated (Agbonlahor *et al.*, 1982; Echeverria *et al.*, 1981), the U.S. Association of Food and Drug Officials labelled *A. hydrophila* as one of eight newly emerging foodborne pathogens in 1983 (Buchanan, 1983). Buchanan (1983) stated that *Aeromonas* appears to have similar fitness to *Salmonella*, and that good hygiene should be sufficient to control the organism.

Buchanan elaborated further and warned that the primary concern for this pathogen is a situation of consumption of a food commodity that was previously stored under refrigeration for a considerable time without further cooking (Buchanan, 1983). For example, Toule and Murphy (1978) had reported that *A. hydrophila*, along with *Pseudomonas putida*, was responsible for spoilage of cooked chicken stored under variable refrigeration conditions. Environmental swabs of the kitchen area at the catering department from the University of Surrey, which originally supplied the cooked chicken, revealed *A. hydrophila* contamination of the work surface, knife, plate, water faucet, knife sharpener and refrigerator, indicating several points of cross-contamination. Myers *et al.* (1982) confirmed the observation of survival and growth of *Aeromonas* at refrigeration temperatures (4°C), as the organism interfered with detection of *Yersinia enterocolitica* using low-temperature enrichment.

Detection of *Aeromonas* from several refrigerated food commodities has been reported. Palumbo *et al.* (1985) found that *Aeromonas* could be isolated from 10 different market fish species, shrimp, scallops, oysters, crab, raw milk, chicken, and ground beef, veal, lamb, and pork, from retail samples from Philadelphia, PA. Fresh sausage products were the only food matrix from which *Aeromonas* was not isolated and failed to grow in at 5°C. The authors speculated that high salt content of this commodity may be the inhibitory factor preventing survival and proliferation. For all other foods, abundance of *Aeromonas* spp. increased after seven days of refrigerated storage. It should be noted that aerobic plate count numbers also increased, indicating that *Aeromonas* spp. are capable of competitive psychrotrophic growth in most foods. Further, *Aeromonas* spp. accounted for a significant portion of the total microbial community in several of the foods following refrigerated storage for seven days.

As stated above, *Aeromonas* is commonly isolated from fish and seafood products. Borrell *et al.* (1998) found that 30% (26/83) of shellfish samples were contaminated with *Aeromonas* spp., while, conversely, less than 4% of meat (6/177) and milk (7/179) products were positive. Neyts *et al* (2000a) reported isolation of *Aeromonas* spp. from 26% (7/27) of vegetable samples, 70% (16/23) of poultry, red meat and meat product samples, and 72% (13/18) of shrimp and fish samples, purchased from retail establishments in Flanders, Belgium. Direct plating and enrichment methods were used. Most of the fish and shrimp samples (10/13) were positive by direct plating, while only five samples from meat and two samples from vegetables were positive by this method, indicative of a higher burden of *Aeromonas* from seafood commodities. Recently, Woodring *et al.* (2012) reported that *Aeromonas* spp. were found in 27% (32/120) of uncooked seafood samples from markets of Bangkok, Thailand. Tsai and Chen (1996) also found that 50% (11/22) of oyster samples from retail sources in Taiwan were contaminated by *A. hydrophila* phenospecies. Twenty-two per cent (4/18) of fish samples and 14% (2/14) of shrimp samples were also contaminated. A similar contamination frequency, 58%, was reported for shellfish from New Zealand (Hudson and De Lacy, 1991), and fish, 19%, from Reading, UK (Fricker and Tompsett, 1989).

There have been several reports on *Aeromonas* detection across a wide array of food commodities. Knochel and Jeppesen (Knochel and Jeppesen, 1990) examined 840 food samples of minced meat and fish, fish fillets, deli meat, ready to eat meals, mayonnaise-based salads, ice cream, and whipped cream from retail sources in Copenhagen, Denmark. The authors found 36% of raw meats were contaminated with *Aeromonas* spp., which was lower than the incidence reported by Palumbo *et al.* (1985), 84% (16/19) for seafood, and 80% (16/20) of poultry, pork and red meat. Ibrahim and Mac Rae (Ibrahim and Mac Rae, 1991) reported isolation of *Aeromonas* spp. from retail meats (64%) and raw bulk milk tanks (27%) from Brisbane, Queensland, Australia. Abbey and Etang (1988) reported that nearly 30% of snails from Nigeria were reservoirs for *A. hydrophila* and *A. sobria*. Nagar *et al.* (2011) analysed chicken, fish, and sprouts from retail stores in Mumbai, India. The authors found that, across all foods, 11.7% (18/154) were contaminated with *Aeromonas*, which included four chicken, 12 fish, and two sprout samples. Awan *et al.* (2009) isolated *Aeromonas* from all food sources, that were tested, from Abu Dhabi, UAE, which included fresh and frozen chicken, game birds, pasteurized milk, bakery products, fish, water, frozen mixed vegetables, and baby food. A survey of meat, fish, milk, cheese, water and soil samples from Ankara, Turkey, revealed contamination with *Aeromonas* for all food commodities (Yucel and Erdogan, 2010). And, a survey of suspected food-poisoning samples in Taipei, Taiwan, found an overall incidence of *Aeromonas* of 16%, dominated by *A. caviae* isolates (42/80) (Chang *et al.*, 2008). The foods included seafood (30%), rice foods (28.6%), fruits and vegetables (23%, apple salad and Chinese broccoli), mixed foods (22%; e.g. maize with pork, radish with egg), processed pork (14%), and beverages (8%, spring water and milk tea).

A few studies have examined the presence of *Aeromonas* in vegetables. For example, McMahon and Wilson (McMahon and Wilson, 2001) reported that 34% of organic vegetables analysed were contaminated with *Aeromonas*, whereas, *Salmonella*, *Campylobacter*, *E. coli* O157 and *Listeria* were not found. The contaminated vegetables included lettuce, cucumber, broccoli, celery, cauliflower, turnip, alfalfa sprouts, asparagus, carrots, tomatoes, mushrooms, peppers, and water cress, purchased from supermarkets from Northern Ireland. Surprisingly, in this study, *A. schubertii* was the dominant species (54.5%) across all food isolates, followed by *A. hydrophila* (15%), *A. trota* (15%), *A. caviae* (9%), and *A. veronii* biovar *veronii* (6%). Nyenje *et al.* (2012) reported that 12% of 252 food samples from Alice, South Africa, were contaminated with *Aeromonas*. *A. hydrophila* was detected in 24% (10/42) of vegetables, as well as 17% (7/42) of rice and 14% (6/42) of potatoes analysed.

Detection of *Aeromonas* in water and food

Aeromonas spp. will grow on a wide array of laboratory media, including selective media for the isolation of enterics. Initially, a number of previously described media intended for use in the clinical setting were used for isolation of *Aeromonas* from environmental sources. For example, Hazen *et al.* (1978) used Rimler–Shotts medium, which had been reported to be highly efficient at isolation of mesophilic aeromonads (Hazen *et al.*, 1978; Shotts and Rimler, 1973). At the same time, new media were developed based on differential metabolic traits. Rippey and Cabelli (Rippey and Cabelli, 1979) reported the use of a differential and selective medium (m-*Aeromonas*, or mA, agar) based primarily on the differential fermentation of trehalose, while Kaper *et al.* (1979) described a medium that exploited the decarboxylation of ornithine as the differential characteristic to identify *Aeromonas*. Davis and Sizemore (Davis and Sizemore, 1981) compared all three media described above and found that concurrent isolation of vibrios, *V. alginolyticus* from estuarine and marine samples, and *V. fluvialis* and *V. furnissii* from freshwater sites, was possible and that these species may be misidentified by API 20E test strips (bioMérieux, Marcy l'Etoile, France).

Palumbo *et al.* (1985) reported the development of a new formulation, starch-ampicillin agar (SA agar), which is composed of Difco phenol red agar base medium (Becton Dickinson, Franklin Lakes, NJ, USA) supplemented with starch and ampicillin. Starch hydrolysis is largely a characteristic of aeromonads and vibrios, while most aeromonads are resistant to ampicillin. It was found that pseudomonads may also grow on this medium, but can be differentiated based on amylase production. This medium has been used extensively for detection of *Aeromonas* spp. from foods (Ibrahim and Mac Rae, 1991; Knochel and Jeppesen, 1990). And, although vibrios grow on this medium, they are not of concern since *Aeromonas* spp. are primarily investigated in refrigerated foods, in which vibrios should not survive and proliferate.

Subsequently, Havelaar *et al.* (1987) combined the selectivity of SA agar, namely ampicillin, with the differential ingredient of dextrin-fuchsin-sulphite agar (DFS), dextrin (Schubert, 1967). This medium, ampicillin-dextrin agar (ADA), in turn has become one of the most widely used media formulations in isolating *Aeromonas* spp. from water, as well as food (Borrell *et al.*, 1998; Neyts *et al.*, 2000a). Other less commonly used media include Ryan's *Aeromonas* medium (Holmes and Sartory, 1993), pril–xylose–ampicillin agar (Rogol *et al.*, 1979), Glutamate Starch Penicillin agar amended with Ampicillin and 10 µg/l C-glucose (SGAP-10C) (Huguet and Ribas, 1991); ampicillin Bile salts Inositol Xylose (MIX) agar (Cunliffe and Adcock, 1989), Bile salt Brilliant green Starch agar (BBS) (Nishikawa and Kishi, 1987), and modified Bile salts-Irgasan-Brilliant Green (BIBG) agar (Neyts *et al.*, 2000b). In a comparison of seven media used for detection of *Aeromonas* spp. from water, Villari *et al.* (1999) found a general trend in which media that gave higher recovery of *Aeromonas* spp. from water samples also had the highest levels of background flora. Also, confirmatory biochemical tests are necessary as 10% false positives and 20–40% false negatives were reported among all media.

For detection of *Aeromonas* spp., many researchers have reported increased sensitivity by first enriching environmental samples in alkaline peptone water (APW), prior to subculturing onto selective media (von Graevenitz and Bucher, 1983). Ampicillin can be added to APW to suppress enterics and other microbes, if high levels of background flora or faecal contamination are anticipated. The use of ampicillin in culture media is now questioned, since *A. trota* is susceptible, as well as some isolates of *A. caviae* (Huddleston *et al.*, 2007). Novobiocin is also used, but a higher percentage of *Aeromonas* spp. isolates have been found to be susceptible to this antibiotic, albeit at relatively high MICs. For detection of *Aeromonas* spp. from drinking water, a copper chelating agent should be incorporated into the agar medium for samples collected from sites using copper pipes as aeromonads are particularly sensitive to this metal (Versteegh *et al.*, 1989). And, the vibriostatic agent O/129 can be added to agar plates when recovery of vibrios are expected, although expense may restrict or prohibit its usage (Borrell *et al.*, 1998; Havelaar *et al.*, 1987).

For enumeration of *Aeromonas* spp. from water, the membrane filtration technique is commonly employed (Borrell *et al.*, 1998; Havelaar *et al.*, 1987; Hazen *et al.*, 1978; Knochel and Jeppesen, 1990; Shotts and Rimler, 1973), in combination with one of the selective and differential media described above. In 2001, the US EPA adopted the use of membrane filtration for the detection and monitoring of *Aeromonas* in 'finished' water (USEPA, 2001). The method is adapted from that reported by Havelaar *et al* (1987), using a modification of ADA agar, which also contains vancomycin, as the plating medium, and incubation of the membrane filter on this medium for 24 hours. The method uses an incubation temperature of 35°C, even though several researchers have found lower temperature superior (Villari *et al.*, 1999).

Following presumptive isolation on selective and differential media, biochemical tests are usually employed for confirmation and species identification of *Aeromonas* isolates (See chapter on Taxonomy for more details). Routine subculturing onto blood agar (BA) plates is useful to distinguish beta-haemolytic *Aeromonas* spp. Additionally, the test for cytochrome oxidase, for which aeromonads are positive, can be performed from BA directly, but not from many of the selective media. Most commercial clinical diagnostic systems incorrectly identify *Aeromonas* to the species level. However, the seven test dichotomous key, Aerokey II (Carnahan *et al.*, 1991), is capable of efficiently differentiating all major human clinical isolates of *Aeromonas* spp. (Lamy *et al.*, 2010). Recently, the use of *rpoD* sequence of *Aeromonas* spp. isolates has been reported to be highly discriminatory in species identification (Saavedra *et al.*, 2006; Soler *et al.*, 2004).

Some molecular methods have been developed to detect *Aeromonas* spp. (Khan *et al.*, 2009), but these methods are rarely used in environmental surveillance. Primarily, this is due to the desirability of an isolate to further characterize and study. Also, lack of consensus and validation on candidate molecular targets has prevented the acceptance of one singular PCR protocol or gene target. One particular approach that has been extensively investigated is the development of a pathogen multiplex assay targeting several genera, species, or pathotypes. For example, Zhou *et al.* (2011) reported the development of a DNA microarray of 26 oligonucleotide probes that can detect 11 common pathogens of drinking water, namely *Legionella pneumophila, Klebsiella pneumoniae, P. aeruginosa, A. hydrophila, Salmonella, Shigella, Staphylococcus aureus, V. cholerae, V. parahaemolyticus, Y. enterocolitica* and *Leptospira interrogans.* Similarly, Liu *et al.* (2012) reported the development of a seven-plex PCR Luminex (microsphere-based oligonucleotide capture probes) assay for *Aeromonas, Campylobacter, Salmonella, Shigella, Vibrio* and *Yersinia.* For *Aeromonas,* the aerolysin gene from *A. hydrophila, A. caviae,* and *A. veronii* was used. This assay was designed to be used for diarrheal disease diagnostic purposes. Indeed, the aerolysin/*act* gene is typically the PCR target used for detection of *Aeromonas* in multiplex assays for multiple organisms (Arora *et al.*, 2006; Gugliandolo *et al.*, 2011), but the use of others has been reported, such as the outer membrane protein, *aha* (Shannon *et al.*, 2007; Wang *et al.*, 2009)

A few PCR assays have been developed to detect *Aeromonas* spp. in food and water, and to also identify virulence genes in the same assay. Kingombe *et al.* (2010) and Chang *et al.* (2008) developed a multiplex PCR that simultaneously detects *act, alt,* and *ast,* while Balakrishna *et al.* (2010) used *hlyA, aerA,* the glycerophospholipid-cholesterol acyl transferase (GCAT) gene, and 16S rRNA.

Distribution of *Aeromonas* and virulence traits in food and water

Several researchers have reported trends in terms of species distribution in water and food and with regard to virulence traits. Due to the convoluted history of the taxonomy of *Aeromonas*, caution should be taken in interpreting these findings (see Chapter 2 for more details). For example, many early reports on *Aeromonas* from water, drinking water and food reported isolation or detection of *A. hydrophila* (phenospecies). Because of the taxonomic uncertainty of the genus at the time of these reports, one should consider these reports as a general observation on the distribution of motile mesophilic aeromonads, and not specifically *A. hydrophila* (genomospecies). Likewise, reports which use the three mesophilic species identification scheme of *A. hydrophila, A. caviae*, and *A. sobria*, should be considered as phenospecies or species complexes. Further, the use of phenospecies and hybridization group nomenclature has and still persists in more recent reports, even though taxonomic updates have occurred. However, certain trends are consistently reported and these are presented in this section.

Several reports indicate that *A. caviae* tends to be the dominant species in water bodies with a high level of organic material (Araujo *et al.*, 1991) (Stecchini and Domenis, 1994). *A. caviae* complex and *A. hydrophila* group strains tend to be equally present in less polluted water, while *A. sobria* complex is increasingly isolated in more pristine and brackish water. For example, Millership *et al.* (1986) found that *A. caviae* (complex) was the most commonly isolated phenospecies in untreated water (63%), as well as from human faeces (55%), from a study of 111 clinical and environmental water isolates from London. Conversely, LeChevallier *et al.* (1982) reported that all *Aeromonas* isolates recovered from chlorinated water distribution system in Oregon, USA, which does not receive inputs from industrial or domestic operations into the source waters, were *A. sobria.*

Conversely, some authors have reported different *Aeromonas* population dynamics, which challenges the validity of the trends reported above across all water samples. For example, Borrell *et al.* (1998) found that *A. veronii* biotype *sobria* was the dominant species from environmental water samples (22%), which included lakes, reservoirs, and treated drinking water, as well as clinical samples (intestinal, 30%, and extra-intestinal, 48%). The authors also found that *A. caviae–media* complex members were the most abundant in seawater samples (26%) and from milk (36%). *A. hydrophila* was the second most abundant species from lake and reservoirs, untreated drinking water, and extra-intestinal infections. Hydrological differences from site to site most likely explain these contradictory reports.

Species distribution of *Aeromonas* in foods is more complex. According to Neyts *et al.* (2000a), *A. hydrophila* was the dominant species (59%) among fish, shrimp, meat and poultry isolates, following by *A. caviae* complex (*A. caviae*/*A. media*; 12%). *A. veronii* biovar *sobria* was also isolated from meat and poultry samples, but, in low abundance. Conversely, the authors found that produce samples were dominated by *A. caviae* complex (71%), then *A. hydrophila* (7%), and *A. bestiarum* (5%). This species distribution was in agreement with several other researchers. For example, Kingombe *et al.* (1999) found that *A. caviae* complex isolates were rarely isolated from foods of animal origin, while this species was very commonly isolated from vegetable samples. Additionally, *A. veronii* was rarely isolated among all food samples analysed, which included beef, minced meat, pork, poultry, seafood, fish and vegetables. Knochel and Jeppesen (Knochel and Jeppesen, 1990) found that 86% of 116 food isolates were *A. hydrophila*, followed by *A. caviae* (complex), 9%, then *A. sobria*, 5%. Ibrahim and Mac Rae (1991) also found that *A. hydrophila* group dominated those isolates from meat samples, which included beef, lamb and poultry (72%, 74%, and 75%, respectively). In agreement with other researchers, the authors found that *A. caviae* was a significant species in raw milk samples (35% of *Aeromonas* spp. isolates); however, they did report the isolation of *A. caviae* complex strains in meat samples, (14% of *Aeromonas* spp. isolates).

A few researchers have reported species distributions among food and water isolates that do not present a clear pattern. For example, recently Woodring *et al.* (2012) found that *A. caviae–media* complex was the dominant species, 15% (18/120), isolated from uncooked seafood, while *A. veronii* biovar *sobria* was second most frequent, 13%, followed by *A. hydrophila* group, 10%. In this study, the prevalence of all three groups was not significantly different.

A number of new species of *Aeromonas* spp. have been reported over the last few years. However, description of many of these is limited to a small number of isolates. One exception to this trend is the novel species, *A. aquariorum* (Martinez-Murcia *et al.*, 2008), synonymous with *A. hydrophila* subsp. *dhakensis* (Huys *et al.*, 2002; Martinez-Murcia *et al.*, 2009). This species is closely related to *A. hydrophila*, often mistaken as this organism, but differentiated primarily by arabinose fermentation (Aravena-Roman *et al.*, 2011; Martinez-Murcia *et al.*, 2008). Indeed, Burke *et al.* (1984a) identified this population of *Aeromonas* spp. in clinical and environmental isolates from Perth. Aravena-Roman *et al.* (2011) recently found that a large number, 30%, of *Aeromonas* spp. isolates from human clinical and water samples were *A. aquariorum*.

Several researchers have reported that aeromonads isolated from water and drinking water express virulence traits. Many of these studies were performed to investigate the hypothesis of water to human transmission of pathogenic *Aeromonas* spp. (Khajanchi *et al.*, 2010). For example, Burke *et al.* (1984a) found that 61% (22/36) and 64% (23/36) isolates of *Aeromonas* from untreated water produced enterotoxin and haemolysin, respectively. By comparison, 68% (38/56) of *Aeromonas* isolates from human faecal samples produced enterotoxin and haemolysin. LeChevallier *et al.* (1980) reported that 80% of *A. sobria* phenospecies isolates from a treated water distribution system in Oregon, USA, were cytotoxic.

The same prevalence of cytotoxic and haemolytic properties among *Aeromonas* isolates from food has been reported. For example, Tsai and Chen (Tsai and Chen, 1996) found that 79% of *Aeromonas* spp. isolates from oyster, fish and shrimp retail samples were haemolytic and 92% were cytotoxic. The difference in the activities was attributed to six isolates from oysters, which were cytotoxic, but not haemolytic. Similarly, Mateos *et al.* (1993) found that 100% of isolates from fish (11/11) were both haemolytic and cytotoxic, while the only 56% (5/9) of clinical isolates produced both toxins. And Handfield *et al.* (1996) found that approximately 50% ($n=72$) of food isolates were haemolytic using six species of red blood cells, while 64% ($n=94$) of water isolates were β-haemolytic. Among the same strains, the authors found that 92% for food isolates and 73% for water isolates were cytotoxic and caused haemagglutination. Fricker and Tompsett (1989) reported a higher number of cytotoxic *Aeromonas* spp. isolates from fish, as compared to those from meat and poultry. Interestingly, in many studies, higher toxin production is observed at 28–30°C. Toxins are still produced at 37°C and at low temperatures (5–10°C).

Some general trends with regard to the distribution of virulence traits among *Aeromonas* spp. have been identified. For example, Notermans *et al.* (1986) reported that all *A. hydrophila* group (100%) and most *A. sobria* (63%) isolates from both human faeces (100%) and drinking water (73%) were haemolytic and/or cytotoxic; although frequency of toxin activity was notably decreased for *A. sobria* isolates from drinking water. In contrast, no (0/27) isolates of *A. caviae* complex, from clinical or environmental sources, were haemolytic or cytotoxic. Similarly, Millership *et al.* (1986) reported that no (0/59) *A. caviae* complex isolates from clinical and drinking water samples were toxigenic. Krovacek *et al.* (1992) reported that 100% of *A. hydrophila* group isolates and 70% of *A. sobria* isolates from treated and untreated drinking water from Sweden were haemolytic. Meanwhile, less than 30% of all isolates were reported to be toxigenic.

The distribution of toxin genes as determined by molecular methods, such as PCR and hybridization, present a less clear picture, mostly due to discrepancies brought about by a failure to reconcile gene homologues, such as aerolysin and Act-encoding genes, which have persisted in the literature, and a further failure to account for species-level divergence among toxin homologue gene sequences. Despite this, some general trends can be deciphered. The cytotoxic enterotoxin gene is apparently equally distributed among *A. hydrophila* strains, regardless of their source of isolation and most strains harbour the *hlyA* gene (Castilho *et al.*, 2009). In a survey of *Aeromonas* spp. isolates, Granum *et al.* (1998) found that all strains of *A. hydrophila* analysed ($n = 15$) were haemolytic and cytotoxic and harboured the alt cytotonic enterotoxin, while 13 of 15 also possessed the aerolysin gene. Although the authors only analysed four strains, *A. trota* isolates had a similar toxigenic profile as that of *A. hydrophila* (Granum *et al.*, 1998). Bhowmik *et al.* (2009) analysed 21 *A. hydrophila* isolates from surface waters from Kolkata, India. Only four of these isolates harboured the *act* gene. In comparison, the authors found that two of five clinical isolates possessed this gene. All of the four *act*$^{+}$ environmental isolates caused fluid accumulation in a sealed adult mouse model, while one of these four was not cytotoxic to HeLa cells. Four isolates from each group also possessed the *hlyA* gene; however 71% were found to be haemolytic, using human red blood cells. Further, 90% of the environmental isolates displayed resistance to killing by human serum.

In contrast, *A. caviae* isolates infrequently possess the *hlyA* haemolysin or the *act* (or aerolysin) cytotoxin genes (Granum *et al.*, 1998); those that do, usually come from human sources (Castilho *et al.*, 2009). Similar results were reported by Pablos *et al.* (2009). The authors isolated 35 strains of *Aeromonas* spp. from 132 water samples from drinking water systems from Leon, Spain. All isolates were found to belong to the *A. caviae* complex and all water samples were negative for *hlyA* gene and the *aerA* gene was detected in only six water samples. In later studies on the same set of isolates, Pablos *et al.* (2010) found that two out of 12 were positive for the *aerA* (aerolysin) gene. Conversely, all twelve possessed the *alt*, cytotonic enterotoxin, and *lafA* (lateral flagella) genes (Pablos *et al.*, 2010). Food sources, namely rabbit meat and fish, from the same geographic area, revealed a predominance of *A. hydrophila* complex isolates (14/18), as compared to *A. caviae* (3/18) (Pablos *et al.*, 2011). In contrast to *A. caviae*, all *A. hydrophila* isolates from this study possessed the *alt* gene, and most (13/14) isolates were positive for *aerA* (Pablos *et al.*, 2011). Kingombe *et al.* (1999) found that 64% (127/200) of food and environmental isolates of *Aeromonas* possessed an *act*/*aerA* homologue. In comparison, 50% (36/71) of clinical and 81% (64/79) of fish isolates harboured this genetic determinant. None of the *A. caviae* isolates from food, environment, or fish were positive for *act*/*aerA*, except for those from vegetables (6/21) while 4 out of 10 clinical isolates possessed this gene. Conversely, almost all *A. hydrophila* isolates from food, environmental, and fish sources harboured the *act* gene (Kingombe *et al.*, 1999).

Additionally, most isolates of *A. bestiarum* and *A. veronii* also were positive for the *act/aerA* gene, regardless of source.

Conclusions

The apparent species distribution of *Aeromonas* in water and food sources is most likely a consequence of source-specific traits, such as type of treatment systems for drinking water, and processing, packaging and preservation practices for individual food commodities, that affect the survival and proliferation of individual species, coupled with initial contamination levels, and environmental conditions that favour the proliferation of *Aeromonas* spp., particularly in a competitive manner with other members of the microbial community. One of the characteristics, that is often overlooked, that strongly affects the distribution of *Aeromonas* in food commodities is their psychrotrophic tolerance. While several reports have correlated elevated abundance of aeromonads in surface waters during late summer months to water temperature, several authors have also reported the survival and proliferation of aeromonads at refrigeration temperatures, and the expression of virulence factors at these temperatures.

References

Abbey, S.D., and Etang, B.B. (1988). Incidence and biotyping of *Aeromonas* species from the environment. Microbios. *56*, 149–155.

Abu Kwaik, Y. (1996). The phagosome containing *Legionella pneumophila* within the protozoan *Hartmannella vermiformis* is surrounded by the rough endoplasmic reticulum. Appl. Environ. Microbiol. *62*, 2022–2028.

Agbonlahor, D.E., Shonekan, R.A., Kazak, W.H., and Coker, A.O. (1982). *Aeromonas* food poisoning in Nigeria: a case report. Cent. Afr. J. Med. *28*, 36–38.

Aguilera-Arreola, M.G., Hernandez-Rodriguez, C., Zuniga, G., Figueras, M.J., Garduno, R.A., and Castro-Escarpulli, G. (2007). Virulence potential and genetic diversity of *Aeromonas caviae*, *Aeromonas veronii*, and *Aeromonas hydrophila* clinical isolates from Mexico and Spain: a comparative study. Can. J. Microbiol. *53*, 877–887.

Alam, M., Hasan, N., Sadique, A., Bhuiyan, N., Ahmed, K., Nusrin, S., Nair, G., Siddique, A., Sack, R., Sack, D., *et al.* (2006). Seasonal cholera caused by *Vibrio cholerae* serogroups O1 and O139 in the coastal aquatic environment of Bangladesh. Appl. Environ. Microbiol. 72, 4096–4104.

Al-Bahry, S.N., Mahmoud, I.Y., Al-Khaifi, A., Elshafie, A.E., and Al-Harthy, A. (2009). Viability of multiple antibiotic resistant bacteria in distribution lines of treated sewage effluent used for irrigation. Water Sci. Technol. *60*, 2939–2948.

Anacarso, I., de Niederhausern, S., Messi, P., Guerrieri, E., Iseppi, R., Sabia, C., and Bondi, M. (2012). *Acanthamoeba polyphaga*, a potential environmental vector for the transmission of food-borne and opportunistic pathogens. J. Basic Microbiol. *52*, 261–268.

Angel, M.F., Zhang, F., Jones, M., Henderson, J., and Chapman, S.W. (2002). Necrotizing fasciitis of the upper extremity resulting from a water moccasin bite. South Med. J. *95*, 1090–1094.

Araujo, R.M., Arribas, R.M., and Pares, R. (1991). Distribution of *Aeromonas* species in waters with different levels of pollution. J. Appl. Bacteriol. *71*, 182–186.

Aravena-Roman, M., Chang, B.J., Riley, T.V., and Inglis, T.J. (2011). Phenotypic characteristics of human clinical and environmental *Aeromonas* in Western Australia. Pathology *43*, 350–356.

Arora, S., Agarwal, R.K., and Bist, B. (2006). Comparison of ELISA and PCR vis-a-vis cultural methods for detecting *Aeromonas* spp. in foods of animal origin. Int. J. Food Microbiol. *106*, 177–183.

Awan, M.B., Maqbool, A., Bari, A., and Krovacek, K. (2009). Antibiotic susceptibility profile of *Aeromonas* spp. isolates from food in Abu Dhabi, United Arab Emirates. New Microbiol. 32, 17–23.

Azzopardi, E.A., Azzopardi, S.M., Boyce, D.E., and Dickson, W.A. (2011). Emerging gram-negative infections in burn wounds. J. Burn Care Res. 32, 570–576.

Balakrishna, K., Murali, H.S., and Batra, H.V. (2010). Detection of toxigenic strains of *Aeromonas* species in foods by a multiplex PCR assay. Indian J. Microbiol. *50*, 139–144.

Barbaree, J.M., Fields, B.S., Feeley, J.C., Gorman, G.W., and Martin, W.T. (1986). Isolation of protozoa from water associated with a legionellosis outbreak and demonstration of intracellular multiplication of *Legionella pneumophila*. Appl. Environ. Microbiol. *51*, 422–424.

Barker, J., Brown, M.R., Collier, P.J., Farrell, I., and Gilbert, P. (1992). Relationship between *Legionella pneumophila* and *Acanthamoeba polyphaga*: physiological status and susceptibility to chemical inactivation. Appl. Environ. Microbiol. *58*, 2420–2425.

Barnett, T.C., and Kirov, S.M. (1999). The type IV *Aeromonas* pilus (Tap) gene cluster is widely conserved in *Aeromonas* species. Microb. Pathog. *26*, 77–84.

Barnett, T.C., Kirov, S.M., Strom, M.S., and Sanderson, K. (1997). *Aeromonas* spp. possess at least two distinct type IV pilus families. Microb. Pathog. *23*, 241–247.

Bechet, M., and Blondeau, R. (2003). Factors associated with the adherence and biofilm formation by *Aeromonas* caviae on glass surfaces. J. Appl. Microbiol. *94*, 1072–1078.

Behera, B., Bhoriwal, S., Mathur, P., Sagar, S., Singhal, M., and Misra, M.C. (2011). Post-traumatic skin and soft tissue infection due to *Aeromonas hydrophila*. Indian J Crit. Care Med. *15*, 49–51.

Bhowmik, P., Bag, P.K., Hajra, T.K., De, R., Sarkar, P., and Ramamurthy, T. (2009). Pathogenic potential of *Aeromonas hydrophila* isolated from surface waters in Kolkata, India. J. Med. Microbiol. *58*, 1549–1558.

Bomar, M.T. (1979). [Microbiological aspects of stored hard-boiled eggs]. Z Ernahrungswiss *18*, 233–238.

Borrell, N., Figueras, M.J., and Guarro, J. (1998). Phenotypic identification of *Aeromonas* genomospecies from clinical and environmental sources. Can. J. Microbiol. *44*, 103–108.

Buchanan, R.L. (1983). The 'new' pathogens: an update of selected examples. Paper presented at: 89th Central Atlantic States Association of Food and Drug Officials Annual Meeting (Madison, WI).

Burke, V., Robinson, J., Cooper, M., Beaman, J., Partridge, K., Peterson, D., and Gracey, M. (1984a). Biotyping and virulence factors in clinical and environmental isolates of *Aeromonas* species. Appl. Environ. Microbiol. *47*, 1146–1149.

Burke, V., Robinson, J., Gracey, M., Peterson, D., and Partridge, K. (1984b). Isolation of *Aeromonas hydrophila* from a metropolitan water supply: seasonal correlation with clinical isolates. Appl. Environ. Microbiol. *48*, 361–366.

Carnahan, A.M., Behram, S., and Joseph, S.W. (1991). Aerokey II: a flexible key for identifying clinical *Aeromonas* species. J. Clin. Microbiol. *29*, 2843–2849.

Castilho, M.C., Castro, T.L., Araujo, V.S., Trajano, R.S., Santos, P.A., Pimenta, P.M., Lucheze, K., Melo, J.T., Goncalves, A.M., Nogueira, R.T., *et al.* (2009). High frequency of haemolytic and cytotoxic activity in *Aeromonas* spp. isolated from clinical, food and environmental in Rio de Janeiro, Brazil. Antonie Van Leeuwenhoek *96*, 53–61.

Chang, Y.C., Wang, J.Y., Selvam, A., Kao, S.C., Yang, S.S., and Shih, D.Y. (2008). Multiplex PCR detection of enterotoxin genes in *Aeromonas* spp. from suspect food samples in northern Taiwan. J. Food Prot. *71*, 2094–2099.

Chauret, C., Volk, C., Creason, R., Jarosh, J., Robinson, J., and Warnes, C. (2001). Detection of *Aeromonas hydrophila* in a drinking-water distribution system: a field and pilot study. Can. J. Microbiol. *47*, 782–786.

Clark, J.A., Burger, C.A., and Sabatinos, L.E. (1982). Characterization of indicator bacteria in municipal raw water, drinking water, and new main water samples. Can. J. Microbiol. *28*, 1002–1013.

Cotuk, A., Dogruoz, N., Zeybek, Z., Kimiran-Erdem, A., and Ilhan-Sungur, E. (2005). The effects of *Pseudomonas* and *Aeromonas* strains on *Legionella pneumophila* growth. Ann. Microbiol. *55*, 219–224.

Cunliffe, D.A., and Adcock, P. (1989). Isolation of *Aeromonas* spp. from water by using anaerobic incubation. Appl. Environ. Microbiol. *55*, 2138–2140.

Daood, N. (2008). Bacterial diversity of local and imported mineral bottled water in Syria. Damascus University Journal for BASIC SCIENCES *24*, 61–80.

Davis, J.W., and Sizemore, R.K. (1981). Nonselectivity of Rimler-Shotts Medium for *Aeromonas hydrophila* in Estuarine Environments. Appl. Environ. Microbiol. *42*, 544–545.

Di Bari, M., Hachich, E.M., Melo, A.M.J., and Sato, M.I.Z. (2007). *Aeromonas* spp. and microbial indicators in raw drinking water sources. Braz. J. Microbiol. *38*, 516–521.

Dingsdag, S., and Coleman, N.V. (2012). Bacterial communities on food court tables and cleaning equipment in a shopping mall. Epidemiol. Infect. *141*, 1647–1651.

Echeverria, P., Blacklow, N.R., Sanford, L.B., and Cukor, G.G. (1981). Travellers' diarrhoea among American Peace Corps volunteers in rural Thailand. J. Infect. Dis. *143*, 767–771.

Egorov, A.I., Best, J.M., Frebis, C.P., and Karapondo, M.S. (2011). Occurrence of *Aeromonas* spp. in a random sample of drinking water distribution systems in the USA. J. Water Health *9*, 785–798.

Enos-Berlage, J.L., Guvener, Z.T., Keenan, C.E., and McCarter, L.L. (2005). Genetic determinants of biofilm development of opaque and translucent Vibrio parahaemolyticus. Mol. Microbiol. *55*, 1160–1182.

Faruque, S., Biswas, K., Udden, S., Ahmad, Q., Sack, D., Nair, G., and Mekalanos, J. (2006). Transmissibility of cholera: In vivo-formed biofilms and their relationship to infectivity and persistence in the environment. Proc. Natl. Acad. Sci. U.S.A. *103*, 6350–6355.

Fernandez, M.C., Giampaolo, B.N., Ibanez, S.B., Guagliardo, M.V., Esnaola, M.M., Conca, L., Valdivia, P., Stagnaro, S.M., Chiale, C., and Frade, H. (2000). *Aeromonas hydrophila* and its relation with drinking water indicators of microbiological quality in Argentine. Genetica *108*, 35–40.

Fliermans, C.B., Gorden, R.W., Hazen, T.C., and Esch, G.W. (1977). *Aeromonas* distribution and survival in a thermally altered lake. Appl. Environ. Microbiol. *33*, 114–122.

Fraser, G.M., and Hughes, C. (1999). Swarming motility. Curr. Opin. Microbiol. *2*, 630–635.

Fricker, C.R., and Tompsett, S. (1989). *Aeromonas* spp. in foods: a significant cause of food poisoning? Int. J. Food Microbiol. *9*, 17–23.

Gavin, R., Merino, S., Altarriba, M., Canals, R., Shaw, J.G., and Tomas, J.M. (2003). Lateral flagella are required for increased cell adherence, invasion and biofilm formation by *Aeromonas* spp. FEMS Microbiol. Lett. *224*, 77–83.

Gavriel, A.A., Landre, J.P., and Lamb, A.J. (1998). Incidence of mesophilic *Aeromonas* within a public drinking water supply in north-east Scotland. J. Appl. Microbiol. *84*, 383–392.

Ghenghesh, K.S., Ahmed, S.F., El-Khalek, R.A., Al-Gendy, A., and Klena, J. (2008). *Aeromonas*-associated infections in developing countries. Journal Infect. Dev. Ctries *2*, 81–98.

Gonzalez, C., Gutierrez, C., and Grande, T. (1987). Bacterial flora in bottled uncarbonated mineral drinking water. Can. J. Microbiol. *33*, 1120–1125.

von Graevenitz, A., and Bucher, C. (1983). Evaluation of differential and selective media for isolation of *Aeromonas* and Plesiomonas spp. from human feces. J. Clin. Microbiol. *17*, 16–21.

Grafova, T.I. (1981). [Characteristics of *Aeromonas* isolated from dry and pasteurized milk]. Zh Mikrobiol Epidemiol Immunobiol, 107–108.

Grafova, T.I. (1982). [Aeromonads in food products]. Gig Sanit, 89–91.

Granum, P.E., O'Sullivan, K., Tomas, J.M., and Ormen, O. (1998). Possible virulence factors of *Aeromonas* spp. from food and water. FEMS Immunol. Med. Microbiol. *21*, 131–137.

Gugliandolo, C., Lentini, V., Spano, A., and Maugeri, T.L. (2011). Conventional and molecular methods to detect bacterial pathogens in mussels. Lett. Appl. Microbiol. *52*, 15–21.

Haburchak, D.R. (1996). *Aeromonas hydrophila*: an underappreciated danger to fisherman. Infect. Med. *13*, 893–896.

Handfield, M., Simard, P., Couillard, M., and Letarte, R. (1996). *Aeromonas hydrophila* isolated from food and drinking water: hemagglutination, hemolysis, and cytotoxicity for a human intestinal cell line (HT-29). Appl. Environ. Microbiol. *62*, 3459–3461.

Havelaar, A.H., During, M., and Versteegh, J.F. (1987). Ampicillin-dextrin agar medium for the enumeration of *Aeromonas* species in water by membrane filtration. J. Appl. Bacteriol. *62*, 279–287.

Havelaar, A.H., Versteegh, J.F., and During, M. (1990). The presence of *Aeromonas* in drinking water supplies in The Netherlands. Zentralbl Hyg Umweltmed *190*, 236–256.

Hazen, T.C., Fliermans, C.B., Hirsch, R.P., and Esch, G.W. (1978). Prevalence and distribution of *Aeromonas hydrophila* in the United States. Appl. Environ. Microbiol. *36*, 731–738.

Holmes, P., and Sartory, D.P. (1993). An evaluation of media for the membrane filtration enumeration of *Aeromonas* from drinking-water. Lett. Appl. Microbiol. *17*, 58–60.

Honma, Y., and Nakasone, N. (1990). Pili of *Aeromonas hydrophila*: purification, characterization, and biological role. Microbiol. Immunol. *34*, 83–98.

Huddleston, J.R., Zak, J.C., and Jeter, R.M. (2007). Sampling bias created by ampicillin in isolation media for *Aeromonas*. Can. J. Microbiol. *53*, 39–44.

Hudson, J.A., and De Lacy, K.M. (1991). Incidence of motile aeromonads in New Zealand retail foods. J. Food Protect. *54*, 696–699.

Huguet, J.M., and Ribas, F. (1991). SGAP-10C agar for the isolation and quantification of *Aeromonas* from water. J. Appl. Bacteriol. *70*, 81–88.

Hunter, P.R., and Burge, S.H. (1987). The bacteriological quality of bottled natural mineral waters. Epidemiol. Infect. *99*, 439–443.

Huys, G., Kampfer, P., Albert, M.J., Kuhn, I., Denys, R., and Swings, J. (2002). *Aeromonas hydrophila* subsp. *dhakensis* subsp. nov., isolated from children with diarrhoea in Bangladesh, and extended description of *Aeromonas hydrophila* subsp. *hydrophila* (Chester 1901) Stanier 1943 (approved lists 1980). Int. J. Syst. Evol. Microbiol. *52*, 705–712.

Ibrahim, A., and Mac Rae, I.C. (1991). Incidence of *Aeromonas* and *Listeria* spp. in red meat and milk samples in Brisbane, Australia. Int. J. Food Microbiol. *12*, 263–269.

Italiana, G.U.d.R. (1997). Integration of the evaluation criteria of microbiological characteristics of natural mineral waters (Italian) (Istituto Poligrafico e Zecca dello Stato, Rome, Italy).

Jablonska-Strynkowska, W., and Potapa, M. (1983). [*Aeromonas hydrophila* var. anaerogenes as a probable etiologic factor in food poisoning]. Przegl Epidemiol. *37*, 455–458.

Jagmann, N., von Rekowski, K.S., and Philipp, B. (2012). Interactions of bacteria with different mechanisms for chitin degradation result in the formation of a mixed-species biofilm. FEMS Microbiol. Lett. *326*, 69–75.

Janda, J.M. (1985). Biochemical and exoenzymatic properties of *Aeromonas* species. Diagn. Microbiol. Infect. Dis. *3*, 223–232.

Jjemba, P.K., Weinrich, L.A., Cheng, W., Giraldo, E., and Lechevallier, M.W. (2010). Regrowth of potential opportunistic pathogens and algae in reclaimed-water distribution systems. Appl. Environ. Microbiol. *76*, 4169–4178.

Johnson, L. (2007). Microcolony and biofilm formation as a survival strategy for bacteria. J. Theor. Biol. *251*, 24–34.

Johnson, L.R. (2008). Microcolony and biofilm formation as a survival strategy for bacteria. J. Theor. Biol. *251*, 24–34.

Kalina, G.P. (1977). [*Aeromonas* in food products and its possible role as a pathogen of food poisoning]. Gig Sanit, *8*, 97–100.

Kaper, J., Seidler, R.J., Lockman, H., and Colwell, R.R. (1979). Medium for the presumptive identification of *Aeromonas hydrophila* and Enterobacteriaceae. Appl. Environ. Microbiol. *38*, 1023–1026.

Kaper, J.B., Lockman, H., Colwell, R.R., and Joseph, S.W. (1981). *Aeromonas hydrophila*: ecology and toxigenicity of isolates from an estuary. J. Appl. Bacteriol. *50*, 359–377.

Khajanchi, B.K., Sha, J., Kozlova, E.V., Erova, T.E., Suarez, G., Sierra, J.C., Popov, V.L., Horneman, A.J., and Chopra, A.K. (2009). N-acylhomoserine lactones involved in quorum sensing control the type VI secretion system, biofilm formation, protease production, and *in vivo* virulence in a clinical isolate of *Aeromonas hydrophila*. Microbiology *155*, 3518–3531.

Khajanchi, B.K., Fadl, A.A., Borchardt, M.A., Berg, R.L., Horneman, A.J., Stemper, M.E., Joseph, S.W., Moyer, N.P., Sha, J., and Chopra, A.K. (2010). Distribution of virulence factors and molecular fingerprinting of *Aeromonas* species isolates from water and clinical samples: suggestive evidence of water-to-human transmission. Appl. Environ. Microbiol. *76*, 2313–2325.

Khan, I.U., Loughborough, A., and Edge, T.A. (2009). DNA-based real-time detection and quantification of aeromonads from fresh water beaches on Lake Ontario. J. Water Health *7*, 312–323.

Kingombe, C.I., Huys, G., Tonolla, M., Albert, M.J., Swings, J., Peduzzi, R., and Jemmi, T. (1999). PCR detection, characterization, and distribution of virulence genes in *Aeromonas* spp. Appl. Environ. Microbiol. *65*, 5293–5302.

Kingombe, C.I., D'Aoust, J.Y., Huys, G., Hofmann, L., Rao, M., and Kwan, J. (2010). Multiplex PCR method for detection of three *Aeromonas* enterotoxin genes. Appl. Environ. Microbiol. *76*, 425–433.

Kirov, S.M. (1993). The public health significance of *Aeromonas* spp. in foods. Int. J. Food Microbiol. *20*, 179–198.

Kirov, S.M., and Sanderson, K. (1996). Characterization of a type IV bundle-forming pilus (SFP) from a gastroenteritis-associated strain of *Aeromonas* veronii biovar sobria. Microb. Pathog. *21*, 23–34.

Kirov, S.M., O'Donovan, L.A., and Sanderson, K. (1999). Functional characterization of type IV pili expressed on diarrhoea-associated isolates of *Aeromonas* species. Infect. Immun. *67*, 5447–5454.

Kirov, S.M., Tassell, B.C., Semmler, A.B., O'Donovan, L.A., Rabaan, A.A., and Shaw, J.G. (2002). Lateral flagella and swarming motility in *Aeromonas* species. J. Bacteriol. *184*, 547–555.

Kirov, S.M., Castrisios, M., and Shaw, J.G. (2004). *Aeromonas* flagella (polar and lateral) are enterocyte adhesins that contribute to biofilm formation on surfaces. Infect. Immun. *72*, 1939–1945.

Kleeberger, A. (1977). [Taxonomic studies on aeromonads of milk, water and minced meat (author's transl)]. Z Lebensm Unters Forsch *163*, 44–47.

Knochel, S., and Jeppesen, C. (1990). Distribution and characteristics of *Aeromonas* in food and drinking water in Denmark. Int. J. Food Microbiol. *10*, 317–322.

Krovacek, K., Peterz, M., Faris, A., and Mansson, I. (1989). Enterotoxigenicity and drug sensitivity of *Aeromonas hydrophila* isolated from well water in Sweden: a case study. Int. J. Food Microbiol. *8*, 149–154.

Krovacek, K., Faris, A., Baloda, S., Lindberg, M., Peterz, M., and Mansson, I. (1992). Isolation and virulence profiles of *Aeromonas* spp. from different municipal drinking water supplies in Sweden. Food Microbiol. *9*, 215–222.

Kuhn, I., Allestam, G., Huys, G., Janssen, P., Kersters, K., Krovacek, K., and Stenstrom, T.A. (1997). Diversity, persistence, and virulence of *Aeromonas* strains isolated from drinking water distribution systems in Sweden. Appl. Environ. Microbiol. *63*, 2708–2715.

Kwon, S., Moon, E., Kim, T.S., Hong, S., and Park, H.D. (2011). Pyrosequencing demonstrated complex microbial communities in a membrane filtration system for a drinking water treatment plant. Microbes Environ. *26*, 149–155.

Lamy, B., Laurent, F., Verdier, I., Decousser, J.W., Lecaillon, E., Marchandin, H., Roger, F., Tigaud, S., de Montclos, H., and Kodjo, A. (2010). Accuracy of 6 commercial systems for identifying clinical *Aeromonas* isolates. Diagn. Microbiol. Infect. Dis. *67*, 9–14.

Lan, X., Zhang, X., Kodaira, R., Zhou, Z., and Shimosaka, M. (2008). Gene cloning, expression, and characterization of a second beta-N-acetylglucosaminidase from the chitinolytic bacterium *Aeromonas hydrophila* strain SUWA-9. Biosci. Biotechnol. Biochem. *72*, 492–498.

LeChevallier, M.W., Seidler, R.J., and Evans, T.M. (1980). Enumeration and characterization of standard plate count bacteria in chlorinated and raw water supplies. Appl. Environ. Microbiol. *40*, 922–930.

LeChevallier, M.W., Evans, T.M., Seidler, R.J., Daily, O.P., Merrell, B.R., Rollins, D.M., and Joseph, S.W. (1982). *Aeromonas* sobria in chlorinated drinking water supplies. Microb. Ecol. *8*, 325–333.

Legnani, P., Leoni, E., Soppelsa, F., and Burigo, R. (1998). The occurrence of *Aeromonas* species in drinking water supplies of an area of the Dolomite Mountains, Italy. J. Appl. Microbiol. *85*, 271–276.

Lehtola, M.J., Torvinen, E., Kusnetsov, J., Pitkanen, T., Maunula, L., von Bonsdorff, C.H., Martikainen, P.J., Wilks, S.A., Keevil, C.W., and Miettinen, I.T. (2007). Survival of Mycobacterium avium, *Legionella pneumophila, Escherichia coli,* and caliciviruses in drinking water-associated biofilms grown under high-shear turbulent flow. Appl. Environ. Microbiol. 73, 2854–2859.

Liu, J., Gratz, J., Maro, A., Kumburu, H., Kibiki, G., Taniuchi, M., Howlader, A.M., Sobuz, S.U., Haque, R., Talukder, K.A., *et al.* (2012). Simultaneous detection of six diarrhoea-causing bacterial pathogens with an in-house PCR-luminex assay. J. Clin. Microbiol. *50*, 98–103.

Loy, A., Beisker, W., and Meier, H. (2005). Diversity of bacteria growing in natural mineral water after bottling. Appl. Environ. Microbiol. *71*, 3624–3632.

Lynch, M.J., Swift, S., Kirke, D.F., Keevil, C.W., Dodd, C.E., and Williams, P. (2002). The regulation of biofilm development by quorum sensing in *Aeromonas hydrophila*. Environ. Microbiol. *4*, 18–28.

Maalej, S., Mahjoubi, A., Elazri, C., and Dukan, S. (2003). Simultaneous effects of environmental factors on motile *Aeromonas* dynamics in an urban effluent and in the natural seawater. Water Res. *37*, 2865–2874.

Maciejska-Roczan, K. (1979). [Occurrence of various strains of gram-negative bacteria: *Yersinia, Klebsiella* and *Aeromonas* and their role in the aetiology of food poisoning]. Rocz Panstw Zakl Hig *30*, 217–223.

McMahon, M.A., and Wilson, I.G. (2001). The occurrence of enteric pathogens and *Aeromonas* species in organic vegetables. Int. J. Food Microbiol. *70*, 155–162.

Markov, G., Kirov, G., Lyutskanov, V., and Konadarev, M. (2007). Necrotizing fasciitis and myonecrosis due to *Aeromonas hydrophila*. Wounds *19*, 223–226.

Martinez-Murcia, A.J., Saavedra, M.J., Mota, V.R., Maier, T., Stackebrandt, E., and Cousin, S. (2008). *Aeromonas* aquariorum sp. nov., isolated from aquaria of ornamental fish. Int. J. Syst. Evol. Microbiol. *58*, 1169–1175.

Martinez-Murcia, A., Monera, A., Alperi, A., Figueras, M.J., and Saavedra, M.J. (2009). Phylogenetic evidence suggests that strains of *Aeromonas hydrophila* subsp. dhakensis belong to the species *Aeromonas aquariorum* sp. nov. Curr. Microbiol. *58*, 76–80.

Massa, S., Altieri, C., and D'Angela, A. (2001). The occurrence of *Aeromonas* spp. in natural mineral water and well water. Int. J. Food Microbiol. *63*, 169–173.

Mateos, D., Anguita, J., Naharro, G., and Paniagua, C. (1993). Influence of growth temperature on the production of extracellular virulence factors and pathogenicity of environmental and human strains of *Aeromonas hydrophila*. J. Appl. Bacteriol. *74*, 111–118.

Matrone-Rocha, S., Piveli, R.P., Matte, G.R., Doria, M.C., Dropa, M., Morita, M., Peternella, F.A., and Matte, M.H. (2010). Dynamics of *Aeromonas* species isolated from wastewater treatment system. J. Water Health *8*, 703–711.

Merino, S., Rubires, X., Aguilar, A., and Tomas, J.M. (1997). The role of flagella and motility in the adherence and invasion to fish cell lines by *Aeromonas hydrophila* serogroup O:34 strains. FEMS Microbiol. Lett. *151*, 213–217.

Merino, S., Gavin, R., Vilches, S., Shaw, J.G., and Tomas, J.M. (2003). A colonization factor (production of lateral flagella) of mesophilic *Aeromonas* spp. is inactive in *Aeromonas* salmonicida strains. Appl. Environ. Microbiol. *69*, 663–667.

Merino, S., Shaw, J.G., and Tomas, J.M. (2006). Bacterial lateral flagella: an inducible flagella system. FEMS Microbiol. Lett. *263*, 127–135.

Messi, P., Guerrieri, E., and Bondi, M. (2002). Survival of an *Aeromonas hydrophila* in an artificial mineral water microcosm. Water Res. *36*, 3410–3415.

Millership, S.E., Barer, M.R., and Tabaqchali, S. (1986). Toxin production by *Aeromonas* spp. from different sources. J. Med. Microbiol. *22*, 311–314.

Moritz, M.M., Flemming, H.C., and Wingender, J. (2010). Integration of *Pseudomonas aeruginosa* and *Legionella pneumophila* in drinking water biofilms grown on domestic plumbing materials. Int. J. Hyg. Environ. Health *213*, 190–197.

Myers, B.R., Marshall, R.T., Edmonson, J.E., and Stringer, W.C. (1982). Isolation of pectinolytic *Aeromonas hydrophila* and Yersinia enterocolitica from vacuum-packaged pork. J. Food Protect. *45*, 33–37.

Nadell, C., Xavier, J., Levin, S., and Foster, K. (2008). The evolution of quorum sensing in bacterial biofilms. PLoS Biol. *6*, e14.

Nagar, V., Shashidar, R., and Bandekar, J.R. (2011). Prevalence, characterization, and antimicrobial resistance of *Aeromonas* strains from various retail food products in Mumbai, India. J. Food Sci. *76*, M486–492.

Narasimhan, T.N. (2005). Basic tenets for water management. Ground Water *43*, 1.

Newsome, A.L., Baker, R.L., Miller, R.D., and Arnold, R.R. (1985). Interactions between *Naegleria fowleri* and *Legionella pneumophila*. Infect. Immun. *50*, 449–452.

Neyts, K., Huys, G., Uyttendaele, M., Swings, J., and Debevere, J. (2000a). Incidence and identification of mesophilic *Aeromonas* spp. from retail foods. Lett. Appl. Microbiol. *31*, 359–363.

Neyts, K., Notebaert, E., Uyttendaele, M., and Debevere, J. (2000b). Modification of the bile salts-Irgasan-brilliant green agar for enumeration of *Aeromonas* species from food. Int. J. Food Microbiol. *57*, 211–218.

Nishikawa, Y., and Kishi, T. (1987). A modification of bile salts brilliant green agar for isolation of motile *Aeromonas* from foods and environmental specimens. Epidemiol. Infect. *98*, 331–336.

Notermans, S., Havelaar, A., Jansen, W., Kozaki, S., and Guinee, P. (1986). Production of 'Asao toxin' by *Aeromonas* strains isolated from faeces and drinking water. J. Clin. Microbiol. *23*, 1140–1142.

Nyenje, M.E., Odjadjare, C.E., Tanih, N.F., Green, E., and Ndip, R.N. (2012). Foodborne pathogens recovered from ready-to-eat foods from roadside cafeterias and retail outlets in Alice, Eastern Cape Province, South Africa: Public health implications. Int. J. Environ. Res. Public Health *9*, 2608–2619.

de Oliveira Scoaris, D., Bizerra, F.C., Yamada-Ogatta, S.F., de Abreu Filho, B.A., Ueda-Nakamura, T., Nakamura, C.V., and Filho, B.P.D. (2008a). The occurrence of *Aeromonas* spp. in the bottled mineral water, well water and tap water from the municipal supplies. Braz. Arch. Biol. Technol. *51*, 1049–1055.

de Oliveira Scoaris, D., Colacite, J., Nakamura, C.V., Ueda-Nakamura, T., de Abreu Filho, B.A., and Dias Filho, B.P. (2008b). Virulence and antibiotic susceptibility of *Aeromonas* spp. isolated from drinking water. Antonie Van Leeuwenhoek *93*, 111–122.

O'Toole, G.A., and Kolter, R. (1998a). Flagellar and twitching motility are necessary for *Pseudomonas aeruginosa* biofilm development. Mol. Microbiol. *30*, 295–304.

O'Toole, G.A., and Kolter, R. (1998b). Initiation of biofilm formation in *Pseudomonas fluorescens* WCS365 proceeds via multiple, convergent signalling pathways: a genetic analysis. Mol. Microbiol. *28*, 449–461.

Pablos, M., Rodriguez-Calleja, J.M., Santos, J.A., Otero, A., and Garcia-Lopez, M.L. (2009). Occurrence of motile *Aeromonas* in municipal drinking water and distribution of genes encoding virulence factors. Int. J. Food Microbiol. *135*, 158–164.

Pablos, M., Remacha, M.A., Rodriguez-Calleja, J.M., Santos, J.A., Otero, A., and Garcia-Lopez, M.L. (2010). Identity, virulence genes, and clonal relatedness of *Aeromonas* isolates from patients with diarrhoea and drinking water. Eur. J. Clin. Microbiol. Infect. Dis. *29*, 1163–1172.

Pablos, M., Huys, G., Cnockaert, M., Rodriguez-Calleja, J.M., Otero, A., Santos, J.A., and Garcia-Lopez, M.L. (2011). Identification and epidemiological relationships of *Aeromonas* isolates from patients with diarrhoea, drinking water and foods. Int. J. Food Microbiol. *147*, 203–210.

Palumbo, S.A., Maxino, F., Williams, A.C., Buchanan, R.L., and Thayer, D.W. (1985). Starch-ampicillin agar for the quantitative detection of *Aeromonas hydrophila*. Appl. Environ. Microbiol. *50*, 1027–1030.

Quevedo-Sarmiento, J., Ramos-Cormenzana, A., and Gonzalez-Lopez, J. (1986). Isolation and characterization of aerobic heterotrophic bacteria from natural spring waters in the Lanjaron area (Spain). J. Appl. Bacteriol. *61*, 365–372.

Rabaan, A.A., Gryllos, I., Tomas, J.M., and Shaw, J.G. (2001). Motility and the polar flagellum are required for *Aeromonas* caviae adherence to HEp-2 cells. Infect. Immun. *69*, 4257–4267.

Rippey, S.R., and Cabelli, V.J. (1979). Membrane filter procedure for enumeration of *Aeromonas hydrophila* in fresh waters. Appl. Environ. Microbiol. *38*, 108–113.

Rippey, R.M., and Cabelli, V.J. (1989). Use of thermotolerant *Aeromonas* group for the trophic classification of freshwaters. Water Res. *23*, 1107–1114.

Rogol, M., Sechter, I., Grinberg, L., and Gerichter, C.B. (1979). Pril-xylose-ampicillin agar, a new selective medium for the isolation of *Aeromonas hydrophila*. J. Med. Microbiol. *12*, 229–231.

Ryu, H., Alum, A., and Abbaszadegan, M. (2005). Microbial characterization and population changes in nonpotable reclaimed water distribution systems. Environ. Sci. Technol. *39*, 8600–8605.

Saavedra, M.J., Figueras, M.J., and Martinez-Murcia, A.J. (2006). Updated phylogeny of the genus *Aeromonas*. Int. J. Syst. Evol. Microbiol. *56*, 2481–2487.

Santos, P.G., Santos, P.A., Bello, A.R., and Freitas-Almeida, A.C. (2011). Association of *Aeromonas caviae* polar and lateral flagella with biofilm formation. Lett. Appl. Microbiol. *52*, 49–55.

Sartory, D.P. (2002). *Aeromonas*. In Guidelines for Drinking Water Quality – Addendum: Microbial Agents In Drinking Water, WHO, ed. (WHO, Geneva, Switzerland), pp. 1–12.

Schets, F.M., Italiaander, R., van den Berg, H.H., and de Roda Husman, A.M. (2010). Rainwater harvesting: quality assessment and utilization in The Netherlands. J. Water Health *8*, 224–235.

Schubert, R.H. (1967). [The occurrence of *Aeromonas* in surface waters]. Arch. Hyg. Bakteriol. *150*, 688–708.

September, S.M., Els, F.A., Venter, S.N., and Brozel, V.S. (2007). Prevalence of bacterial pathogens in biofilms of drinking water distribution systems. J. Water Health *5*, 219–227.

Seshadri, R., Joseph, S.W., Chopra, A.K., Sha, J., Shaw, J., Graf, J., Haft, D., Wu, M., Ren, Q., Rosovitz, M.J., *et al.* (2006). Genome sequence of *Aeromonas hydrophila* ATCC 7966T: jack of all trades. J. Bacteriol. *188*, 8272–8282.

Shak, J.R., Whitaker, J.A., Ribner, B.S., and Burd, E.M. (2011). Aminoglycoside-resistant *Aeromonas hydrophila* as part of a polymicrobial infection following a traumatic fall into freshwater. J. Clin. Microbiol. *49*, 1169–1170.

Shannon, K.E., Lee, D.Y., Trevors, J.T., and Beaudette, L.A. (2007). Application of real-time quantitative PCR for the detection of selected bacterial pathogens during municipal wastewater treatment. Sci. Total Environ. *382*, 121–129.

Shimada, T., Sakazaki, R., and Suzuki, K. (1985). Peritrichous flagella in mesophilic strains of *Aeromonas*. Jpn J. Med. Sci. Biol. *38*, 141–145.

Shotts, E.B., Jr., and Rimler, R. (1973). Medium for the isolation of *Aeromonas hydrophila*. Appl. Microbiol. *26*, 550–553.

Sisti, M., Albano, A., and Brandi, G. (1998). Bactericidal effect of chlorine on motile *Aeromonas* spp. in drinking water supplies and influence of temperature on disinfection efficacy. Lett. Appl. Microbiol. *26*, 347–351.

Soler, L., Yanez, M.A., Chacon, M.R., Aguilera-Arreola, M.G., Catalan, V., Figueras, M.J., and Martinez-Murcia, A.J. (2004). Phylogenetic analysis of the genus *Aeromonas* based on two housekeeping genes. Int. J. Syst. Evol. Microbiol. *54*, 1511–1519.

Someya, N., Ikeda, S., Morohoshi, T., Noguchi Tsujimoto, M., Yoshida, T., Sawada, H., Ikeda, T., and Tsuchiya, K. (2011). Diversity of culturable chitinolytic bacteria from rhizospheres of agronomic plants in Japan. Microbes Environ. *26*, 7–14.

Stampi, S., Zanetti, F., Crestani, A., and De Luca, G. (2000). Occurrence and seasonal variation of airborne gram negative bacteria in a sewage treatment plant. New Microbiol. *23*, 97–104.

Stecchini, M.L., and Domenis, C. (1994). Incidence of *Aeromonas* species in influent and effluent of urban wastewater purification plants. Lett. Appl. Microbiol. *19*, 237–239.

Stojek, N.M., and Dutkiewicz, J. (2011). Co-existence of *Legionella* and other Gram-negative bacteria in potable water from various rural and urban sources. Ann. Agric. Environ. Med. *18*, 330–334.

Stratev, D., Vashin, I., and Rusev, V. (2012). Prevalence and survival of *Aeromonas* spp. in foods – a review. Revue Méd Vét *163*, 486–494.

Szollosy, E., Marjai, E., and Lantos, J. (1974). Bacterial contamination and sparing heat treatment of mother's milk. Acta Microbiol. Acad. Sci. Hung. *21*, 319–325.

Temmerman, R., Vervaeren, H., Noseda, B., Boon, N., and Verstraete, W. (2006). Necrotrophic growth of *Legionella pneumophila*. Appl. Environ. Microbiol. *72*, 4323–4328.

Toule, G., and Murphy, O. (1978). A study of bacteria contaminating refrigerated cooked chicken; their spoilage potential and possible origin. J. Hyg. (Lond.) *81*, 161–169.

Trouwborst, T. (1992). Government policy with regard to occurrence of *Aeromonas* in drinking-water (Dutch). In *Aeromonas* in Drinking-Water: Occurrence, Control and Significance (Dutch), van der Kooij, D. ed. (Kiwa, Nieuwegein, Netherlands), pp. 95–104.

Tsai, G.J., and Chen, T.H. (1996). Incidence and toxigenicity of *Aeromonas hydrophila* in seafood. Int. J. Food Microbiol. *31*, 121–131.

USEPA (2001). Method 1605: *Aeromonas* in Finished Water by Membrane Filtration using Ampicillin-Dextrin Agar with Vancomycin (ADA-V), EPA-821-R-01-034 (USEPA, Washington, DC).

USEPA (2005). Drinking Water Contaminant Candidate List 2; Final Notice (Federal Register), pp. 9071–9077.

USEPA (2006). *Aeromonas*: Human Health Criteria Document (USEPA, Washington, DC).

USEPA (2009). Drinking Water Contaminant Candidate List 3-Final (Federal Register), pp. 51850–51862.

USEPA (2012). Guidelines for Water Reuse, EPA/600/R-12/618 (USEPA, Washington, DC).

Vandewalle, J.L., Goetz, G.W., Huse, S.M., Morrison, H.G., Sogin, M.L., Hoffmann, R.G., Yan, K., and McLellan, S.L. (2012). Acinetobacter, *Aeromonas* and Trichococcus populations dominate the microbial community within urban sewer infrastructure. Environ. Microbiol. *14*, 2538–2552.

Versteegh, J.F., Havelaar, A.H., Hoekstra, A.C., and Visser, A. (1989). Complexing of copper in drinking water samples to enhance recovery of *Aeromonas* and other bacteria. J. Appl. Bacteriol. *67*, 561–566.

Villari, P., Pucino, A., Santagata, N., and Torre, I. (1999). A comparison of different culture media for the membrane filter quantification of *Aeromonas* in water. Lett. Appl. Microbiol. *29*, 253–257.

Villari, P., Crispino, M., Montuori, P., and Boccia, S. (2003). Molecular typing of *Aeromonas* isolates in natural mineral waters. Appl. Environ. Microbiol. *69*, 697–701.

Vitkov, M. (1974). [Psychrophilic organisms in poultry meat]. Vet. Med. Nauki *11*, 85–92.

Wadowsky, R.M., Wilson, T.M., Kapp, N.J., West, A.J., Kuchta, J.M., States, S.J., Dowling, J.N., and Yee, R.B. (1991). Multiplication of Legionella spp. in tap water containing *Hartmannella vermiformis*. Appl. Environ. Microbiol. *57*, 1950–1955.

Wang, H.B., Wang, D.C., Bi, Z.Q., and Kan, B. (2009). [Development of TaqMan real-time PCR in detection of *Aeromonas hydrophila*]. Zhonghua Yu Fang Yi Xue Za Zhi *43*, 611–614.

Warburton, D.W. (1993). A review of the microbiological quality of bottled water sold in Canada. Part 2. The need for more stringent standards and regulations. Can. J. Microbiol. *39*, 158–168.

Warburton, D.W., Dodds, K.L., Burke, R., Johnston, M.A., and Laffey, P.J. (1992). A review of the microbiological quality of bottled water sold in Canada between 1981 and 1989. Can. J. Microbiol. *38*, 12–19.

Warburton, D.W., McCormick, J.K., and Bowen, B. (1994). Survival and recovery of *Aeromonas hydrophila* in water: development of methodology for testing bottled water in Canada. Can. J. Microbiol. *40*, 145–148.

Watnick, P.I., Fullner, K.J., and Kolter, R. (1999). A role for the mannose-sensitive hemagglutinin in biofilm formation by *Vibrio cholerae* El Tor. J. Bacteriol. *181*, 3606–3609.

Watnick, P.I., Lauriano, C.M., Klose, K.E., Croal, L., and Kolter, R. (2001). The absence of a flagellum leads to altered colony morphology, biofilm development and virulence in *Vibrio cholerae* O139. Mol. Microbiol. 39, 223–235.

WHO (2006). Wastewater use in agriculture. In WHO Guidelines for the Safe Use of Wastewater, Excreta, and Greywater, vol. 2 (WHO Press, Geneva, Switzerland).

Woodring, J., Srijan, A., Puripunyakom, P., Oransathid, W., Wongstitwilairoong, B., and Mason, C. (2012). Prevalence and antimicrobial susceptibilities of Vibrio, salmonella, and *Aeromonas* isolates from various uncooked seafoods in Thailand. J. Food Prot. 75, 41–47.

Ye, L., and Zhang, T. (2011). Pathogenic bacteria in sewage treatment plants as revealed by 454 pyrosequencing. Environ. Sci. Technol. *45*, 7173–7179.

Yucel, N., and Erdogan, S. (2010). Virulence properties and characterization of aeromonads isolated from foods of animal origin and environmental sources. J. Food Prot. 73, 855–860.

Zhou, G., Wen, S., Liu, Y., Li, R., Zhong, X., Feng, L., Wang, L., and Cao, B. (2011). Development of a DNA microarray for detection and identification of *Legionella pneumophila* and ten other pathogens in drinking water. Int. J. Food Microbiol. *145*, 293–300.

Index

www.ingramcontent.com/pod-product-compliance
Lightning Source LLC
LaVergne TN
LVHW051050210926
848486LV00001B/61